MW00748728

About Island Press

Since 1984, the nonprofit organization Island Press has been stimulating, shaping, and communicating ideas that are essential for solving environmental problems worldwide. With more than 800 titles in print and some 40 new releases each year, we are the nation's leading publisher on environmental issues. We identify innovative thinkers and emerging trends in the environmental field. We work with world-renowned experts and authors to develop cross-disciplinary solutions to environmental challenges.

Island Press designs and executes educational campaigns in conjunction with our authors to communicate their critical messages in print, in person, and online using the latest technologies, innovative programs, and the media. Our goal is to reach targeted audiences—scientists, policymakers, environmental advocates, urban planners, the media, and concerned citizens—with information that can be used to create the framework for long-term ecological health and human well-being.

Island Press gratefully acknowledges major support of our work by The Agua Fund, The Andrew W. Mellon Foundation, Betsy & Jesse Fink Foundation, The Bobolink Foundation, The Curtis and Edith Munson Foundation, Forrest C. and Frances H. Lattner Foundation, G.O. Forward Fund of the Saint Paul Foundation, Gordon and Betty Moore Foundation, The Kresge Foundation, The Margaret A. Cargill Foundation, The Overbrook Foundation, The S.D. Bechtel, Jr. Foundation, The Summit Charitable Foundation, Inc., V. Kann Rasmussen Foundation, The Wallace Alexander Gerbode Foundation, and other generous supporters.

The opinions expressed in this book are those of the author(s) and do not necessarily reflect the views of our supporters.

Climate Change in the Northwest

Implications for Our Landscapes, Waters, and Communities

Climate Change in the Northwest

Implications for Our Landscapes, Waters, and Communities

COORDINATING LEAD EDITORS

Meghan M. Dalton
Oregon Climate Change Research Institute,
Oregon State University

Philip W. Mote
Oregon Climate Change Research Institute,
Oregon State University

Amy K. Snover
Climate Impacts Group, University of Washington

ISLANDPRESS

Washington | Covelo | London

Copyright © 2013 Oregon Climate Change Reasearch Institute

All rights reserved under International and Pan-American Copyright Conventions. Reproduction of this report by electronic means for personal and noncommercial purposes is permitted as long as proper acknowledgement is included. Users are restricted from photocopying or mechanical reproduction as well as creating derivative works for commercial purposes without the prior written permission of the publisher.

ISLAND PRESS is a trademark of the Center for Resource Economics.

✪ Printed on recycled, acid-free paper

Suggested Citation: Dalton, M.M., P.W. Mote, and A.K. Snover [Eds.]. 2013. *Climate Change in the Northwest: Implications for Our Landscapes, Waters, and Communities*. Washington, DC: Island Press.

Editor Contact:
Meghan M. Dalton: mdalton@coas.oregonstate.edu (541) 737-3081
Philip W Mote: pmote@coas.oregonstate.edu (541) 737-5694
Amy K. Snover: aksnover@uw.edu (206) 221-0222

Keywords: Climate change, energy supply, climate variability, water supply, environmental management, solar variability, National Climate Assessment, energy consumption, water treatment, heating, cooling, adaptation, mitigation, renewable energy, oil production, thermal electrics, future risk management

Manufactured in the United States of America

10 9 8 7 6 5 4 3 2 1

Front Cover Images:
(cityscape) W. Spencer Reeder; Location: City of Seattle, Washington waterfront
(shellfish) W. Spencer Reeder; Location: Shi Shi beach, Olympic National Park, Washington coast
(river) Philip W. Mote; Location: North Fork of the Willamette River near Oakridge, Oregon
(mountain) Robert Campbell; Location: Mt. Broken Top in the Three Sisters Wilderness Area, Oregon
(forest) Susan Capalbo; Location: Ten-Mile Creek Road, Yachats, Oregon
(pasture) Kate Painter; Location: a farm near Colfax, Washington

About This Series

This report is published as one of a series of technical inputs to the Third National Climate Assessment (NCA) report. The NCA is being conducted under the auspices of the Global Change Research Act of 1990, which requires a report to the President and Congress every four years on the status of climate change science and impacts. The NCA informs the nation about already observed changes, the current status of the climate, and anticipated trends for the future. The NCA report process integrates scientific information from multiple sources and sectors to highlight key findings and significant gaps in our knowledge. Findings from the NCA provide input to federal science priorities and are used by U.S. citizens, communities and businesses as they create more sustainable and environmentally sound plans for the nation's future.

In fall of 2011, the NCA requested technical input from a broad range of experts in academia, private industry, state and local governments, non-governmental organizations, professional societies, and impacted communities, with the intent of producing a better informed and more useful report. In particular, the eight NCA regions, as well as the Coastal and the Ocean biogeographical regions, were asked to contribute technical input reports highlighting past climate trends, projected climate change, and impacts to specific sectors in their regions. Each region established its own process for developing this technical input. The lead authors for related chapters in the Third NCA report, which will include a much shorter synthesis of climate change for each region, are using these technical input reports as important source material. By publishing this series of regional technical input reports, Island Press hopes to make this rich collection of information more widely available.

This series includes the following reports:

Climate Change and Pacific Islands: Indicators and Impacts
Coastal Impacts, Adaptation, and Vulnerabilities
Great Plains Regional Technical Input Report
Climate Change in the Midwest: A Synthesis Report for the National Climate Assessment
Climate Change in the Northeast: A Sourcebook
Climate Change in the Northwest: Implications for Landscapes, Waters, and Communities
Oceans and Marine Resources in a Changing Climate
Climate of the Southeast United States: Variability, Change, Impacts, and Vulnerability
Assessment of Climate Change in the Southwest United States
Climate Change and Infrastructure, Urban Systems, and Vulnerabilities: Technical Report for the US Department of Energy in Support of the National Climate Assessment
Climate Change and Energy Supply and Use: Technical Report for the US Department of Energy in Support of the National Climate Assessment

Electronic copies of all reports can be accessed on the Climate Adaptation Knowledge Exchange (CAKE) website at *www.cakex.org/NCAreports*. Printed copies are available for sale on the Island Press website at *www.islandpress.org/NCAreports*.

Report Author Team

John T. Abatzoglou
Department of Geography, University of Idaho

Jeffrey Bethel[*]
College of Public Health and Human Sciences, Oregon State University

Susan M. Capalbo[#]
Department of Applied Economics, Oregon State University

Jennifer E. Cuhaciyan
Idaho Department of Water Resources

Meghan M. Dalton
Oregon Climate Change Research Institute, Oregon State University

Sanford D. Eigenbrode[*#]
Regional Approaches to Climate Change - Pacific Northwest Agriculture,
Plant, Soil, and Entomological Sciences, University of Idaho

Patty Glick[#]
National Wildlife Federation, Pacific Regional Center

Oliver Grah
Nooksack Indian Tribe

Preston Hardison
Tulalip Natural Resources, Treaty Rights Office

Jeffrey A. Hicke
Department of Geography, University of Idaho

Jennie Hoffman
EcoAdapt

Laurie L. Houston
Department of Applied Economics, Oregon State University

Jodi Johnson-Maynard
Plant, Soil, and Entomological Sciences, University of Idaho

Ed Knight
Swinomish Indian Tribal Community

Chad Kruger
Center for Sustaining Agriculture and Natural Resources, Washington State University

Kenneth E. Kunkel
Cooperative Institute for Climate and Satellites, North Carolina State University
NOAA National Climatic Data Center

Jeremy S. Littell[*#]
US Geological Survey, Alaska Climate Science Center

Kathy Lynn[*]
Pacific Northwest Tribal Climate Change Project, University of Oregon

Philip W. Mote[*#]
Oregon Climate Change Research Institute, Oregon State University

Jan A. Newton
Applied Physics Laboratory, University of Washington

Beau Olen
Department of Agricultural and Resource Economics, Oregon State University

Steven Ranzoni
College of Public Health and Human Sciences, Oregon State University

Rick R. Raymondi[*#]
Idaho Department of Water Resources

W. Spencer Reeder[*][#]
Cascadia Consulting Group

Amanda Rogerson
Pacific Northwest Tribal Climate Change Project, University of Oregon
Peter Ruggiero
College of Earth Ocean and Atmospheric Sciences, Oregon State University

Sarah L. Shafer
US Geological Survey, Geosciences and Environmental Change Science Center

Amy K. Snover[*][#]
Climate Impacts Group, University of Washington

Patricia Tillmann
National Wildlife Center, Pacific Regional Center

Carson Viles
Pacific Northwest Tribal Climate Change Project, University of Oregon

Paul Williams
Suquamish Tribe

[*] Northwest Report Chapter Lead Author
[#] Third NCA Northwest Chapter Lead Author

About this Report

Climate Change in the Northwest: Implications for Our Landscapes, Waters, and Communities is a report aimed at assessing the state of knowledge about key climate impacts and consequences to various sectors and communities in the Northwest United States. This report draws on two recent state climate assessments in Washington in 2009 (Washington State Climate Change Impacts Assessment; http://cses.washington.edu/cig/res/ia/waccia) and in Oregon in 2010 (Oregon Climate Assessment Report; occri.net/ocar) and a wealth of additional literature and research prior to and after these state assessments. As an assessment, this report aims to be representative (though not exhaustive) of the key climate change issues as reflected in the growing body of Northwest climate change science, impacts, and adaptation literature available at this point in time.

This report process co-evolved with the process to produce the Northwest chapter of the Third National Climate Assessment (NCA), specifically through a shared risk framework to identify key risks of climate change facing the Northwest. Beginning with a workshop in December 2011, scientists and stakeholders from all levels and types of organizations from all over the Northwest engaged in a discussion and exercise to begin the process of ranking climate risks according to likelihood of occurrence and magnitude of consequences. The risks considered were previously identified in the Oregon Climate Change Adaptation Framework. A summary of the workshop was submitted as a technical input to the NCA (http://downloads.usgcrp.gov/NCA/Activities/northwestncariskframingworkshop.pdf). This initial risk exercise was continued by the lead author team of the Northwest chapter of the Third NCA resulting in several informal white papers that were (1) condensed and synthesized into the Northwest chapter of the Third NCA and (2) expanded on and added to forming the present report.

We anticipate that this report will serve as (1) an updated resource for scientists, stakeholders, decision makers, students, and interested community members on current climate change science and key impacts to sectors and communities in Oregon, Washington, and Idaho; (2) a resource for adaptation planning, (3) a more detailed, foundational report supporting the key findings presented in the Northwest chapter of the Third NCA; and (4) a resource directing readers to the wealth of climate literature in the Northwest as cited in each chapter.

Organization of This Report

This report begins with an overview of the Northwest's varied natural and human systems (Chapter 1) followed by a description of observed and projected physical climate changes for the Northwest (Chapter 2), which together provides a context for understanding climate impacts within our geographically diverse region. The remainder of the report is organized by sectors of economic and cultural importance that are especially vulnerable to impacts of climate change. Key climate impacts and their consequences as well as adaptation measures and gaps in knowledge are described for freshwater resources and ecosystems (Chapter 3), coastal communities and ecosystems (Chapter 4), forest ecosystems (Chapter 5), agriculture (Chapter 6), human health (Chapter 7), and tribal communities (Chapter 8).

Partners

The production of this report was led jointly by representatives from the Pacific Northwest Climate Impacts Research Consortium (CIRC) and the University of Washington's Climate Impacts Group (CIG). Partners include the following federal, state, tribal, private, non-profit, university, and other organizations represented by the author team: National Oceanic and Atmospheric Administration (NOAA) National Climatic Data Center, US Geological Survey, US Department of Interior Alaska Climate Science Center, Idaho Department of Water Resources, Cascadia Consulting Group, National Wildlife Federation, EcoAdapt, Oregon State University, University of Idaho, University of Washington, Washington State University, University of Oregon, Nooksack Indian Tribe, Tulalip Natural Resources, Swinomish Indian Tribal Community, and the Suquamish Tribe.

Acknowledgments

Funding for the development of this report was provided by the NOAA Climate Program Office Regional Integrated Sciences and Assessment (RISA) program for the Pacific Northwest Climate Impacts Research Consortium (CIRC) (Grant #: NA10OAR431028). Additional funding was provided by the US Department of Interior Northwest Climate Science Center (Grant #: G1OAC00702). We recognize the support from all the organizations represented by the author teams. The editors and authors wish to thank the 27 reviewers for their time and effort providing thoughtful comments and suggestions that ultimately improved this report. The editors also wish to thank Kim Carson (Oregon Climate Change Research Institute) for her logistical and administrative assistance and Rachel Calmer (Oregon Climate Change Research Institute) for her editorial assistance. We also acknowledge the expert skill of Robert Norheim (geospatial analyst and cartographer with the Climate Impacts Group at the University of Washington) who produced all of the maps in this report.

Contents

CHAPTER 8 NORTHWEST TRIBES: CULTURAL IMPACTS AND ADAPTATION RESPONSES

Executive Summary

AUTHORS
Meghan M. Dalton, Jeffrey Bethel, Susan M. Capalbo, J. E. Cuhaciyan,
Sanford D. Eigenbrode, Patty Glick, Laurie L. Houston, Jeremy S. Littell, Kathy Lynn,
Philip W. Mote, Rick R. Raymondi, W. Spencer Reeder, Sarah L. Shafer, Amy K. Snover

Chapter 1 Introduction: The Changing Northwest

The Northwest's climatic, ecological, and socioeconomic diversity set the stage for a diverse array of climate impacts, many of which will be united by their dependence on availability of water and other natural resources. (Section 1.1)

Nestled between the Pacific Ocean and the Rocky Mountains, the Northwest (NW, fig. 1.1) experiences relatively wet winters and dry summers, with locations west of the Cascade Range considerably wetter than the sometimes desert-like conditions on the east side. In addition, the thousands of miles of NW coastline support a variety of coastal environments. On the whole, the Northwest's diverse climate and landscape make it one of the most ecologically rich areas in the United States, a feature that has been integral to sustaining the region's economy, culture, and way of life. NW tribes have cultural, social,

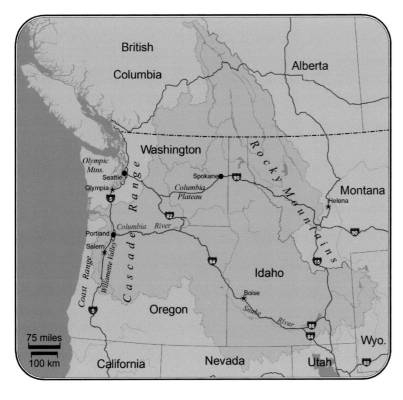

Figure 1.1 The Northwest, comprising the states of Washington, Oregon and Idaho and including the Columbia River basin (shaded).

and spiritual traditions that are inseparable from the landscape and environmental conditions on and beyond reserved tribal lands. The region's water resources and seasonality of snow accumulation and melt shape the migration of iconic salmon and steelhead; growth and distribution of forests; and availability of water for drinking, irrigation, and hydropower production, among many other uses. Land ownership, population distribution, economic and cultural dependence on natural resources, current ecological conditions, and patterns of resource use will substantially shape the regional and local consequences of a changing climate.

Key regionally consequential risks in the Northwest include impacts of warming on watersheds where snowmelt is important, coastal consequences of sea level rise combined with other stressors, and the cumulative effects of fire, insects, and disease on forest ecosystems. (Section 1.2)

This report focuses on the major drivers of regional climate change and impacts on systems of high regional and local importance. Three key issues of concern were identified through a qualitative risk assessment that evaluated the relative likelihood and consequences of climate change impacts for the region's economy, infrastructure, natural systems, and human health. These are: impacts of warming on snow accumulation and melt and their effects on regional hydrology and related systems; coastal consequences of sea level rise combined with other drivers of change, including river flooding, coastal storms and changes in the coastal ocean, and the cumulative effects of climate change on fire, insects, and tree diseases in forest ecosystems. In addition to these three risk areas, this report focuses on three climate-sensitive sectors of regional importance: agriculture, human health, and NW tribes. Regionally-identified risks are complemented with discussion of locally-specific risks and vulnerabilities.

This assessment of climate change in the Northwest reveals a familiar story of climate impacts, but highlights new details at multiple scales considering multiple interacting drivers of change and vulnerabilities resulting from human choices throughout time. (Section 1.3.1)

The findings presented in this report largely confirm over fifteen years of research, but add new details regarding how impacts are likely to vary across the region. Analyzing climate impacts at local to regional scales and how impacts vary between natural and managed systems is essential to ensure a complete picture of projected climate impacts on the region and development of appropriate adaptive responses. Considering multiple drivers of change and their interactions is also necessary as some of the largest impacts can occur when multiple drivers align and some individual drivers of change can offset each other. Past and present human choices and actions are a large determinant of current social and ecological vulnerability to climate; understanding these causal linkages and adjusting relevant choices and actions could help reduce future climate vulnerability.

The Northwest has been a leader in applied regional climate impacts science since the 1990s, and the region's resource managers, planners, and policy makers have been early engagers in climate change issues. This report provides a solid foundation for

identifying challenges posed by climate change in order to assist adaptation efforts throughout the region. (Section 1.3.2)

Climate change adaptation focuses on adjusting existing practices in order to reduce negative consequences and take advantage of opportunities. Adaptation begins with identifying and characterizing the problem posed by climate change, a goal this report aims to serve. It then proceeds with identifying, assessing, and selecting alternative actions, and ultimately implementing, monitoring, and evaluating the selected actions. Many federal, state, local, and tribal entities in the Northwest are already engaged in various stages of climate change adaptation, including state-level climate change response strategies; however, adaptation is not yet wide-spread and few efforts have moved beyond planning to implementation.

Chapter 2 Climate: Variability and Change in the Past and the Future

Variations in solar output, volcanic eruptions, and changes in greenhouse gases all contribute to the energy balance at the top of the atmosphere, which influences global surface temperature fluctuations and changes over time. (Section 2.1)

Global surface temperature is governed by the balance at the top of the atmosphere between incoming and reflected solar radiation and outgoing infrared radiation, or heat, radiated from the Earth. Clouds and certain gases in the atmosphere (e.g., water vapor, CO_2, methane, ozone, etc.) absorb some of Earth's radiated energy reducing the amount escaping to space. Changes in these infrared–absorbing gases (or more commonly, greenhouse gases) force a change in the energy balance of the climate system, with CO_2 changes being the dominant factor. Other important factors include changes in solar output and volcanic eruptions. Variations in solar output are partially responsible for changes in the past climate, but play a small role in climate changes today. Large volcanic eruptions act to cool the Earth for a few years afterward as tiny sunlight-reflecting particles spread throughout the upper atmosphere.

Climate variability and change in the Northwest is influenced by both global and local factors, such as the El Niño-Southern Oscillation and mountain ranges. (Section 2.2)

More important than global changes in the Earth's energy balance for understanding regional and local climate variability and change are the natural variability of atmospheric and ocean circulation and effects of local topography. NW climate variability is dominated by the interaction between the atmosphere and ocean in the tropical Pacific Ocean responsible for El Niño and La Niña. During El Niño, winter and spring in the Northwest have a greater chance of being warmer and drier than normal. The complex topography of the Northwest, which includes the Coast, Cascade, and Rocky Mountain ranges, results in large changes in temperature and precipitation over relatively short distances.

During 1895–2011, the Northwest warmed approximately 0.7 °C (1.3 °F) while precipitation fluctuated with no consistent trend. (Section 2.2)

For the last 30 years, temperatures averaged over the Northwest have generally exceeded the 20th century average. During 1895–2011, the Northwest warmed by about 0.7 °C (1.3 °F). Year-to-year fluctuations in precipitation averaged over the Northwest have been slightly larger since 1970 compared with the previous 75 years, with some of the wettest and driest years occurring in the most recent 40 years. However, there has not been a clear overall increase or decrease in average precipitation over the 20th century. The observed changes in temperature include contributions from both natural climate variability and human influences. Seasonal trends in temperature, while influenced by fluctuations in atmospheric circulation patterns, are consistent with expected changes from human activities.

The frequency of extreme high nighttime minimum temperatures increased in the Northwest during 1901–2009, but observed changes in extreme precipitation are ambiguous. (Section 2.3)

Confidently detecting changes in extreme events is challenging. During 1901–2009, the number of extreme high nighttime minimum temperatures increased in the Northwest, but other extreme temperature measures showed no clear change. Observed changes in extreme precipitation are ambiguous in most areas, with some increases and some decreases, and depend on the specific type of extreme precipitation event examined. Changes are most pronounced in western Washington where most measures show increases of 10–20%.

State-of-the-art global and regional climate modeling provides a consistent basis for understanding projections of future climate and related impacts in the Northwest. (Section 2.4)

Coordinated global and regional climate modeling approaches provide a framework for understanding uncertainty associated with model projections of future climate. Three such modeling frameworks are the Coupled Model Intercomparison Project phases 3 and 5 (CMIP3/5), the North American Regional Climate Change Assessment Program (NARCCAP), and regional climateprediction.net (regCPDN) with spatial resolutions ranging from 300 to 25 km (186 to 15 mi). All three datasets are generally consistent in the broad story of projected future NW climate.

The Northwest is expected to experience an increase in temperature year-round with more warming in summer and little change in annual precipitation, with the majority of models projecting decreases for summer and increases during the other seasons. (Section 2.4.1)

Over the period from 1950–1999 to 2041–2070, CMIP5 models project NW mean annual warming of 1.1 °C to 4.7 °C (2 °F to 8.5 °F), with the lower end possible only if greenhouse gas emissions are significantly reduced (RCP4.5 scenario; fig. 2.5 a). All models project warming of at least 0.5 °C (0.9 °F) in every season. Projected warming is greater during the summer with increases ranging from 1.9 °C to 5.2 °C (3.4 °F to 9.4 °F) for the very high growth scenario (RCP8.5). Annual average precipitation is projected to change by about +3% with individual models ranging from –4.7% to +13.5%. For every season, some models project decreases and others increases, although for summer more models project decreases than increases, with the largest projected change of about –30%

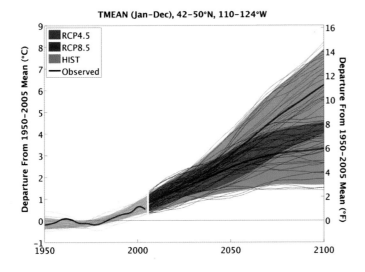

TMEAN (Jan–Dec), 42–50°N, 110–124°W

Figure 2.5. (a) Observed (1950–2011) and simulated (1950–2100) regional mean annual temperature for selected CMIP5 global models for the RCP4.5 and RCP8.5 scenarios.

by 2041–2070. In addition, the models that project the largest warming in summer also tend to project the largest precipitation decreases.

Measures of temperature and precipitation extremes are projected to increase in the Northwest. (Section 2.4.2)

Climate models are unanimous that measures of heat extremes will increase and measures of cold extremes will decrease. Averaged over the Northwest, NARCCAP results project that in the period averaged over 2041 to 2070 there will be more days above maximum temperature thresholds and fewer days below minimum temperature thresholds compared with the 1971–2000 average. For example, the number of days greater than 32 °C (90 °F) increases by 8 days (± 7), and the number of days below freezing decreases by 35 days (± 6). Future changes in precipitation extremes are more certain than changes in total seasonal precipitation. The number of days with greater than 1 in (2.5 cm) of precipitation is projected to increase by 13% (± 7%) and the 20-year and 50-year return period extreme precipitation events are projected to increase 10% (-4 to +22%) and 13% (-5 to +28%), respectively, by mid-century.

Chapter 3 Water Resources: Implications of Changes in Temperature and Precipitation

Changes in precipitation and increasing air temperatures are already having, and will continue to have, significant impacts on hydrology and water resources in the Northwest. (Section 3.1)

Such climate changes will alter streamflow magnitude and timing, water temperatures, and water quality. Hydrologic impacts will vary by watershed type. Snow-dominant watersheds are projected to shift toward mixed rain-snow conditions, resulting in earlier and reduced spring peak flow, increased winter flow, and reduced late-summer flow; mixed rain-snow watersheds are projected to shift toward rain-dominant conditions; and rain-dominant watersheds could experience higher winter streamflows if winter precipitation increases, but little change in streamflow timing (fig. 3.3). Such hydrologic impacts have important consequences for reservoir systems, hydropower production, irrigated agriculture, floodplain and municipal drinking water infrastructure, freshwater aquatic ecosystems, and water-dependent recreation.

Reduced snowpack and shifts in streamflow seasonality due to climate change pose an additional challenge to reservoir system managers as they strive both to minimize flood risk and to satisfy warm season water demands. (Section 3.2.1)

Reservoir systems in the Northwest rely heavily on the ability of snowpack to act as additional water storage. During the snowmelt season, reservoir managers face the challenge of simultaneously maximizing water storage for summer water supply and maintaining sufficient space for capturing floodwaters to minimize downstream flood risk. Earlier snowmelt and peak flow means that more water will run off when it is not needed for human uses and that less water will be available to help satisfy early summer water demand. Flood risk may decrease in some basins and will likely increase in others.

The Columbia River Basin, whose reservoir storage capacity is much smaller than its annual flow volume, is ill-equipped to handle the projected shift to earlier snowmelt and peak flow timing and will likely be forced to pass much of these earlier flows out of the system, under current operating rules. With reservoir drawdown starting earlier in the year, managers would be faced with complex tradeoffs between multiple objectives; namely, hydropower, irrigation, instream flow augmentation for fish, and flood control.

Due to earlier peak streamflow, summer hydropower generation is projected to decline, but winter hydropower generation may increase. (Section 3.3.2)

Hydropower production provides two-thirds of the region's electricity and the Northwest produces 40% of all US hydropower. The shifts in streamflow timing caused by reduced snowpack and earlier snowmelt will reduce the opportunity for hydropower generation in the late spring and summer. In one study, summer hydropower production is projected to decline by about 15% by 2040, while winter hydropower production may

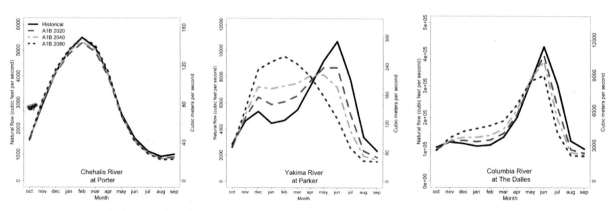

Figure 3.3. Simulated monthly streamflow hydrographs for the historical baseline (1916–2006 average, black) and the 2020s (blue), 2040s (yellow), and 2080s (red) under the SRES-A1B scenario of continued emissions growth peaking at mid-century (after Elsner et al. 2010) for three representative watershed types in the Northwest, namely rain dominant (Chehalis River at Porter, *top*), mixed rain-snow (Yakima River at Parker, *center*), and snowmelt dominant (Columbia River at The Dalles, *bottom*).

slightly increase (4%) compared to 1917–2006 levels. Further reductions in hydropower generation may also result from climate change adaptation for other competing water management objectives; for example, flood control and instream flow augmentation for fish.

Reduced water supply combined with increased water demands in the summer could lead to water shortages, reducing the proportion of irrigable cropland and the value of agricultural production. (Section 3.3.1)

Irrigated agriculture is the largest consumptive water user in the Columbia River Basin and poses the greatest demands on regional reservoir systems. Warmer, drier summers and a longer growing season may increase those demands. A case study in the Yakima River Basin projects the more frequent occurrence of conditions in which senior water right holders experience shortage. Water shortages could impact the proportion of cropland able to be irrigated during the growing season and lead to substantially reduced value of agricultural production; however, certain producer strategies may mitigate the shortage. Some evidence also suggests that increased atmospheric CO_2 concentrations may benefit water use efficiency in plants, possibly mitigating potential effects of drought.

Floodplain and municipal water supply infrastructure are vulnerable to projected increases in extreme precipitation and flood risk. (Section 3.3.3, 3.3.4)

Increases in extreme precipitation and flooding are expected, though changes in flood risk depend on the type of basin. Warmer winter temperatures and increased precipitation variability have already increased winter flood risk in mixed rain-snow basins in Washington and Oregon. Developed areas in floodplains may be particularly vulnerable to the increased flood risk, depending on flood control capacity. Water management may be stressed also by more frequent temperature extremes, warmer stream temperatures, lower summer flows, and the projected increase in municipal water demands. State and local government agencies in the Northwest are building strategies to address issues around how climate and hydrological change affects municipal water supply and use.

Changes in hydrologic flow regimes and warming stream and lake temperatures pose significant threats to aquatic ecosystems and are expected to alter key habitat conditions for salmon and other aquatic species. (Section 3.3.5)

Hydrologic changes in streamflow may harm the spawning and migration of salmon and trout species. Continued warming of stream and lake temperatures may also affect the health of and the extent of suitable habitat for many other aquatic species. Salmonids and other species that currently live in conditions near the upper range of their thermal tolerance are particularly vulnerable to higher stream temperatures, increasing susceptibility to disease and rates of mortality. Upstream migration for thermally-stressed species may be impeded by changes in channel structure from altered low-flow regimes. Reduced glacier area and volume over the long-term, which is projected for the future in the North Cascades, may challenge Pacific salmonids in those streams in which glacier melt comprises a significant portion of streamflow, although the consequences of glacial loss are not well quantified.

Water-dependent recreational activities will be affected by dry conditions, reduced snowpack, lower summer flows, impaired water quality, and reduced reservoir storage. (Section 3.3.6)

The sport fishing industry is likely to be affected by climate change effects on native fish including Pacific salmon. Mid-elevation ski resorts, located near the freezing elevation, will be the most sensitive to decreased snow, increased rain, and earlier spring snowmelt. The shortened ski-season will not only affect skiers, but the livelihood of local communities that are dependent on snow-recreation.

Chapter 4 Coasts: Complex Changes Affecting the Northwest's Diverse Shorelines

Sea level along the Northwest coast is projected to rise 4–56″ (9–143 cm) by 2100, with significant local variations. (Section 4.2)

Global mean sea level rose 0.12 in/year (3.1 mm/year) during 1993–2012, and there is high confidence that global sea level will continue to rise throughout the 21st century and beyond. Many local and regional factors modify the global trend in the Northwest. The active tectonics underlying western Oregon and Washington cause uplift in some locations, such as the Olympic Peninsula, at nearly the same rate as sea level rise resulting in little observed local sea level change, whereas subsidence in other locations leads to larger local sea level rise. End-of-the-century sea level rise projections along the NW coast range from 4 to 56 in (9–143 cm) relative to the year 2000, with variation in local factors adding to or subtracting from this range in different locations. Increasing wave heights in recent decades may have been a dominant factor in the observed increased frequency of coastal flooding along the outer coast. Regional sea levels can rise up to 12 in (30.5 cm) during an El Niño event, compounding impacts of sea level rise, but it is unknown whether and how El Niño-Southern Oscillation (ENSO) intensity and frequency may change in the future.

Increasingly acidified waters hinder the ability of some marine organisms to build shells and skeletons, which could alter key ecological processes, threatening our coastal marine ecosystems, fisheries, and aquaculture. (Sections 4.3, 4.5.3)

Anthropogenic additions of CO_2, seasonal coastal upwelling, and inputs of nutrients and organic matter combine to produce some of the most acidified marine waters worldwide along our coast; conditions in estuaries can reduce pH even further. Decreased abundance of shell forming species, many of which are highly vulnerable to ocean acidification, may alter the abundance and composition of other marine species. A simulation of ocean acidification impacts on the shelled species at the base of the marine food web resulted in a 20–80% decline of commercially important groundfish such as English sole. The rate at which mussels and oysters form shells is projected to decline by 25% and 10%, respectively, by the end of the century, and oyster larval growth rates are slower under low pH levels. Some species, such as sea grasses, may actually benefit from increased ocean acidification. Because of the serious implications of ocean acidification for marine species, several recent research initiatives have focused on identifying the impacts of ocean acidification in the Northwest.

Ocean temperatures off the Northwest coast have increased in the past and, though highly variable, are likely to increase in the future, causing shifts in distribution of marine species and contributing to more frequent harmful algal blooms. (Sections 4.4, 4.5.2)

Future increases in ocean temperature will continue to be highly variable and will affect the distribution of marine species found in NW coastal waters. Cooling of the eastern equatorial Pacific and ENSO-related changes in wind over the North Pacific may moderate warming of the northeast Pacific. Near coastal sea surface temperature (SST) varies by about 4–6 °C (7–11 °F) annually and is influenced by local coastal upwelling and downwelling and other weather and oceanographic-related factors. The range and abundance of Pacific Coast marine fish, birds, and mammals vary from year-to-year and serve as important indicators for potential fish species' responses to future climate change. For example, Pacific mackerel and hake are drawn to warmer coastal waters during El Niño events. One study found that long-term climate change, rather than climate variability, was the predominant factor in observed changes in the breeding and abundance of several seabird species in the California Current System. Blue whale and California sea lion habitats are projected to decrease over the 21st century, while northern elephant seal habitat is projected to increase. Increases in SST also contribute to more frequent and extended incidences of harmful algal blooms, increasing risks associated with paralytic shellfish toxins.

Coastal marine ecosystems in the Northwest provide important habitat for a diverse range of species. Coastal changes, such as sea level rise, erosion, and saltwater intrusion, could lead to loss or decline of some habitats, with impacts varying along the coast. (Section 4.5.1, Fig. 4.2.b)

Coastal wetlands, tidal flats, and beaches in low-lying areas with limited opportunity to move upslope (either by migrating inland or directly upwards by accumulating sediment) are highly vulnerable to sea level rise and coastal erosion, threatening the loss of key habitats and supported species. Significant beach erosion has occurred in north-central Oregon, where local sea levels have been rising, whereas southern Oregon beaches, where local sea levels have not risen, are relatively stable. Beach erosion increasingly exposes upland habitat to extreme tides and storm surges, affecting, for example, haulout sites used by harbor seals for resting, breeding, and rearing pups. Coastal freshwater marsh and swamp habitats are projected to convert to salt or transitional marsh due to increasing saltwater inundation, reducing the extent of tidal flats and estuarine and outer coast beaches and affecting associated species, such as shorebirds and forage fish. Sea level rise could reduce the extent of certain coastal marshes and riparian habitat used by juvenile Chinook salmon as they transition between freshwater and ocean life stages. Potential increases in surface and groundwater salinity, due to sea level rise, may affect coastal plant and animal species unable to tolerate such increases. Some coastal habitats may be able to accommodate moderate rates of sea level rise by migrating inland, provided that there are no barriers such as dikes and seawalls.

Sea level rise and flooding will affect Northwest coastal transportation infrastructure, though the degree of potential impacts will vary. (Section 4.6.1)

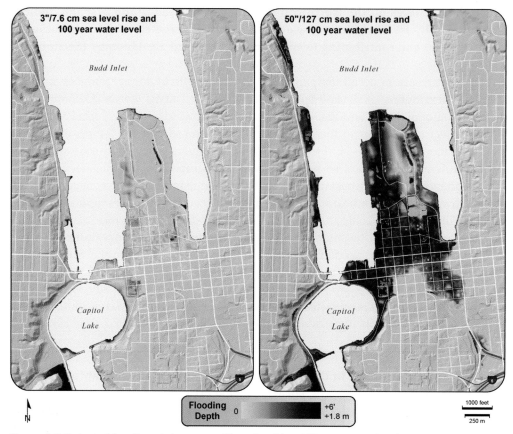

Figure 4.6 Projected flooding of downtown Olympia with a 100-year water level and 127 cm (50 in) of sea level rise. Redrawn from Coast and Harbor Engineering (2011).

About 2800 miles of roads in Washington and Oregon coastal counties are in the 100-year floodplain. The Washington State Department of Transportation assessed the climate change vulnerability of state-owned transportation infrastructure, identifying some outer coast and low-lying highways near Puget Sound that may face long-term inundation from 2 ft (0.6 m) of sea level rise. Most major state highways in Washington are situated high enough to experience only temporary closures. Highways near the mouth of the Columbia River near Astoria, Oregon, are also at risk. Inundation of low-lying secondary transportation routes in many coastal areas of the Northwest will very likely worsen and has the potential to temporarily cut off access to some communities during high tide and storm events.

Northwest coastal cities face multiple climate impacts and risks, including sea level rise, erosion, and flooding. Some local governments are evaluating and preparing for climate-related risks and vulnerabilities. (Section 4.6.2, Box 4.1)

The City of Seattle is assessing the vulnerability of its infrastructure to sea level rise and storm surge and is developing adaptation options. The City of Olympia is similarly examining areas of future exposure to inundation in the downtown core under various sea level rise and creek flooding scenarios (fig. 4.6), examining engineering and regulatory responses, and incorporating sea level rise response in their comprehensive planning process. The City of Anacortes has examined risks to their water treatment facility from projected increases in river flooding and resultant increases in sediment loading. The Swinomish Indian Tribal Community has examined a wide range of climate vulnerabilities and corresponding adaptation strategies and is incorporating assessment findings into ongoing regulatory and economic development efforts.

Climate driven changes in ocean conditions may have important economic impacts on marine fisheries, including shellfish aquaculture and fish landings. (Section 4.7.1)

Marine and coastal resources, particularly marine fisheries, provide communities in the Northwest with numerous economic benefits. The response of fish species to climate change will vary, so there may be both positive and negative economic impacts on commercial and recreational fisheries. Shellfish aquaculture, which provides many jobs and 49% and 72% of the commercial fishing landing value in Oregon and Washington, respectively, is threatened by ocean acidification. Climate–driven changes in the distribution, abundance, and productivity of key commercial species in Oregon and Washington could impact landings and revenues, which averaged around $275 million per year from 2000 to 2009.

Chapter 5 Forest Ecosystems: Vegetation, Disturbance, and Economics

The spatial distribution of suitable climate for many important Northwest tree species and vegetation types may change considerably by the end of the 21st century, and some vegetation types, such as subalpine forests, will probably become extremely limited. (Section 5.2)

Climate change is likely to affect the distribution, growth, and function of NW forests. Tree growth responses to future climate change will vary both within the region and in time with climate variability, but some locations are likely to experience higher growth (e.g., higher elevations) whereas other areas are likely to experience reduced growth (e.g., the lower elevation eastern parts of the Cascade Range). Forests limited by water availability will likely experience longer, more severe water-limitation under projected warming and reduced warm-season precipitation, resulting in decreased tree growth. Forests limited by energy or temperature will likely experience increased growth, depending on water availability. Area climatically favorable for Douglas-fir is projected to decrease by 32% by the 2060s in Washington in one study, but another study suggests that Douglas-fir may be able to balance loss of climate suitability at lower elevations with increases at higher elevations. Sub-alpine tree species are projected to decline and have limited potential to migrate upslope, resulting in potential loss of these high-elevation habitats, affecting associated wildlife and biodiversity. Vulnerability to disturbances is expected to increase in most forests.

Grasslands in some areas may expand under warmer and drier conditions, while sagebrush steppe habitat may transition to other vegetation (woodland or even forest)

depending on the amount and seasonality of precipitation change. Increased fire activity and expansion of invasive species will also determine the response of these systems to climate change. (Box 5.1)

Grassland and shrubland systems have already declined through land use and management changes, and the effect of future climate change will vary. Grass-dominated prairies and oak savannas in western parts of the Northwest are adapted to periodic drought and may expand under future warmer and drier conditions. Sagebrush steppe systems and associated species are sensitive to altered precipitation patterns and may decline, being replaced by woodland and forest vegetation. Expansion of new and current invasive species, both native (e.g., western juniper) and non-native (e.g., yellow starthistle), will influence the response of grassland and shrubland systems to climate change. Many grassland and shrubland systems are adapted to frequent fires, but projected increases in future fire activity threaten fire intolerant shrubs and the greater sage-grouse that depend on them for feeding, nesting, and protection.

The cumulative effects of climate change on disturbances (fire, insects, tree disease), and the interactions between them, will dominate changes in forest landscapes over the coming decades. (Sections 5.3, 5.3.4)

Large areas have been affected by disturbances in recent years (fig. 5.7), and climate change is expected to increase the probability of disturbance. The interaction between multiple disturbances (insect or disease outbreaks and wildfires) will heighten impacts on forests. The forests that establish after disturbance will depend on disturbance, climate, and other conditions that affect forest processes, though cumulative effects will vary. At least in the first half of the 21st century, climate change impacts on plant productivity, life history, and distribution are likely to be secondary to disturbance in terms of the area affected and risk presented to human values via altered forest ecosystem services.

Fire activity in the Northwest is projected to increase in the future in response to warmer and drier summers that reduce the moisture of existing fuels, facilitating fire. One study estimated that the regional area burned per year will increase by roughly 900 sq. mi. by the 2040s. (Section 5.3.1)

Climate influences both vegetation growth prior to the fire season and short-term vegetation moisture during the fire season, which influence fire-season activity. Fire activity in most NW forests tends to increase with higher summer temperature and lower summer precipitation. In one study, regional area burned is projected to increase by 0.3, 0.6, and 1.5 million acres by the 2020s, 2040s, and 2080s, respectively. Years with abnormally high area burned may become more frequent in the future: the chance of a given year being what was historically a "high" fire year is projected to increase by up to 30% for non-forested systems, 19% for the western Cascade Range, and 76% for the eastern Cascade Range. Greater fire severity is expected as increases in extreme events, particularly droughts and heat waves, will likely increase fire activity in the Northwest.

Recent mountain pine beetle and other insect outbreaks were facilitated by higher-than-average temperatures and drought stress, and the frequency and area of such

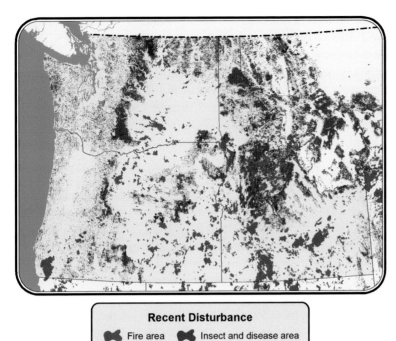

Figure 5.7. Areas of recent fire and insect disturbance in the Northwest.

Recent Disturbance
Fire area   Insect and disease area

outbreaks is expected to increase, particularly in high-elevation forests. Certain forest diseases, such as Swiss needle cast in Douglas-fir, are also expected to increase in the future. (Sections 5.3.2, 5.3.3)

Insect life-stage development and mortality rates are influenced by temperature, and drought can cause host trees to be more vulnerable to insects, leading to higher tree mortality. The frequency and area of mountain pine and spruce beetle outbreaks is expected to increase with future warming in the Northwest, particularly in high-elevation forests that are typically too cold to support the insect. Climate also influences the range and survival of forest pathogens, but the climate-disease relationship is unclear for many diseases and depends on pathogen-host interaction. Higher average temperatures and increased spring precipitation in the Oregon Coast Range have contributed to an increase in severity and distribution of Swiss needle cast in Douglas-fir, which is projected to have a greater impact in the future.

While the Northwest's forest economy is sensitive to climate changes, federal and state policies governing management and harvest have and will continue to impact the net returns to this sector, and the magnitudes of the impacts from policy changes and from climate change are difficult to separate. (Section 5.4.1)

The sustainability, net returns, and long-term future of the forest economy depend on the interaction of climate factors and management practices and policies. In the Northwest, while yields may increase due to a more favorable set of climate changes, leading to increased timber production, timber markets may be adversely impacted because of declining global prices and reduced net returns to timber producers. Timber yield losses due to regional insect and disease outbreaks and wildfires could also offset any potential economic benefit from increased growth in the Northwest. Furthermore, increasing severity and intensity of Swiss needle cast affecting the commercially and culturally important Douglas-fir could pose a threat to the NW timber industry west of the Cascade Range; the dominant commercial species east of the Cascade Range, ponderosa pine, is increasingly affected by mountain pine beetle and other insect and disease attacks, decreasing growth and yield.

Tourism and recreation on publicly owned lands (about two-thirds of Northwest forests) are important economically and socially in the region and may be affected by climate change. (Section 5.4.1.3)

Although no specific studies have been conducted on the NW economy, national scale estimates suggest forest recreation revenue losses of $650 million by 2060. Given the extent of forested and recreational land in the Northwest, along with projected increased risk of wildfire and decreased snowpack, impacts on the NW recreational economy will likely be negative. In the short-term, summer recreational opportunities in publicly owned forest land could increase due to lengthening of the high-use summer season, while winter recreational opportunities may decline. The local economies in drier regions of the Northwest could experience economic losses because of forest closures from wildfires.

Forest ecosystem services, such as flood protection or water purification, and goods, such as species habitat or forest products, add wealth to society and will be affected by climate change. (Section 5.4.1.4)

Valuing changes in these ecosystem goods and services is based on demand for these services. Changes in the demand of these services is influenced by many factors including land development, water demands, and air pollution, which all interact with climate change, making it difficult to isolate the impact of climate change on the value of ecosystem goods and services. However, values of some ecosystem goods and services in the Northwest have been estimated: water purification function of forests ($3.2 million per year); erosion control in the Willamette Valley ($5.5 million per year); cultural and aesthetic uses ($144 per household per year); and endangered species habitat ($95 per household per year).

Northwest forest ecosystems that will be affected by climate change support many species of fish and wildlife whose abundance and distribution may also be affected. (Section 5.4.2)

Wolverines and pika are particularly vulnerable to projected loss of alpine and subalpine habitat provided by snow cover and high-elevation tree species. Changes in fire regime could negatively impact old-growth habitat species, such as marbled murrelets and northern spotted owls, and affect stream temperatures and riparian vegetation important for spawning and juvenile bull trout. Some species, such as the northern flicker and hairy woodpecker, may thrive with more frequent fires. The effects of climate change may exacerbate existing stressors to natural systems.

Chapter 6 Agriculture: Impacts, Adaptation, and Mitigation

Agriculture is important to the Northwest's economy, environment, and culture. Our region's diverse crops depend on adequate water supplies and temperature ranges, which are projected to change during the 21st century. (Sections 6.1, 6.2, 6.3)

Agriculture contributes 3% of the Northwest's gross domestic product, crop and pastureland comprise about one-quarter of NW land area, and farming and ranching have been a way of life for generations. Wheat, potatoes, tree fruit, vineyards, and over 300 minor crops, as well as livestock grazing and confined animal feeding operations such as beef and dairy, depend on adequate supply of water and temperature ranges. Higher temperatures and altered precipitation patterns throughout the 21st century may benefit some cropping systems, but challenge others. Vulnerabilities differ among agricultural sectors, cropping systems (fig. 6.3), and location. Climate changes could alter pressure from pests, diseases, and invasive species. Available studies specifically examining climate change and NW agriculture are limited, and have focused on major commodities.

Projected climate changes will have mixed implications for dryland crops. Warmer, drier summers increase risks of heat and drought stress. At the same time, warmer winters could be advantageous for winter wheat and other winter crops, and increases in atmospheric CO_2 can improve yields at least until mid-century (Section 6.4.1.1)

Dryland cereal-based cropping systems occur mainly in the semiarid portion of central Washington and the Columbia Plateau of northeastern Oregon and northern Idaho. Winter wheat may benefit from warmer winters, but drier summers may delay fall planting of this crop. Increased winter precipitation could hamper spring wheat planting, but could also mitigate projected reductions in summer precipitation. Taking into account the beneficial effects of atmospheric CO_2, winter wheat yields are projected to increase 13–25% while spring wheat yields are projected to change by –7% to +2% by mid-21st century across several locations in Washington.

Irrigated crops are vulnerable to higher temperatures and projected water shortages from increasing demands and reduced supplies; potato yields are generally projected to increase with increasing atmospheric CO_2 to mid-century and decline to levels similar to or substantially below current yields by end of century. (Section 6.4.1.2)

The rivers of the Columbia and Klamath Basins provide irrigation water for surrounding agricultural areas that receive low summer and annual precipitation. Irrigation demands are expected to increase in the summertime with warmer temperatures, while water supplies are likely to be reduced, which could exacerbate water shortages in some areas, potentially reducing yields of irrigated wheat, potatoes, sugarbeets, forages, corn, tree fruit, and vegetable crops. Potatoes, grown under irrigation primarily in central Washington and the Snake River valleys of Idaho, are projected to experience yield losses from higher temperature, but when considering CO_2 fertilization, losses may only be 2–3% by the end of the century. Some studies project higher losses of up to 40% in Boise, Idaho.

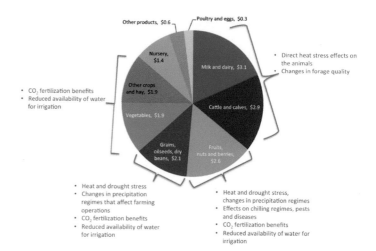

Figure 6.3. Northwest agricultural commodities with market values shown in $ (billion) in 2007. Potential effects of climate change on these sectors, if any have been projected, are shown.

Warmer winters could adversely affect tree fruits dependent on chilling for fruit set and quality. Tree fruits, most of which are produced with irrigation, are vulnerable to projected reduction in water supplies. Increased CO_2 may offset these effects; irrigated apple production is projected to increase 9% by the 2040s. (Section 6.4.2.1)

Payette County, Idaho, the Willamette Valley in Oregon, and central Washington are home to major tree-fruit production that requires irrigation and adequate chilling periods. Projected warmer temperatures that disrupt chilling requirements could hamper production of some existing tree fruits while allowing new cold-sensitive varieties to be grown. Under warming, irrigated apple production is projected to decrease by 3% in the 2040s, but increase by 9% when CO_2 fertilization is included. In addition, early budding from warmer spring temperatures could put trees more at risk to damage by frost. Tree fruits are water-intensive crops, making them vulnerable to projected reduced water supplies in some locations.

Northwest wine regions are already seeing an increase in the length of the frost-free period and warmer temperatures, which could adversely impact this growing industry. (Section 6.4.2.2)

Wine grapes are primarily grown in western Oregon and the Columbia River Basin. Each wine grape varietal has an optimal growing-season temperature range. Warmer temperatures could shift which varieties are produced in specific locations and alter wine quality. While some varietals, such as Pinot Noir and Pinot Gris (dominant grapes grown in Oregon), may experience temperatures in excess of optimal thresholds by mid-century, other varietals may become viable or more favorable in Oregon and Washington, although the cost of replacing long-lived vines must be considered.

Warming may reduce the productivity and nutritional value of forage in rangelands and pastures, though alfalfa production may increase as long as water is available. Higher temperatures can affect animal health, hampering milk production and beef cattle growth. (Section 6.4.3)

Grazing lands provide important ecosystem services. A warming climate may reduce productivity and nutritional value in rangelands located in warmer, drier climates while benefiting those in wetter environments. As long as water is not limiting, alfalfa production may increase in the Northwest under warmer temperatures and higher CO_2 concentrations. Climate change in rangeland systems may alter pressure from invasive species leading to degradation. Decreased availability, nutritional quality, and digestibility of forages, projected under higher CO_2 concentrations, may adversely affect livestock. Increased temperatures and extreme heat days can also affect animal health. Warmer temperatures can reduce milk production and decrease the rate of beef cattle growth, reducing the economic value of these products.

Agriculture is both impacted by and contributes to climate change. There are opportunities to reduce Northwest agriculture's contribution to climate change. (Box 6.1)
Opportunities to mitigate emissions in the Northwest include reducing tillage (which increases carbon storage in the soil), improving nitrogen fertilization efficiency to limit nitrous oxide production and release to the atmosphere (nitrous oxide is a greenhouse gas), and capturing methane emissions from manure. Mitigation strategies may have co-benefits that help with adaptation, sustainability and profitability of farming.

Northwest agriculture may be well positioned to adapt autonomously to climate changes due to the flexible nature of agriculture in responding to variable weather conditions and the relatively moderate projected impacts for the Northwest region. (Section 6.5)
Inherent adaptability varies by cropping system, with diversified systems potentially more adaptable than semi-arid inland wheat production and rangeland grazing. Agriculture's adaptive capacity is constrained by availability and time required for transitioning to new varieties, risk aversion among farmers, water availability in irrigation-dependent regions, and some economic, environmental, and energy policies. Partnerships and investments between public and private sectors have helped ensure agriculture remained strong in the preceding century and will be essential in the future.

Chapter 7 Human Health: Impacts and Adaptation

While the potential health impact of climate change is low for the Northwest relative to others parts of the United States, key climate-related risks facing our region include heat waves, changes in infectious disease epidemiology, river flooding, and wildfires. (Section 7.1)
Climate change in the Northwest will have implications for all aspects of society, including human health. Communities in the Northwest will experience the effects of climate change differently depending on existing climate and varying exposure to climate-related risks. While vulnerability remains relatively low in the Northwest, the negative impacts of climate change outweigh any positive ones. Increasing temperatures, changing precipitation patterns, and the possibility of more extreme weather could increase morbidity and mortality due to heat-related illness, extreme weather hazards, air pollution and allergenic disease exacerbation, and emergence of infectious diseases.

Average temperatures and heat events are projected to increase in the Northwest with an expected increase in incidence of heat-related illness and death (Section 7.2.1)

Heat-related deaths in the US have increased over the past few decades. In Oregon, the hottest days in the 2000s had about three times the rate of heat-related illness compared with days 10 °F (5.6 °C) cooler. Warmer temperatures and more extreme heat events are expected to increase the incidence of heat-related illnesses (e.g., heat rash, heat stroke) and deaths. One study projected up to 266 excess deaths among persons 65 and older in 2085 in the greater Seattle area compared to 1980–2006. Outdoor workers are especially vulnerable to heat-related illnesses.

People in the Northwest are threatened by projected increases in the risk of extreme climate-related hazards such as winter flooding and drought. (Section 7.2.2)

Wintertime flood-risk is likely to increase in mixed rain-snow basins in Washington and Oregon due to increased temperatures and, potentially, increased winter precipitation. Decreased summer precipitation and temperature-driven loss of snowpack can lead to more frequent drought conditions in the Northwest, leading to human health impacts due to food insecurity and associated wildfires. Drought can reduce global food supply and increase food prices, threatening food insecurity, especially for the poor and those living in rural areas of the Northwest. The 2012 US drought, one of the most extensive in 25 years with an estimated loss of up to $7–$20 billion, resulted in disaster declarations across the country, including counties in Oregon and Idaho.

Climate change can have a negative impact on respiratory disorders due to longer and more potent pollen seasons, increases in ground-level ozone, and more wildfire particulate matter (Section 7.2.3, 7.2.2)

Extended growing periods due to increased temperature can lengthen the pollen season and increase pollen production. Greater CO_2 concentrations can also heighten the potency of some pollens such as ragweed, found throughout the Northwest. A relatively small increase in ozone is expected for the Northwest (fig. 7.2) compared to other regions of the US, but increased ground-level ozone, or air pollution, can exacerbate asthma symptoms and lead to a higher risk of cardio-pulmonary death. Excess deaths due to ground-level ozone between 1997–2006 and mid-21st century are projected to increase from 69 to 132 and 37 to 74, respectively, in King and Spokane counties in Washington under a scenario of continued emissions growth (SRES-A2). The Northwest is expected to experience more burned acres during the wildfire season, releasing more particulate matter into the air. Wildfire risk is greatest east of the Cascade Range, but all population centers in the region are at risk of poor air quality from drifting smoke plumes, which could exacerbate respiratory disease.

Changes in climate can potentially impact the spread of vector-borne, water-borne, and fungal diseases, raising the risk of exposure to infectious diseases. (Section 7.2.4)

Longer, drier, warmer summers in the Northwest may have a significant impact on the incidence of arboviruses, such as West Nile virus. Higher ocean and estuarine temperatures in the Northwest have the potential to increase the number of *Vibrio parahaemolyticus* infections from eating raw oysters or other shellfish. Anticipated increases in

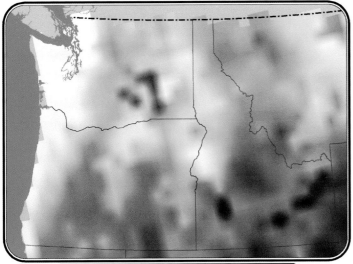

Figure 7.2. Change in summer averaged daily maximum 8-hr ozone mixing ratios (ppbv) between a future case (2045–2054) and base case (1990–1999) based on future climate from a model forced with the continued growth emissions scenario (SRES-A2). Changes in ground-level ozone are due to global and local emissions, changes in environmental conditions and urbanization, and increasing summer temperatures. Adapted from Chen et al. (2009).

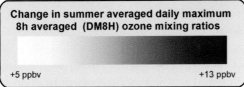

Change in summer averaged daily maximum 8h averaged (DM8H) ozone mixing ratios

+5 ppbv   +13 ppbv

precipitation and subsequent flooding have the potential to wash animal intestinal pathogens into drinking water reservoirs and recreational waters, potentially increasing the risk of *Cryptosporidium* outbreaks. The emergence of the fungus *Cryptococcus gattii* in the Northwest in the early 2000s may have some relationship to climate change.

Longer harmful algal blooms increase the risk of paralytic shellfish and domoic acid poisoning in humans. (Section 7.2.5)

The frequency, intensity, and duration of harmful algal blooms appear to be increasing globally, but the exact relationship to climate change is unknown. In the Puget Sound, rising water temperatures promote earlier and longer lasting harmful algal blooms, which can cause paralytic shellfish and domoic acid poisoning in humans who consume infected shellfish.

Climate change may affect mental health and well-being. (Section 7.2.6)

Like physical health impacts, there are direct and indirect mental health impacts of climate change. Extreme weather events can cause mental distress, and even the threat of a climate event, the uncertainty of the future, or the loss of control over a situation can result in feelings of depression or helplessness.

Public health practitioners and researchers in the Northwest are actively engaging local communities regarding adaptation measures for climate change. Additional efforts

are needed to engage a greater number of communities and build our understanding of how climate change will affect human health. (Sections 7.3, 7.4)

Public health officials, universities, and state agencies in the Northwest are engaged in numerous adaptation activities to address the potential impact of climate change on human health by developing public health adaptation resources, integrating planning at various government levels, and creating programs to monitor and respond to public health issues. Even some local health departments are creating their own climate change adaptation plans. In order to better understand the full impact of climate change on human health and for communities to effectively adapt, several needs must be addressed including accurate surveillance data on climate-sensitive health and environmental indicators.

Chapter 8 Northwest Tribes: Cultural Impacts and Adaptation Responses

Northwest tribes are intimately connected to the land's resources, and are tied to their homelands by law as well as by culture. The impacts of climate change will not recognize geographic or political boundaries. (Sections 8.1, 8.2)

Climate change will have complex and profound effects on tribal resources, cultures, and economies. In ceding lands and resources to the US, tribes were guaranteed the rights to hunt, fish, and gather on their usual and accustomed places both on and off reservation lands (fig. 8.2). Climate change could potentially affect these treaty-protected rights. For example, treaty-protected fish and shellfish populations may become threatened or less accessible to tribes due to climate change. Treaty water rights could also be affected by climate change through changes to water quantity and quality that affect salmon and other fisheries.

Reduced snowpack and shifts in timing and magnitude of precipitation and runoff could significantly affect culturally and economically important aquatic species, such as salmon. (Section 8.3.1, Box 8.1)

Salmon are culturally and economically significant to inland and coastal tribes throughout the Northwest. Spring Chinook salmon that spawn in the Nooksack River watershed, for example, are especially important to the Nooksack Indian Tribe for ceremonial, commercial, and subsistence uses. Past land-use practices have resulted in loss of fish habitat in the Nooksack watershed; observed changes in climate, such as decreased summer flows, increased stream temperatures, and higher peak winter flows, exacerbate the existing stressors that affect the migration and spawning of Chinook and other Pacific salmonids. Continued climate change will further challenge salmonid survival, highlighting the need for effective restoration strategies that consider both existing stressors and those added by climate change.

Increasing ocean acidification, hypoxia, and warmer air and water temperatures threaten many species of fish and shellfish widely used by tribes. (Section 8.3.2)

In the Puget Sound, fish and shellfish harvests are primary sources of income for tribal members. The health of these fisheries depends on how they are managed and the

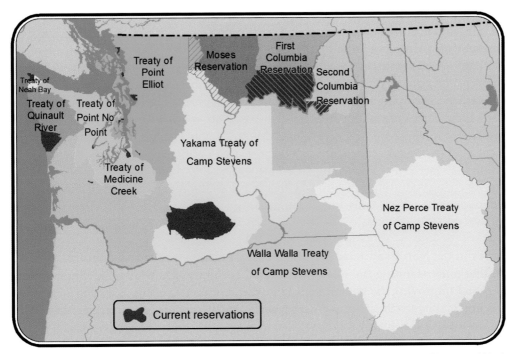

Figure 8.2. Treaty Ceded Lands. Washington State Historic Tribal Lands (Tribal Areas of Interest. Washington Department of Ecology)

health of the waters and ecosystems they inhabit. Decreasing pH is already associated with observed declines in the abundance and mean size of mussels from Tatoosh Island on the Makah Reservation in Washington. Warmer air temperatures have led to a decrease in the vertical extent of the California mussel in the Strait of Juan de Fuca.

Tribal coastal infrastructure and ecosystems are threatened by sea level rise, storm surge, and increasing wave heights. (Section 8.3.3)

Rising seas threaten culturally important areas of coastal tribes' homelands, such as burial grounds and traditional fishing and shellfish gathering areas, as well as infrastructure in low-lying areas. Small coastal reservations may face tension between allowing coastal habitat to shift inland (to limit habitat loss from sea level rise) and maintaining space for land-based needs and infrastructure.

Changes in forest ecosystems and disturbances will affect important tribal resources. (Section 8.3.4)

Projected changes in large-scale tree distribution across the Northwest, including those already occurring such as northward and elevational migration of temperate forests, will affect resources and habitats that are important for the cultural, medicinal, economic, and community health of tribes. Compounding impacts from forest disturbances, including wildfires and insect outbreaks, also pose a threat to traditional foods, plants, and wildlife that tribes depend on.

There are numerous tribes in the region pro-actively addressing climate change and bridging opportunities with non-tribal entities to engage in climate change research, assessments, plans, and policies. (Section 8.4)

There are many tribes in the Northwest pro-actively addressing climate change through a myriad of efforts. The Swinomish Indian Tribal Community showed early innovation in developing a tribal climate change impacts assessment and adaptation plan. The Tulalip Tribes are taking an ecosystem-based approach to understand and address interrelated changes in local ecosystems due to climate change. The Suquamish Tribe is engaged in federal, state, and academic research partnerships to study the effects of pH on crab larvae and is creating an online database to direct teachers to high quality climate change education materials. Other tribes in the region have initiated efforts to reduce greenhouse gas emissions through energy efficiency, renewable energy sources, and carbon sequestration.

Tribes in the Northwest have identified climate change needs and opportunities, including understanding the role of traditional ecological knowledge in climate initiatives and improving the government-to-government relationship. (Section 8.5)

Vulnerability to climate change and tribal adaptation strategies require explicit attention because of the unique social, legal, and regulatory context for tribes. It will be important for future climate research and policies to consider how reserved rights, treaty rights, and tribal access to cultural resources will be affected by climate change, potential species and habitat migration, and implementation of adaptation and mitigation strategies. Traditional knowledge can inform tribal and non-tribal understanding of how climate change may impact tribal resources and traditional ways of life. Strengthening government-to-government relationships is important in order to protect tribal rights and resources in the face of climate change, as is effective communication, collaboration, and federal-tribal partnerships.

Chapter 1

Introduction
The Changing Northwest

AUTHORS
Amy K. Snover, Patty Glick, Susan M. Capalbo

Human influences on climate, already apparent at the global and continental scales (IPCC 2007), are projected to alter the climate, ecology and economy of the Northwest (NW). Despite large natural variations, changes in regional temperature, snowpack, snow-melt timing, and river flows have already been observed that are consistent with expected human-caused trends (Mote 2006; Pierce et al. 2008; Stewart et al. 2005; Hidalgo et al. 2009; Luce and Holden 2009). With 21st century rates of global and regional warming projected to be at least double those observed in the 20th century (IPCC 2007; Mote and Salathé 2010; see Chapter 2), these changes are expected to continue even as new changes emerge.

Climate change is projected to alter environmental conditions across the region, affecting the Northwest's natural resource base and changing habitat conditions for fish and wildlife. The regional consequences of climate change will pose new risks to health, safety, and personal property, alter the reliability of transportation interconnections, and drive changes in local and regional economies. More fundamentally, these changes mean that many of the climatic assumptions inherent in decisions, infrastructure, and policies across the Northwest—from where to build, to what to grow where, to how to manage variable water resources to meet multiple needs—will become increasingly incorrect.

Many of the changes set in motion are unavoidable, caused by greenhouse gases already emitted (Solomon et al. 2009), though they may be temporarily obscured by the Northwest's highly variable climate (Hawkins and Sutton 2009; Deser et al. 2012). What risks will a changing climate bring for the region as a whole and for specific sectors and locations? What strategies are emerging for evaluating and altering management of regional water and energy supplies, infrastructure, transportation, health, and ecological and agricultural systems to address these risks? To what extent is the region preparing?

This report synthesizes currently available information to provide answers to these questions. It focuses on impacts that matter for the region as a whole, chosen with an eye toward the likely major drivers of regional change and consequences of highest regional and local importance. It is an assessment of existing knowledge that builds on and augments previous assessments (e.g., Climate Impacts Group 2009, Oregon Climate Change Research Institute 2010) and draws on a wealth of resources from local government and state agency reports to academic peer-reviewed journal articles. It is intended to be a resource for preparing the Northwest for climate change.

While we can do our best to discern the most likely consequences of climate change for NW ecosystems and communities, the ultimate consequences of the changes now in motion remain partially contingent on future societal actions and choices. Whether the consequences of the climate impacts outlined in this report are severe or mild depends in part on the degree to which regional social, economic, and infrastructural systems are adjusted to align with the changing climate, and the degree to which natural systems are provided with the room, flexibility, and capacity to respond. The regional consequences of climate change will also be strongly shaped by past choices—of what to build where, what to grow where—and by the laws, institutions, and procedures that shape how natural resources are managed and allocated, risks from natural hazards are identified, and trade-offs among conflicting objectives resolved.

This chapter sets the stage for the detailed, sector-specific examination of climate risks that follow. It provides an introduction to the physical, ecological, and economic characteristics of today's Northwest, describes the risk assessment methods used to prepare this report, comments on common themes about future change that cross all sectors, and describes the current state of regional preparation for climate change.

Figure 1.1
The Northwest, comprising the states of Washington, Oregon and Idaho and including the Columbia River basin (shaded).

1.1 Regional Introduction: The Physical, Ecological, and Social Template

Bordering Canada and the Pacific Ocean, comprising the states of Washington, Oregon, and Idaho, and including boundary-spanning watersheds like the Columbia River basin (fig. 1.1), the Northwest is a region defined in large part by its landscape and abundant natural resources. With craggy shorelines, volcanic mountains, and high sage deserts, the Northwest's complex and varied topography contributes to the region's rich climatic, ecological, and social diversity. Natural resources—timber, fisheries, productive soils, and plentiful water—remain important to the region's economy and strong connections to the environment are common. These regional characteristics set the stage for current and future regional climate vulnerabilities.

1.1.1 LANDSCAPE AND CLIMATE

Lying between the Pacific Ocean and the Rocky Mountains and punctuated by the Cascade and Olympic mountain ranges, the Northwest experiences a Mediterranean-type climate with relatively wet winters and dry summers. The mountains enhance winter precipitation, with some of the wettest locations in North America found on the west slopes of the Olympic Mountains where annual precipitation over 16.4 feet (5 meters) of water equivalent, supports the region's dramatic coastal temperate rainforest. In contrast, the lee side of the Cascade Range is much drier, with desert-like conditions occurring on the high plateau of the interior Columbia Basin where annual precipitation can be less than 8 inches (20.3 cm) (Jackson and Kimmerling 1993; see fig. 2.1).

With 453 miles (729 km) of coastline and 4,436 miles (7139 km) of tidal shoreline (including Puget Sound and the Columbia River estuary; US Department of Commerce et al. 2009), the NW coast spans seven degrees of latitude. Coastal mountains, strong Pacific currents, and diverse coastal landforms—including rocky shores, hilly headlands and sandy beaches, broad coastal plains, and barrier beaches and dunes—create varied and diverse coastal environments next to some of the most productive coastal waters in the world (Good 1993).

1.1.2 ECOSYSTEMS, SPECIES, AND HABITATS

Together, Washington, Oregon, and Idaho constitute one of the most ecologically rich areas in the United States, reflecting the region's topographically induced climatic diversity. The region contains diverse species and habitats, ranging from the sage grouse and pygmy rabbits that rely on the shrub steppe habitats of southern Idaho and the Columbia Plateau, to the subalpine fir and mountain hemlock forests of the Cascade and Olympic Mountains; from iconic trout, salmon, and steelhead that spawn in lakes and streams across the region, to the seabirds, orca whales, and shellfish that inhabit the coastal and marine waters of Puget Sound and the Pacific Ocean. The 'California Current', running along the Pacific West Coast from southern British Columbia to southern California, brings cooler marine waters southward and is linked to an upwelling of nutrient-rich sub-surface waters that supports abundant seabirds, marine mammals, and fisheries, including Dungeness crab, Pacific sardines, Chinook salmon, albacore tuna, and halibut.

Although these diverse ecological resources have been integral to sustaining the region's economy, culture, and way of life, human activities have significantly altered many NW ecosystems, causing habitat fragmentation, degradation and loss, and species decline, and for NW tribes, significant cultural losses (see section 1.1.4). For example, Oregon's Willamette Valley, now among the state's most densely populated areas, retains only about 4% of the prairies and savannas that covered 49% of the area at the time of Euro-American settlement (Hulse et al. 2002). Over 70% of the Northwest's original old-growth conifer forest has been lost, mainly through logging and other development (Strittholt et al. 2006). Coastal habitat degradation is significant in Washington's Puget Sound, where over 50% of the central Puget Sound shoreline has been modified (Washington Biodiversity Council 2007) and three quarters of saltwater marsh habitat has been eliminated (Puget Sound Partnership 2012). Construction of dams and reservoirs has altered natural streamflow patterns on many of the region's rivers, one of several factors contributing to the rapid decline of NW wild salmon populations (Cone and Ridlington 1996; NRC 1996; Lichatowich 1999), resulting in extinction of several salmon populations and the listing of 19 species of salmon and steelhead as threatened or endangered under the Endangered Species Act. In all, the region has 71 species of plants and animals listed under the Act (FWS 2013) and dozens of invasive plants, animals, and insects, causing an array of ecological challenges (Ray 2005; Eissinger 2009; Eastern Forest Environmental Threat Assessment Center 2013; USGS 2013; EDDMaps 2013).

As is the case across the nation, protecting the region's wildlife and natural habitats has been a challenge in the face of growing pressures from urban and industrial development, agriculture, and natural resource extraction. Climate change is projected to exacerbate and intensify many of these existing problems, resulting in new sets of impacts and stressors on NW ecological systems.

1.1.3 POPULATION AND ECONOMY

The region's population is concentrated west of the Cascades, with the region's major urban centers and about 60% of the region's 12 million residents along the Interstate 5 corridor in Washington's Puget Sound lowlands and Oregon's Willamette Valley (fig. 1.2; OR: US Census Bureau 2010b, ID: US Census Bureau 2010a, WA: Washington OFM 2010). With the exception of a handful of interior population centers, the largest being Spokane, Washington (population: 208,916, US Census Bureau 2010d) and Boise, Idaho (population: 205,671, US Census Bureau 2010c), the remainder of the region has relatively low population density of about 44 persons per square mile (17 per square hectare) (US Census Bureau 2010a, 2010b; Washington OFM 2010). Washington and Oregon's (Pacific Ocean) coastal counties are also sparsely populated; the largest town on the Oregon Coast is Coos Bay, population 30,000 (Foushee 2010). During the last several decades, the Northwest has undergone population and economic growth at nearly twice the national rate. The NW population has nearly doubled since 1970 (Foushee 2010) and is expected to grow nearly 50% in the next three decades (Oregon Office of Economic Analysis 2013; Washington OFM 2012; US Census Bureau 2012).

Low population density in much of the Northwest reflects the relatively high percentage of land that is mountainous, in public ownership (fig. 1.3), and/or in agricultural

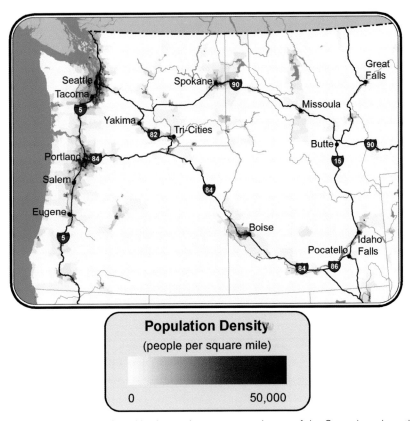

Figure 1.2 The population of the Northwest is concentrated west of the Cascades, along the Interstate 5 corridor in the Puget Sound lowlands of Washington and the Willamette Valley of Oregon. Data source: US Census Bureau, www.census.gov, accessed May 2, 2013.

usage. The fraction of land in public (federal and state) ownership is about 70% in Idaho, 50% in Oregon, and 38% in Washington. With over 31 million hectares (76 million acres) in federal ownership, the US Forest Service and Bureau of Land Management are the region's the major landowners (Pease 1993). Approximately 24% of the land area of Idaho, Oregon, and Washington states is devoted to agricultural crops or rangeland and pastureland (US Department of Agriculture Census of Agriculture 2010), predominantly in the interior Columbia Basin and the Willamette Valley (see fig. 5.1 and fig. 6.2).

Along the Northwest's diverse coastline, regional economic centers are juxtaposed with diverse habitats and ecosystems that support thousands of species of fish and wildlife, with commercial fish and shellfish landings valued at $480 million in 2011 (National Marine Fisheries Service 2012). The shores of Puget Sound alone contain forests, farms, commercial shellfish beds, American Indian tribal lands, urban landscapes, military installations, wetlands, bluffs, and beaches. Communities involved in marine fishing and harvesting are found along both the outer coast and the inner tidal shoreline (including Washington's Puget Sound and the region's largest inland

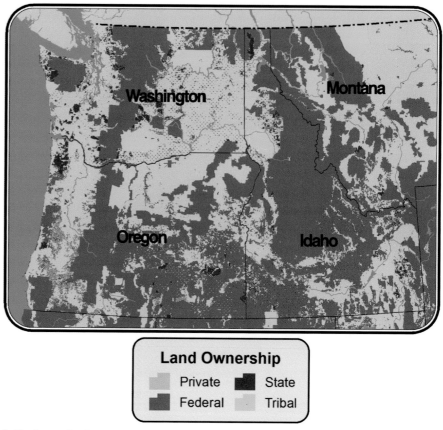

Figure 1.3 Northwest land ownership. Data sources: State lands–Interior Columbia Basin Ecosystem Management Plan, ICBEMP.gov; Federal and tribal lands–USGS, National Atlas, NationalAtlas.gov.

waterway, the Columbia River) where the region's only major metropolitan "coastal" cities are located.

From the standpoint of the region's total economy, the natural resources sectors continue to decline in importance, compared to the major regional economic drivers of software, telecommunications, aerospace, biotech, manufacturing, transportation, and defense. Even in many rural areas, the economic contribution of natural resource sectors has waned. For example, although agriculture, timber, and fisheries remain components of Oregon's coastal economies, at 2%, 9% and 5% of total earned personal income, respectively, in 2003, they were dwarfed by the 46% of income deriving from investments and transfers (social security and other government assistance), primarily from retirees (OCZMA 2006). In absolute numbers, the value of the natural resources-related components of the NW economy remains large. The forest industry contributes $12.7 billion to Oregon's economy each year (Oregon Forest Resources Institute 2012) and 15% of Washington's manufacturing jobs (Washington State Department of Natural Resources 2007), while Idaho's wood and paper industries account for nearly one-fifth of all the labor income generated in the state (Idaho Forest Products Commission 2012).

Agriculture remains a significant contributor to regional and rural economies and cultures, and a major regional employer; agricultural commodities constituted 3% of the region's GDP, i.e., $17 billion (US Department of Agriculture Census of Agriculture 2010).

Although the demographic implications of climate change remain highly speculative (e.g., the likelihood of significant in-migration of climate refugees), climate change will clearly affect regional and local economies, through its influence on not only regional natural resources, but maintenance and repair of public infrastructure and private

---

Box 1.1

Assessing the Economic Impacts of Climate Change: A Commentary and Challenge

When it comes to thinking about the regional impacts of climate change, the big unanswered questions include: How big is the problem? What will climate change cost regional and local economies? How much could these costs be reduced by adaptation actions and/or policy interventions?

Answers to these questions are essential for characterizing rational and effective adaptive responses and policies, and for prioritizing adaptation efforts and investments. But relatively few answers are available, at either the regional or national levels. Why isn't there more information on the costs likely to be associated with a changing climate? What do we need in order to develop more comprehensive estimates?

Understanding and quantifying the economic implications of a changing climate are complex and challenging, requiring information about the magnitude of local impacts of climate change, including information on changes in natural assets and services, on behavioral responses to these climate-induced changes, and on changes that may occur beyond the region of focus. For example, information about changes in water availability and timing as a result of a changing climate is criti-cal, but must be supplemented with information about the likelihood and economic feasibility of behavioral responses (adaptation alternatives) in order to quantify the economic dimensions of the change. For completeness, information about how climate impacts and behavioral responses shape regional and global markets and costs should

also be included, but few assessments include the types of systems modeling and projections this would require.

Across the board, valuation of climate-induced changes is made more difficult by the expected heterogeneity of climate impacts; because impacts will vary across different physical and ecological systems, and across different sectors and industries, the economic valuation of those impacts will vary. Some sectors and geographic areas may benefit, while others will be adversely affected.

But there's an even bigger challenge. Not only must assessments of the net economic costs of climate change consider impacts that will differentially affect regional economic activity, via changes in the production of goods, in the rates of infrastructure damage and loss, etc., they should also include impacts that will affect the provision of ecosystem services, commonly referred to as changes in natural capital.

Recognizing that the environment provides a range of services that have value to humans today and in the future leads to a characterization of the environment as a form of natural capital or natural assets, and thus, similar to other forms of capital such as financial wealth, education, physical infrastructure, it generates current and future flows of income or benefits. However, unlike more familiar types of capital, our environmental assets are dynamic and understanding natural rates of growth are critical for understanding

Box 1.1 (Continued)

trade-offs over time. And as Barbier and Markandya (2013) note, regardless of whether or not there exists a recognized market for the services from the environmental asset, the asset's value is the discounted net present value of these income or benefit flows. Assessing how these future income or benefit flows change with climate change is fundamental to quantifying the full range of economic impacts. If the value of natural capital is enhanced, the changes due to climate would benefit society, and conversely, if the value is diminished, then the changes would adversely impact society. Under the latter case, society may be able to adapt to these declines to lessen the negative impacts.

The challenge, then, is to understand and track the magnitude of the changes to our environmental and natural assets that are projected to occur in the future, to link the impacts of those changes on the ecosystem services and the resulting net present values of the benefits from these assets, and to quantify the opportunities (and costs of these actions) that may be available to partially offset any projected disinvestments in the environmental asset. The valuation challenge for natural capital stems fundamentally from the lack of direct markets (and prices) for the services of these assets, and limited information regarding the changes in these assets over time. But ignoring the changes in these net present values in our assessments of climate change, whether by default or ignorance, is paramount to assuming that the value of these changes are zero. Underpricing (or zero-pricing) these assets will result in management decisions that overuse and thus degrade the stock of natural capital.

So where do we stand? A limited number of empirical studies have focused on quantifying the economic market impacts. Less has been done to use existing economic information to assess the non-market and non-consumptive uses of ecosystems or to project the long-term implications of climate change for natural capital stocks. Many studies sidestep the issue of projecting a behavioral response to climate changes and quantify economic impacts based on business-as-usual scenarios. While these scenarios provide useful information, they should be viewed as an upper bound on "costs" or impacts of climate change: these are the resulting impacts assuming that people do not adapt or respond to different economic or biophysical scenarios, and they tend to ignore any spatial variations in the impacts. Existing economic assessment studies in Washington and Oregon (Climate Leadership Initiative 2006, 2009, 2010) have taken this approach, likely overstating the economic costs associated with conventionally measured economic activity in the region and at the same time ignoring the additional costs of the continued disinvestment in the region's natural capital.

Economists, ecologists, and other scientists have made some progress in addressing the challenges associated with costing climate impacts; the information in this report is testimony to these efforts. However, assessing the (non-market) value of changes in these environmental services is essential for finding the desired balance among conservation, sustainability, and development over time. Applying sound economic principles and values to these changes rightly conveys to society that these services contribute to our well-being and that the disinvestments are real. As noted by Heal (1998), "we are coming to realize, in part through the process of losing them, that environmental assets are key determinants of the quality of life . . ."

property, impacts on regional transportation and interconnectivity, and less tangible impacts on non-market ecosystem services and environmental amenities. Many questions remain about the overall economic consequences of climate change; current understanding is highlighted in subsequent chapters and some of the challenges associated with quantifying these costs explored in box 1.1.

1.1.4 NORTHWEST TRIBES

Indigenous peoples have lived in the region for thousands of years, developing cultural and social customs that revolve around traditional foods and materials and a spiritual tradition that is inseparable from the environment. Today, 43 federally-recognized American Indian tribes have reserved lands within the region (see fig. 8.1). Each has a unique history and relationship with the landscape and environment of the Northwest, and yet many are united by their connection to the plants, animals, and habitats of the region.

Throughout history, tribes have maintained geographically bounded rights to natural resources and heritage that occur both on their reservations and on off-reservation lands (Gates 1955; Ovsak 1994). These off-reservation, "usual and accustomed", areas stretch across the region (see fig. 8.2).

Northwest tribes can have a significant role in natural resource management beyond management of tribal reserved lands. Government-to-government relationships between tribes and state or federal resource management agencies enable consulting on agency management plans and, in some instances, co-management of natural resources. For fisheries, this can involve collaboration in setting conservation goals and harvest limits, in-season management, monitoring and assessment, and hatchery management. For example, state and tribal representatives participate in the Pacific Fishery Management Council that sets annual fisheries levels for groundfish and salmon fishing in federal waters from 3 to 200 miles (~5 to 320 km) off the coasts of Washington, Oregon, and California.

Climate change has the potential to affect tribal hunting, fishing, and gathering rights through changes in tribally important species and habitat on a wide variety of lands. As a result, NW tribes are becoming increasingly involved in climate change research, assessment, and adaptation efforts (see Chapter 8).

1.1.5 A REGION SHAPED BY WATER

The seasonal cycle of water availability—winter delivery of rain and snow, spring snowmelt, and relatively dry summers—has shaped the region's ecosystems, economies, and infrastructure, affecting what grows where, who lives where, and which strategies are employed for water management and use. As climate change alters these patterns—shifting the balance between rain and snow and altering streamflow timing—all of the Northwest's snowmelt-dependent systems could be affected.

Snow accumulates in mountains, melting in spring to power both the region's rivers and economy, creating enough hydropower (40% of national total) to supply about two-thirds of the region's electricity (NWPCC 2012) and export 2 to 6 million megawatt hours/month (EIA 2011). Many manufacturing industries, including timber, paper, and food processing are located in the Northwest because of its relatively inexpensive hydro-power and provide hundreds of thousands of jobs, while more than 100,000 port jobs depend on river commerce (NW RiverPartners 2013). Snowmelt waters crops in the dry interior, helping the region produce tree fruit (#1 in the world) and almost $17 billion worth of agricultural commodities including 55%, 15%, and 11% of US potato, wheat, and milk production respectively (USDA 2012a, 2012b). Irrigated agriculture represents over 90% of the consumptive water use in the Columbia River Basin (Washington State Department of Ecology 2011) and 21%, 27%, and 48% of the cropland in Washington, Oregon, and Idaho, respectively, is irrigated (US Department of Agriculture Economic Research Service 2012).

Seasonal water patterns shape the region's flora and fauna, including iconic salmon and steelhead, whose seasonal migration timing is linked to streamflow timing patterns. Water availability is a major controlling factor in the forests of the Northwest, which cover 47% of the landscape (Smith et al. 2009); even in the western Cascade Range forests are limited by summer water availability. The great rivers, lakes, streams, and wetlands in the Northwest provide habitat for fish and wildlife, support transportation, commercial fisheries, and agriculture, and are an essential part of the region's outdoor traditions. In many basins, however, existing water supply is overallocated, leading to shortages and conflicts among objectives and uses during current low flow years; these difficulties are expected to worsen as the climate warms (Hamlet 2011; Miles et al. 2000).

The combination of past climate and previous human choices has shaped the ecosystems, communities, and economies of today's Northwest. The current structure and composition of NW forests, for example, reflects the combination of forests' non-equilibrium response to the varying climate of the Holocene and the changes caused by human activities across the landscape, including logging, development, introduction or suppression of fire, etc. The Northwest of tomorrow will be shaped by the combination of this legacy, today's and tomorrow's choices, and the non-stationary climate of the 21st century.

1.2 A Focus on Risk

As the following chapters show, the regional impacts of climate change are numerous and complex, as a result of the region's physical, social, and ecological heterogeneity. Recognizing that this diversity makes cataloging *all* projected climate impacts impractical, this report was born of an effort to focus on impacts that matter most for the region as a whole.

Written to augment the synthesis developed for the Northwest chapter of the Third National Climate Assessment, this report is grounded in the National Assessment's risk framework approach. While a quantitative comparative risk assessment across the sectors and issues of importance in the Northwest is beyond the scope of this effort, qualitative risk assessment was helpful in focusing the content of both this report and the Northwest chapter of the National Assessment. This process involved evaluating the relative consequences of each projected impact of climate change for the region's economy, infrastructure, natural systems, and the health of NW residents.[1] The likelihood of each impact was qualitatively ranked. Together, these rankings allowed identification of the impacts posing the highest risk, i.e., *likelihood x consequence*, to the region as a whole. Each impact's qualitative scorings for likelihood and consequence were reassessed multiple times by the authors, both individually and as a team, to ensure inter-consistency of scores across risks and sectors.

The resultant key *regionally consequential* risks are those deriving from warming-related impacts in watersheds where snowmelt is an important contributor to flow; coastal consequences of the combined impact of sea level rise and other climate-related drivers; and changes in forest ecosystems. This report therefore focuses on the implications of these risks for water resources, coastal systems, and forest ecosystems. In addition, we focus on three additional climate-sensitive sectors of significance to the region—agriculture; human health vulnerabilities and threats; and NW tribal communities, resources, and values. Under this approach, some important issues cut across multiple chapters, like climate impacts on NW salmon (see Chapter 3, box 3.1).

For all sectors, the focus on risks of importance to the region's overall economy, ecology, built environment, and health is complemented by discussion of the local specificity of climate impacts, vulnerabilities, and adaptive responses, recognizing that impacts of negligible consequence to the region as a whole may sometimes have very significant local consequences. Finally, a focus on risks leads to a stronger focus in this report on negative than positive impacts of climate change. This is consistent with the existing climate impacts literature as well as reflecting our prioritization of assessment to support loss reduction over identification of potential opportunities.

Much has been written about the uncertainties associated with climate change projections, from the range of possible futures represented by alternate greenhouse gas emission scenarios (e.g., Nakićenović et al. 2000, Moss et al. 2010), to the range and variability in resultant global, regional, and local climate change (e.g., IPCC 2007, Hawkins and Sutton 2009, Deser et al. 2012), to the uncertainty in physical and biological impacts and human responses (e.g., Littell et al. 2011). Although it might be tempting to try to base a cross-cutting, cross-sectoral assessment, such as this, on a unified set of climate change projections (e.g., for all reported analyses to be based on the same assumptions about future greenhouse gas emissions), and all changes reported for the same future time periods, the wide-ranging and evolving nature of climate and climate impacts science precludes such consistency. Instead, this report relies on the "ensemble of opportunity", that is, the suite of currently available impact analyses. For example, projected impacts described in subsequent chapters derive from analyses using scenarios based on a variety of greenhouse gas emission scenarios, i.e., SRES-A1FI and RCP8.5: "very high growth", SRES-A2: "continued growth", SRES-A1B: "continued growth peaking at mid-century", SRES-B1 and RCP4.5: "substantial reductions" (Nakicenovic et al. 2000; Moss et al. 2010). Reflecting the lag time between availability of climate model runs and of related impact analyses, only the climate chapter presents results from the latest

---

1 This evaluation began in December 2011, when scientists and stakeholders from all levels and types of organizations from across the Northwest engaged in a discussion and exercise to rank climate risks according to likelihood of occurrence and magnitude of consequences (Dalton et al. 2012).

global climate model runs, developed as part of the Coupled Model Intercomparison Project phase 5, which are being synthesized in the Intergovernmental Panel on Climate Change's 2013 fifth assessment report. To support intercomparison of findings, we compare those new projections to the earlier projections upon which the analyses in the remaining chapters are based (see figs. 2.5, 2.6, 2.7). The careful reader desiring detailed intercomparisons will appreciate the attention paid throughout this report to providing the scenario origin, i.e., emissions scenario, time period, reference period, and GCM used, for each result reported.

Finally, this volume focuses almost exclusively on one side of the climate change issue, that is the projected impacts and requirements for *adaptation* to climate change, largely neglecting regional contributions to the drivers of climate change, such as greenhouse gas emissions. Virtually all of the sectors covered in this report have important linkages to greenhouse gas emissions, including synergies and trade-offs with potential emission reduction strategies—from implications of increased wildfire risk for regional carbon fluxes (Raymond and McKenzie 2012) to recent challenges incorporating wind-generated electricity into a transmission system flooded with peak season hydropower generation (Behr 2011; BPA 2012). With the exception of agriculture, where we briefly discuss relationships between greenhouse gas emissions and management practices, we leave the topics of greenhouse gas emissions and regional efforts to control or reduce them for other assessment efforts.

1.3 Looking Toward the Future

Projected regional warming (see Chapter 2) and sea level rise (see Chapter 4) are expected to bring new conditions to the Northwest, many of which will be different from those for which regional infrastructure and natural resources policies were intended, and those recently experienced by regional ecosystems. The resultant altered patterns of water supply and demand would challenge NW water resources management, agriculture, and ecosystems from fish to forests (see Chapter 3). Coastal habitat and ecosystems, infrastructure and communities are expected to experience ongoing reshaping of the physical and ecological environment caused by climate changes on both land and sea (see Chapter 4), while the combined risk of fires, insects, and diseases could cause significant forest mortality and long-term transformation of NW forest landscapes (see Chapter 5). The agricultural sector is expected to experience mixed impacts, with some sectors and locations benefiting from the projected changes, others sustaining losses, and new opportunities arising (see Chapter 6). While the projected human health impact of climate change is low for the Northwest, relative to other parts of the United States, key climate-related risks facing our region include extreme heat waves, changes in infectious disease epidemiology, river flooding, and wildfires (see Chapter 7). Climate change will have complex and profound effects on the lands, resources, and economies of NW tribes, and on tribal homelands, traditions, and cultural practices that have relied on native plant and animal species since time immemorial (see Chapter 8). Although many of these changes may be obscured in the near term by natural variations in climate, they will become increasingly apparent over time, especially those driven by regional warming.

Spending time in freshwater, coastal, and marine environments, NW salmon will experience the impacts of a changing climate through a wide variety of impact pathways. With their ecological, cultural, and economic importance to the region, and legal protection for some populations, climate impacts on salmon will resonate across many elements of today's Northwest, affecting management and allocation of water resources, treaty obligations to NW tribes and tribal cultural practices, coastal and inland ecosystems, and local economies (see Chapters 3, 4, 8).

1.3.1 COMMON THEMES IN A CHANGING CLIMATE

Familiar Story, New Details. Many of the projections described here may sound familiar. Indeed, the research reviewed for this assessment largely confirms previous projections and analyses, painting the same broad picture of climate impacts on the Northwest that has been described for over fifteen years (e.g., Snover et al. 1998, Mote et. al, 1999). Recent work, however, provides more detailed insight into how impacts are likely to vary from place to place, and from system to system, within the region (e.g., Hamlet et al. 2013, WSDOT 2011). And as efforts increase to apply information about climate change, new knowledge gaps become apparent. The following chapters identify some of these gaps; orienting future research towards filling them would enhance the knowledge base necessary to support regional adaptation.

No "One-Size-Fits-All": Understanding and Preparing for Climate Change Requires Analysis at Multiple Scales. It is increasingly recognized that there is no one-size fits all answer to the question of "what are the implications of climate change for the Northwest?" and that climate impacts will vary within any particular sector or issue area within the region. The extent to which projected higher summer temperatures and lower streamflows in NW streams stresses resident and migratory coldwater fish will depend, for example, on whether and how river flow is managed. This clearly differs between natural and regulated rivers, but will also differ among each broad type; in regulated systems, for example, as a function of available water storage, operating rules, and other demands on the system.

As a result, analysis at multiple scales is essential to ensure completeness. Recognition of commonalities and differences within the Northwest must be included in any effort to develop adaptation strategies over large areas, by state governments and federal landowners, for example. The chapters that follow provide both a broad, region wide examination of climate change implications and insight into the finer scale details of where, how, and why impacts projected for each sector differ from that overarching picture.

Interacting Drivers of Change Must be Considered. Projecting likely climate impacts requires consideration of the combined effect of multiple climate impact pathways and other interacting drivers of change, whether political, economic, social, or ecological. Piecemeal assessment, focusing on individual drivers in isolation, can cause underestimation of risk, since the largest—or sometimes simply different—impacts can occur when multiple drivers align. The specific locations within the City of Olympia, Washington identified as being most at risk to climate change, for example, are different when the combined drivers of high creek flows, high tide levels, storm waves, and sea level rise are considered, compared to when flooding risk is assessed due to sea level rise alone (see Chapter 4). Piecemeal assessment can also cause underestimation of climate risks, when individual drivers would offset each other. For some plants, for example, the beneficial "fertilizing" effect of higher atmospheric concentrations of CO_2 can temporarily offset the negative effects of increased temperatures.

Our Choices Shape Our Vulnerability. The degree to which regional climate change impacts "matter", that is, cause significant or lasting economic, ecologic, or social damage, depends as much on human choices and actions as it does on the rate and magnitude of climate change. This includes choices about where to locate assets or activities that determine exposure to climate impacts, such as the 2899 miles (4665 km) of Washington, Oregon, and Idaho highways and railroads currently located within the 100-year floodplain (MacArthur et al. 2012). It includes choices that affect a system's sensitivity to climate change, like how much of a buffer to maintain in existing systems. Fully- or over-appropriated basins (such as the Yakima; Vano et al. 2012) will be sensitive to a reduction in spring and summer streamflow, while such changes may seem irrelevant to watersheds with abundant supply compared to demands, like those supplying the major urban areas of Puget Sound (Cascade Water Alliance 2012). In many cases, human actions will also determine the capacity of regional systems to adapt to climate change; with 830 miles (1336 km) of Puget Sound coastline already armored by dikes, seawalls and other structures, and more being added each year (Puget Sound Partnership 2012), how many of the basin's remaining coastal wetlands and intertidal habitat will be able to migrate landward in response to sea level rise? Identifying how and why human actions shape natural and social vulnerability to climate change can provide insights useful for reducing those vulnerabilities.

1.3.2 CLIMATE CHANGE ADAPTATION IN THE NORTHWEST

There are two categories of potential response to human-caused climate change. *Mitigation* efforts aim to reduce the magnitude of climate change that occurs by decreasing the causes of that change, e.g., by reducing greenhouse gas emissions. *Adaptation* efforts focus on addressing the consequences of a changing climate, e.g., adjusting practices, processes, or structures of systems to reduce the negative consequences of climate change and, where relevant, take advantage of new opportunities (Adger et al. 2005). These adjustments may be proactive (i.e., in anticipation of projected impacts) and/or reactive (i.e., in response to impacts) and can include both actions intended to reduce impacts and those intended to build capacity for reducing impacts (Whitely Binder 2010). Although appearing to some as an avenue to consider only if mitigation efforts become insufficient, the need for adaptation is becoming more widely recognized (Moser 2009).

If preparing for climate change is a rational adaptive cycle (fig. 1.4, Moser and Ekstrom 2010), it begins with awareness and characterization of the problem, moves into a planning phase, in which objectives are defined, alternatives identified, assessed and selected, and proceeds to a management phase during which actions are implemented, monitored, and progress evaluated, leading to adjustments if necessary. Doing this well, given the uncertainty in present and future conditions, suggests the need for an iterative, evolutionary approach that allows adjustment over time (Brunner and Nordgren

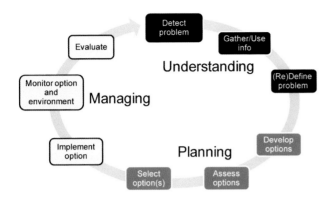

Figure 1.4 The phases and associated components of climate change adaptation, as an iterative cycle. Most adaptation efforts within the Northwest are within the "understanding" or "planning" phases; few have moved into "management". Figure source: Moser and Ekstrom (2010).

2012). Though widely recognized that planning and implementation are rarely sequential, this model is a useful organizing framework for describing the current state of climate change adaptation in the Northwest.

Tremendous progress has been made to identify climate change consequences of practical concern, and adaptation efforts can be found across the region, with some entities beginning to test new strategies (analysis, partnerships, management approaches) for dealing with climate risks. However, adaptation is not yet wide-spread and the preponderance of effort in the region remains focused on the initial steps of awareness raising, problem identification and, to some extent, planning (fig. 1.4); few examples of implementation can be found with which to begin to evaluate effective responses (Hansen et al. 2012).

The scientific synthesis provided in this report provides a solid foundation for identifying the challenges posed by climate change, i.e., identifying the changes projected and their implications for the sector, system, location or community of interest. Its detail reflects the extraordinary amount of scientific effort devoted within the Northwest to understanding potential local impacts. With climate impacts expertise now evident in nearly every academic research institution in the region and in many state, federal, and tribal science and resource management agencies and non-profit organizations, the Northwest has been a leader in applied regional climate impacts science since the mid 1990s (e.g., Chatters et al. 1991, Franklin et al. 1991, Lettenmaier et al. 1995, 1996, Mantua et al. 1997, Francis and Hare 1997, Snover et al. 1998, 2003, Mote et al. 1999, Zolbrod and Peterson 1999), and is now relatively rich in localized climate change information, assessments, tools, and resources (e.g., Hamlet et al. 2013, Snover et al. 2007, Climate Impacts Group 2009, Oregon Climate Change Research Institute 2010, Macarthur et al. 2012). Due, in part, to efforts such as these, NW resource managers, planners and policy makers were early engagers in climate change issues (e.g., Oregon Task Force on Global Warming 1990, Canning 1991, Craig 1993, King County 2007) and continue to lead by example. Following is a brief synopsis of relevant efforts at various levels of jurisdiction within the Northwest.

Local: NW public water utilities were among the first natural resource management agencies in the region to consider climate change impacts (e.g., Palmer and Hahn 2002, Palmer et al. 2004, Palmer 2007) and several have since organized nationally to provide input into climate change research priorities and develop adaptation strategies (Barsugli et al. 2009). Numerous cities and counties (e.g., King County, Seattle, Olympia, Snohomish, and Port of Bellingham (Washington); Portland, Multnomah County, and Eugene, (Oregon)) have assessed climate risks, developed response strategies, and/ or implemented adaptive actions at various levels and for various sectors within local government.

State: Both Washington and Oregon have developed state level climate change response strategies (State of Oregon 2010; Washington State Department of Ecology 2012) aligned with commissioned assessments of climate change impacts on sectors of inter-est (Oregon Climate Change Research Institute 2010; Climate Impacts Group 2009). These set out overarching objectives across all issue areas, and are intended to inform the development of more targeted plans by state agencies and local jurisdictions. Follow-on efforts include the development of agency-specific analyses of climate change risks and vulnerabilities (WA State Departments of Transportation [WSDOT 2011] and Emergency Management, OR Department of Transportation, OR Public Health Authority), adaptation plans (WA Dept of Natural Resources), regulatory or planning guidance (WA Ecology and Office of the Insurance Commissioner; Leurig and Dlugolecki 2013) and public health community engagement and training (OR Public Health Authority).

Federal: Consistent with President Obama's 2009 Executive Order (E.O. 13514), which required federal agency adaptation planning, NW federal entities are incorporating climate change information in assessment and planning and developing innovative approaches to integrating risks across issue areas and actors. Recent examples include a US Forest Service/National Parks Service partnership to "increase awareness of climate change, assess the vulnerability of cultural and natural resources; and incorporate climate change adaptation into current management of [over 6 million acres of] federal lands in the North Cascades region" (Raymond et al. 2013), an Environmental Protection Agency pilot project to consider how projected climate change impacts could be incorporated into stream temperature standards and influence restoration plans (EPA 2013) and a collaboration among Columbia River Basin water management agencies to develop climate and hydrology datasets for use in long-term planning in preparation for the renegotiation of the Columbia River Treaty with Canada (USBR et al. 2011).

Tribal: Numerous NW tribes have begun addressing adaptation. Among these, the Swinomish Indian Tribal Community is a national leader in evaluating tribal climate change vulnerabilities and adaptation needs from a multi-risk, multi-sector, multi-timescale perspective (Swinomish Indian Tribal Community 2009, 2010). Other tribes addressing climate change risks include the Nez Perce, the Coquille, and the Port Gamble S'Klallam and Jamestown S'Klallam Tribes (see Chapter 8).

1.4 Conclusion

Implicit assumptions about past and future climatic conditions underlie many plans, policies, and management strategies. Decisions about how much timber to harvest, fish to catch, or water to store in reservoirs include implicit expectations about how fast forests regenerate, how many fish will return next year, and when the rains will start in the fall, all of which are influenced by climate. Similarly, most infrastructure construction decisions and associated management policies—from highway location and culvert sizing to dam construction and water rights decisions—contain embedded expectations about climate risks, based on qualitative or quantitative assessment of past climatic conditions. As we look toward the future, the key question in front of us is: How will the region meet the additional challenges climate change will bring? By identifying, assessing, and preparing for the potential risks? By re-examining and, where necessary, adjusting our infrastructure, plans, policies, and operating procedures to function successfully under new and changing conditions? Or by proceeding as before, using the past as a guide to the future and basing decisions on assumptions about the future that are becoming increasingly incorrect?

Acknowledgments

The authors acknowledge the NCA NW chapter author team (Sanford D. Eigenbrode, Jeremy S. Littell, Philip W. Mote, Rick R. Raymondi, and W. Spencer Reeder) for their contributions, and Elisabet Eppes (University of Washington) for research assistance.

References

Adger, W. N., N. W. Arnell, and E. L. Tompkins. 2005. "Successful Adaptation to Climate Change across Scales." *Global Environmental Change* 15: 77-86.

Barbier, E., and A. Markandya. 2013. *A New Blueprint for a Green Economy*. Routledge Publishers, London, United Kingdom and New York, NY, USA.

Barsugli, J., C. Anderson, J. B. Smith, J. M. Vogel. 2009. "Options for Improving Climate Modeling to Assist Water Utility Planning for Climate Change." Water Utility Climate Alliance White Paper. http://www.wucaonline.org/assets/pdf/pubs_whitepaper_120909.pdf.

Beechie, T. J., E. Buhle, M. Ruckelshaus, A. Fullerton, and L. Hoslinger. 2006. "Hydrologic Regime and the Conservation of Salmon Life History Diversity." *Biological Conservation* 130: 560-572.

Bottom, D. L., C. A. Simenstad, J. Burke, A. M. Baptista, D.A. Jay, K. K. Jones, E. Casillas, and M. H. Schiewe. 2005. "Salmon at River's End: The Role of the Estuary in the Decline and Recovery of Columbia River Salmon." US Department of Commerce, NOAA Technical Memo NMFS-NWFSC-68.

BPA (Bonneville Power Administration). 2012. "BPA Proposes Resolution to Electricity Oversupply." BPA Fact Sheet DOE/BP-4397.

Brunner, R., and J. Nordgren. 2012. "Climate Adaptation as an Evolutionary Process: A White Paper Based on the Kresge Grantees and Practitioners Workshop on Climate Change Adaptation." Portland, OR. http://kresge.org/library/climate-adaptation-evolutionary-process-white-paper.

Callahan, B. M., E. L. Miles, and D. L. Fluharty. 1999. "Policy Implications of Climate Forecasts for Water Resources Management in the Pacific Northwest." *Policy Sciences* 32: 269-293.

Canning, D. J. 1991. "Sea Level Rise in Washington State: State-of-the-Knowledge, Impacts, and Potential Policy Issues." Shorelands and Coastal Zone Management Program, Washington Department of Ecology. Olympia, WA.

Cascade Water Alliance. 2012. "Transmission and Supply Plan." Bellevue, WA.

Chatters, J. C., D. A. Neitzel, M. J. Scott, and S. A. Shankle. 1991. "Potential Impacts of Global Climate Change on Pacific Northwest Spring Chinook Salmon (Oncorhynchus Tshawytscha): An Explanatory Case Study." *NW Environmental Journal* 7: 71-92.

Climate Impacts Group. 2009. "The Washington Climate Change Impacts Assessment," eds. M. McGuire Elsner, J. Littell, and L. Whitely Binder. Center for Science in the Earth System, Joint Institute for the Study of the Atmosphere and Oceans, University of Washington, Seattle, WA. http://www.cses.washington.edu/db/pdf/wacciareport681.pdf.

Climate Leadership Initiative. 2006. "Impacts of Climate Change on Washington's Economy: A Preliminary Assessment of Risks and Opportunities." The Washington Economic Steering Committee and the Climate Leadership Initiative, Institute for a Sustainable Environment, University of Oregon. Eugene, OR. http://www.ecy.wa.gov/pubs/0701010.pdf.

Climate Leadership Initiative. 2009. "An Overview of Potential Economic Costs to Washington of a Business-as-Usual Approach to Climate Change." A Report from the Program on Climate Economics, Climate Leadership Initiative, Institute for a Sustainable Environment, University of Oregon. Eugene, OR. http://www.ecy.wa.gov/climatechange/docs/021609 _ClimateEconomicsImpactsReport.pdf.

Climate Leadership Initiative. 2010. "Additional Analysis of the Potential Economic Costs to the State of Washington of a Business as-Usual Approach to Climate Change: Lost Snowpack Water Storage and Bark Beetle Impacts." A Report from the Program on Climate Economics, Climate Leadership Initiative, Institute for a Sustainable Environment, University of Oregon. Eugene, OR. http://www.ecy.wa.gov/climatechange/docs/20101230 _EconomicAnalysis.pdf.

Coast and Harbor Engineering. 2011. "City of Olympia Engineered Response to Sea Level Rise." Prepared for the City of Olympia Public Works Department, Planning and Engineering. Olympia, WA. http://olympiawa.gov/community/sustainability/~/media/Files/PublicWorks /Sustainability/Sea%20Level%20Rise%20Response%20Technical%20Report.ashx.

Cone, J. and S. Ridlington (eds.). 1996. "The Northwest Salmon Crisis: A Documentary History." Oregon State University Press, Corvallis, OR.

Craig, D. 1993. "Preliminary Assessment of Sea Level Rise in Olympia, Washington: Technical and Policy Implications." Policy and Program Development Division, Public Works Department. Olympia, WA.

Dalton, M., P. Mote, J. A. Hicke, D. Lettenmaier, J. Littell, J. Newton, P. Ruggiero, and S. Shafer. 2012. "A Workshop in Risk-Based Framing of Climate Impacts in the Northwest: Implementing the National Climate Assessment Risk-Based Approach." Technical Input Report to the Third National Climate Assessment. http://downloads.usgcrp.gov/NCA/Activities /northwestncariskframingworkshop.pdf.

Deser, C., A. S. Phillips, V. Bourdette, and H. Teng. 2012. "Uncertainty in Climate Change Projections: The Role of Internal Variability." *Climate Dynamics* 38: 527-546. doi: 10.1007 /s00382-010-0977-x.

DeVries, P. E. 1997. "Riverine Salmonids Egg Burial Depths: Review of Published Data and Implications for Scour Studies." *Canadian Journal of Fisheries and Aquatic Sciences* 54: 1685–1698.

Eastern Forest Environmental Threat Center. 2013. "Idaho, Oregon, and Washington Threats." http://forestthreats.org/threatsummary.

EDDMaps. 2013. "Status of Invasive Plants in Idaho, Oregon, and Washington." http://www .eddmaps.org.

EIA. 2011. "A Quarter of California's Energy Comes From Outside the State." *Today in Energy Newsletter*. http://www.eia.gov/todayinenergy/detail.cfm?id=4370.

Eissinger, A. 2009. "Marine Invasive Species Identification Guide for the Puget Sound Area." Puget Sound Marine Invasive Species Volunteer Monitoring Program.

EPA (Environmental Protection Agency). 2013. "Helping to Protect Wild Salmon." *Science Matters Newsletter Current Issue: Climate Change Research.*

Executive Order Number 13,514, 3 C.F.R. 13514. 2010. http://www.whitehouse.gov/assets /documents/2009fedleader_eo_rel.pdf.

Foushee, A. 2010. "Identifying Ecological Indicators of Climate Change and Land Use: Impacts to a Coastal Watershed." University of Vermont Field Naturalist Program.

Francis, R. C., and S. R. Hare. 1997. "Regime Scale Climate Forcing of Salmon Populations in the Northwest Pacific - Some New Thoughts and Findings." In *Estuarine and Ocean Survival of Northeastern Pacific Salmon*, edited by R. L. Emmett and M. H. Schiewe, 113-128. US Department of Commerce, NOAA Technical Memo NMFS-NWFSC-29, 313 pp.

Franklin, J. F., F. J. Swanson, M. E. Harmon, D. A. Perry, T. A. Spies, V. H. Dale, A. McKee, W. L. Ferrell, J. E. Means, S. V. Gregory, J. D. Lattin, T. D. Schowalter, and D. Larsen. 1991. "Effects of Global Climatic Change on Forests in Northwestern North America. *NW Environmental Journal* 7: 233-254.

FWS. 2013. "Species Reports: Listings and Occurrences for Idaho, Oregon, and Washington." http://www.ecos.fws.gov.

Gates, C. 1955. "The Indian Treaty of Point No Point." *Pacific Northwest Quarterly* 46 (2): 52-58.

Good, J. W. 1993. "Ocean Resources." In *Atlas of the Pacific Northwest*, edited by P. L. Jackson and A. J. Kimerling, 110-121. Oregon State University Press, Corvallis, OR.

Hamlet, A. F. 2011. "Assessing Water Resources Adaptive Capacity to Climate Change Impacts in the Pacific Northwest Region of North America." *Hydrology and Earth System Sciences* 15: 1427-1443. doi: 10.5194/hess-15-1427-2011.

Hamlet, A. F., and D. P. Lettenmaier. 1999. "Effects of Climate Change on Hydrology and Water Resources in the Columbia River Basin." *Journal of the American Water Resources Association* 35 (6): 1597-1623.

Hamlet, A. F., M. M. Elsner, G. Mauger, S. Y. Lee, and I. Tohver. 2013. "An Overview of the Columbia Basin Climate Change Scenarios Project: Approach, Methods, and Summary of Key Results." *Atmosphere-Ocean* (accepted).

Hansen, L., R. M. Gregg, V. Arroyo, S. Ellsworth, L. Jackson and A. Snover. 2012. "The State of Adaptation in the United States." Report for the MacArthur Foundation. http://www .georgetownclimate.org/sites/default/files/The%20State%20of%20Adaptation %20in%20the%20United%20States.pdf.

Hawkins, E., and R. Sutton. 2009. "The Potential to Narrow Uncertainty in Regional Climate Predictions." *Bulletin of the American Meteorological Society* 90: 1095-1107. doi: 10.1175/2009BAMS2607.1.

Heal, G. 1998. *Valuing the Future: Economic Theory and Sustainability*. Columbia University Press, New York, NY.

Hidalgo, H. G., T. Das, M. D. Dettinger, D. R. Cayan, D. W. Pierce, T. P. Barnett, G. Bala, A. Mirin, A. W. Wood, and C. Bonfils. 2009. "Detection and Attribution of Streamflow Timing Changes to Climate Change in the Western United States." *Journal of Climate* 22: 3838-3855. doi: 10.1175/2009JCLI2470.1.

Hulse, D., J. Branscomb, J. G. Duclos, S. Gregory, S. Payne, D. Richey, H. Deraborn, L. Ashkenas, P. Minear, J. Christy, E. Alverson, D. Diethelm, and M. Richmond. 2002. *Willamette River*

Basin Planning Atlas: Trajectories of Environmental and Ecological Change. Oregon State University Press, Corvallis, OR.

Idaho Forest Products Commission. 2012. "Economics." http://www.idahoforests.org/money 1.htm.

IPCC (Intergovernmental Panel on Climate Change). 2007. "Climate Change 2007: The Physical Science Basis." Contribution of Working Group I to Solomon, S., D. Qin, M. Manning, Z. Chen, M. Marquis, K. B. Averyt, M. Tignor and H. L. Miller (eds.). "The Fourth Assessment Report of the Intergovernmental Panel on Climate Change." Cambridge University Press, Cambridge, United Kingdom and New York, NY, USA.

ISAB (Independent Scientific Advisory Board). 2007. "Climate Change Impacts on Columbia River Basin Fish and Wildlife." Northwest Power and Conservation Council, Portland, OR.

Jackson, P. L. and A. J. Kimerling (eds.). 2003. *Atlas of the Pacific Northwest.* Oregon State University Press, Corvallis, OR. 152 pp.

King County. 2007. 2007 Climate Plan. King County, WA. http://your.kingcounty.gov/exec /news/2007/pdf/ClimatePlan.pdf.

Lettenmaier, D. P., S. M. Fisher, R. N. Palmer, S. P. Millard, J. P. Hughes, and J. C. Sias. 1995. Water Management Implications of Global Warming: 2. The Tacoma Water Supply System." Report to the US Army Corps of Engineers, Institute for Water Resources. Fort Belvoir, VA.

Lettenmaier, D. P., D. Ford, S. M. Fisher, J. P. Hughes, and B. Nijssen. 1996. "Water Management Implications of Global Warming: 4. The Columbia River Basin." Report to the US Army Corps of Engineers, Institute for Water Resources. Fort Belvoir, VA.

Leurig, S., and A. Dlugolecki. 2013. "Insurer Climate Risk Disclosure Survey: 2012 Findings and Recommendations." *Ceres.* http://www.ceres.org/resources/reports/naic-report.

Lichatowich, J. 1999. *Salmon without Rivers.* Island Press, Washington, D.C.

Littell, J. S., D. McKenzie, B. K. Kerns, S. Cushman, and C. G. Shaw. 2011. "Managing Uncertainty in Climate-Driven Ecological Models to Inform Adaptation to Climate Change." *Ecosphere* 2 (9): 102. doi: 10.1890/ES11-00114.1.

Luce, C., and Z. Holden. 2009. "Declining Annual Streamflow Distributions in the Pacific Northwest United States, 1948–2006." *Geophysical Research Letters* 36: L16401. doi: 10.1029/2009GL039407.

MacArthur, J., P. Mote, J. Ideker, M. Figliozzi, and M. Lee. "Climate Change Impact Assessment for Surface Transportation in the Pacific Northwest and Alaska." 2012. Washington State Department of Transportation Research Report WA-RD 772.1.

Mantua, N. J., I. Tohver, and A. F. Hamlet. 2010. "Climate Change Impacts on Streamflow Extremes and Summertime Stream Temperature and Their Possible Consequences for Freshwater Salmon Habitat in Washington State." *Climatic Change* 102: 187-223. doi: 10.1007/ s10584-010-9845-2.

Mantua, N. J., S. R. Hare, Y. Zhang, J. M. Wallace, and R. C. Francis. 1997. "A Pacific Interdecadal Climate Oscillation with Impacts on Salmon Production." *Bulletin of the American Meteorological Society* 78(6): 1069-1079.

Miles, E. L., A. K. Snover, A. F. Hamlet, B. Callahan, and D. Fluharty. 2000. "Pacific Northwest Regional Assessment: the Impacts of Climate Variability and Climate Change on the Water Resources of the Columbia River Basin." *Journal of the American Water Resources Association* 36: 399-420.

Moser, S. C. 2009. "Good Morning America! The Explosive Awakening of the US to Adaptation." NOAA, Charleston, SC and California Energy Commission, Sacramento, CA.

Moser, S. C. and J. Ekstrom. 2010. "A Framework to Diagnose Barriers to Climate Change Adaptation. *Proceedings of the National Academy of Sciences* 107 (51): 22026-22031. doi: 10.1073/pnas.1007887107.

Moss, R. H., J. A. Edmonds, K. A. Hibbard, M. R. Manning, S. K. Rose, D. P. van Vuuren, T. R. Carter, S. Emori, M. Kainuma, T. Kram, G. A. Meehl, J. F. B. Mitchell, N. Nakicenovic, K. Riahi, S. J. Smith, R. J. Stouffer, A. M. Thomson, J. P. Weyant, and T. J. Wilbanks. 2010. "The Next Generation of Scenarios for Climate Change Research and Assessment." *Nature* 463: 747-756. doi: 10.1038/nature08823.

Mote, P. W. 2006. "Climate-Driven Variability and Trends in Mountain Snowpack in Western North America." *Journal of Climate* 19: 6209-6220. doi: 10.1175/JCLI3971.1.

Mote, P. W., and E. P. Salathé. 2010. "Future climate in the Pacific Northwest." *Climatic Change* 102: 29-50. doi: 10.1007/s10584-010-9848-z.

Mote, P. W., D. J. Canning, D. L. Fluharty, R. C. Francis, J. F. Franklin, A. F. Hamlet, M. Hershman, M. Holmberg, K. N. Ideker, W. S. Keeton, D. P. Lettenmaier, L. R. Leung, N. J. Mantua, E. L. Miles, B. Noble, H. Parandvash, D. W. Peterson, A. K. Snover, and S. R. Willard. 1999. "Impacts of Climate Variability and Change, Pacific Northwest." National Atmospheric and Oceanic Administration, Office of Global Programs, and JISAO/SMA Climate Impacts Group, Seattle, WA. 110 pp.

Nakićenović, N., O. Davidson, G. Davis, A. Grubler, T. Kram, E. L. La Rovere, B. Metz, T. Morita, W. Pepper, H. Pitcher, A. Sankovski, P. Shukla, R. Swart, R. Watson, and Z. Dadi. 2000. *Intergovernmental Panel on Climate Change Special Report on Emissions Scenarios.* Cambridge University Press, Cambridge, United Kingdom.

National Marine Fisheries Service. 2012. "Annual Commercial Landings Statistics." Accessed December 2012. http://www.st.nmfs.noaa.gov/commercial-fisheries/commercial-landings/annual-landings/index.

Northwest Power and Conservation Council. 2012. "Hydroelectricity in the Columbia River Basin." http://www.nwcouncil.org/energy/powersupply/dams/hydro.htm.

Northwest RiverPartners. 2013. "Facts Tell the Story of NW Hydro and Dams." *Current Reflections* 83: 1-2.

NRC (National Research Council). 1996. "Upstream: Salmon and Society in the Pacific Northwest." Committee on Protection and Management of Pacific Northwest Anadromous Salmonids, Board on Environmental Studies and Toxicology, Commission on Life Sciences. National Academy Press, Washington, DC.

OCZMA (Oregon Coastal Zone Management Association). 2006. "A Demographic and Economic Description of the Oregon Coast: 2006 Update." Newport, OR. http://www.oczma.org/pdfs/ED%20landscape%20report2003_1.pdf.

Oregon Climate Change Research Institute. 2010. "Oregon Climate Assessment Report," eds. K. D. Dello and P. W. Mote. College of Oceanic and Atmospheric Sciences, Oregon State University, Corvallis, OR. http://www.occri.net/OCAR.

Oregon Forest Resources Institute. 2012. "The 2012 Forest Report: An Economic Assessment of Oregon's Forest and Wood Products Manufacturing Sector." http://oregonforests.org/sites/default/files/publications/pdf/OFRI_Forest_Report_2012_0.pdf.

Oregon Office of Economic Analysis. 2013. "Long-term Oregon State's County Population Forecast, 2010-2050." http://www.oregon.gov/DAS/OEA/Pages/demographic.aspx#Long_Term_County_Forecast.

Oregon Task Force on Global Warming. 1990. "Report to the Governor and Legislature: Part One: Possible Impacts on Oregon from Global Warming." Oregon Department of Energy.

Orr, J. C., V. J. Fabry, O. Aumont, L. Bopp, S. C. Doney, R. A. Feely, A. Gnanadesikan, N. Gruber, A. Ishida, F. Joos, R. M. Key, K. Lindsay, E. Maier-Reimer, R. Matear, P. Monfray, A. Mouchet, R. G. Najjar, G. K. Plattner, K. B. Rodgers, C. L. Sabine, J. L. Sarmiento, R. Schlitzer, R. D. Slater, I. J. Totterdell, M. F. Weirig, Y. Yamanaka, and A. Yool. 2005. "Anthropogenic Ocean Acidification over the Twenty-First Century and its Impact on Calcifying Organisms." *Nature* 437: 681-686. doi: 10.1038/nature04095.

Ovsak, C. M. 1994. "Reaffirming the Guarantee: Indian Treaty Rights to Hunt and Fish Off-Reservation in Minnesota." *William Mitchell Law Review* 20 (4): 1177.

Palmer, R. N. 2007. "Final Report of the Climate Change Technical Committee." A Report Prepared by the Climate Change Technical Subcommittee of the Regional Water Supply Planning Process. Seattle, WA.

Palmer, R. N., and M. A. Hahn. 2002. "The Impacts of Climate Change on Portland's Water Supply: An Investigation of Potential Hydrologic and Management Impacts on the Bull Run System." Report Prepared for the Portland Water Bureau. University of Washington, Seattle, WA. 139 pp.

Palmer, R. N., E. Clancy, N. T. VanRheenen, and M. W. Wiley. 2004. "The Impacts of Climate Change on the Tualatin River Basin Water Supply: An Investigation into Projected Hydrologic and Management Impacts." Department of Civil and Environmental Engineering, University of Washington, Seattle, WA. 91 pp.

Pearcy, W. G. 1992. "Ocean Ecology of North Pacific Salmonids." University of Washington Press, Seattle, WA.

Pease, J. R. 1993. "Land Use and Ownership." In *Atlas of the Pacific Northwest*, edited by P. L. Jackson and A. J. Kimerling, 31-39. Oregon State University Press, Corvallis, OR.

Pierce, D. W., T. P. Barnett, H. G. Hidalgo, T. Das, C. Bonfils, B. D. Santer, G. Bala, M. D. Dettinger, D. R. Cayan, and A. Mirin. 2008. "Attribution of Declining Western US Snowpack to Human Effects." *Journal of Climate* 21: 6425-6444. doi: 10.1175/2008JCLI2405.1.

Population Research Center. 2010. "2010 Census Profiles." http://www.pdx.edu/prc.

Puget Sound Partnership. 2012. "2012 State of the Sound: A Biennial Report on the Recovery of Puget Sound." http://www.psp.wa.gov/sos.php.

Ray, G. 2005. "Invasive Marine and Estuarine Animals of the Pacific Northwest and Alaska." Aquatic Nuisance Species Research Program, ERDC/TN ANSRP-05-6.

Raymond, C. L., and D. McKenzie. 2012. "Carbon Dynamics of Forests in Washington, USA: 21st Century Projections Based on Climate-Driven Changes in Fire Regimes." *Ecological Applications* 22: 1589-1611.

Raymond, C. L., D. L. Peterson, and R. M. Rochefort. 2013. "Climate Change Vulnerability and Adaptation in the North Cascades Region, Washington." General Technical Report PNW-GTR-xxx. US Department of Agriculture, Forest Service, Pacific Northwest Research Station, Portland, OR. In press. Draft available from: http://northcascadia.org/pdf/DRAFT_raymond_et_al_NCAP.pdf.

Sarachik, E. S. 2000. "The Application of Climate Information." *Consequences* 5: 27-36.

Smith, W. B., P. D. Miles, C. H. Perry, and S. A. Pugh. 2009. "Forest Resources of the United States, 2007." General Technical Report WO-78. US Department of Agriculture, Forest Service, Washington Office, Washington, DC. 336 pp. http://www.fs.fed.us/nrs/pubs/gtr/gtr_wo78.pdf.

Snover, A. K., A. F. Hamlet, and D. P. Lettenmaier. 2003. "Climate Change Scenarios for Water Planning Studies: Pilot Applications in the Pacific Northwest." *Bulletin of the American Meteorological Society* 84 (11): 1513-1518. doi: 10.1175/BAMS-84-11-1513.

Snover, A. K., E. L. Miles, and B. Henry. 1998. "OSTP/USGCRP Regional Workshop on the Impacts of Global Climate Change on the Pacific Northwest: Final Report." NOAA Climate and Global Change Program Special Report No. 11.

Solomon, S., G. Plattner, R. Knutti, and P. Friedlingstein. 2009. "Irreversible Climate Change Due to Carbon Dioxide Emissions." *Proceedings of the National Academy of Sciences* 106: 1704-1709. doi: 10.1073/pnas.0812721106.

State of Oregon. 2010. "The Oregon Climate Change Adaptation Framework." http://www .oregon.gov/ENERGY/GBLWRM/docs/Framework_Final_DLCD.pdf?ga=t.

Stewart, I. T., D. R. Cayan, and M. D. Dettinger. 2005. "Changes Toward Earlier Streamflow Timing across Western North America." *Journal of Climate* 18: 1136-1155. doi:10.1175 /JCLI3321.1.

Stöckle, C. O., R. L. Nelson, S. Higgins, J. Brunner, G. Grove, R. Boydston, M. Whiting, and C. Kruger. 2010. "Assessment of Climate Change Impact on Eastern Washington Agriculture." *Climatic Change* 102: 77-102.

Strittholt, J. R., D. A. DellaSala, and H. Jiang [Abstract]. 2006. "Status of Mature and Old-Growth Forests in the Pacific Northwest." *Conservation Biology* 20 (2): 363-374.

Swinomish Indian Tribal Community. 2009. "Swinomish Climate Change Initiative: Impact Assessment Technical Report." http://www.swinomish-nsn.gov/climate_change/Docs/SITC _CC_ImpactAssessmentTechnicalReport_complete.pdf.

Swinomish Indian Tribal Community. 2010. "Swinomish Climate Change Initiative: Climate Adaptation Action Plan." http://www.swinomish-nsn.gov/climate_change/Docs/SITC_CC _AdaptationActionPlan_complete.pdf.

US Census Bureau. 2010a. "Idaho QuickFacts." quickfacts.census.gov/qfd/states/16000 .html.

US Census Bureau. 2010b. "Oregon QuickFacts." http://quickfacts.census.gov/qfd/states/41000 .html.

US Census Bureau. 2010c. "State & County QuickFacts: Boise City, Idaho." quickfacts.census .gov/qfd/states/16/1608830.html.

US Census Bureau. 2010d. "State & County QuickFacts: Spokane (city), Washington." quickfacts .census.gov/qfd/states/53/5367000.html.

US Census Bureau. 2012. "Population Projections to 2060 by Selected Age Groups and Sex (Idaho)." http://quickfacts.census.gov/qfd/states/16000lk.html.

US Department of Agriculture Census of Agriculture. 2010. "Production Fact Sheet." http: //www.agcensus.usda.gov/Publications/2007/Online_Highlights/Fact_Sheets /Production/.

US Department of Agriculture Economic Research Service. 2012. "Data Products State Fact Sheets: Washington, Oregon, and Idaho." http://www.ers.usda.gov/data-products/state -fact-sheets/.

US Department of Commerce, NOAA (National Oceanic and Atmospheric Administration), and National Marine Fisheries Service. 2009. "Fishing Communities of the United States 2006." NOAA Technical Memorandum NMFS-F/SPO-98. https://www.st.nmfs.noaa.gov/st5 /publication/communities/CommunitiesReport_ALL.pdf.

USBR (Department of the Interior Bureau of Reclamation), USACE (Army Corps of Engineers), and BPA (Bonneville Power Administration). 2011. "Climate and Hydrology Datasets for Use in the River Management Joint Operating Committee (RMJOC) Agencies' Longer-Term Planning Studies: Part IV – Summary." http://www.usbr.gov/pn/programs/climatechange /reports/finalpartIV-0916.pdf.

USDA. 2012a. "Crop Production 2011 Summary." US Department of Agriculture, National Agricultural Statistics Service. http://usda01.library.cornell.edu/usda/current/CropProdSu/CropProdSu-01-12-2012.pdf.

USDA. 2012b. "Milk Production, Disposition, and Income, 2011 Summary." US Department of Agriculture, National Agricultural Statistics Service. http://usda01.library.cornell.edu/usda/current/MilkProdDi/MilkProdDi-04-25-2012.pdf.

USGS. 2013. "Nonindigenous Aquatic Species: Idaho, Oregon, and Washington." http://nas.er.usgs.gov.

Vano, J. A., M. Scott, N. Voisin, C. Stöckle, A. F. Hamlet, K. E. B. Mickelson, M. M. Elsner, and D. P. Lettenmaier. 2010. "Climate Change Impacts on Water Management and Irrigated Agriculture in the Yakima River Basin, Washington, USA." *Climatic Change*, doi: 10.1007/s10584-010-9856-z.

Washington Biodiversity Council. 2007. "Washington's Biodiversity Status and Threats." http://www.biodiversity.wa.gov.

Washington OFM (Office of Financial Management). 2010. "Population Density." http://www.ofm.wa.gov/pop/popden/.

Washington OFM (Office of Financial Management). 2012. "Forecast of the State Population by Age and Sex, 2010-2040." http://www.ofm.wa.gov/pop/stfc/default.asp.

Washington State Department of Ecology. 2011. "Columbia River Basin Long-Term Water Supply and Demand Forecast." Publication No. 11-12-011. https://fortress.wa.gov/ecy/publications/publications/1112011.pdf.

Washington State Department of Ecology. 2012. "Preparing For a Changing Climate: Washington State's Integrated Climate Response Strategy." http://www.ecy.wa.gov/pubs/1201004.pdf.

Washington State Department of Natural Resources. 2007. "The Future of Washington Forests." Washington State Department of Natural Resources. http://www.dnr.wa.gov/ResearchScience/Topics/ForestResearch/Pages/futureofwashingtonsforest.aspx.

Whitely Binder, L. C., J. Krencicki Barcelos, D. B. Booth, M. Darzen, M. M. Elsner, R. A. Fenske, T. F. Graham, A. F. Hamlet, J. Hodges-Howell, J. E. Jackson, C. Karr, P. W. Keys, J. S. Littell, N. J. Mantua, J. Marlow, D. McKenzie, M. Robinson-Dorn, E. A. Rosenberg, C. Stöckle, and J. A. Vano. 2010. "Preparing for Climate Change in Washington State." *Climatic Change* 102: 351-376. doi: 10.1007/s10584-010-9850-5.

WSDOT (Washington State Department of Transportation). 2011. "Climate Impacts Vulnerability Assessment." Report to the Federal Highway Administration. http://www.wsdot.wa.gov/NR/rdonlyres/B290651B-24FD-40EC-BEC3-EE5097ED0618/0/WSDOTClimateImpactsVulnerabilityAssessmentforFHWAFinal.pdf.

Zolbrod, A. N., and D. L. Peterson. 1999. "Response of High-Elevation Forests in the Olympic Mountains to Climatic Change." *Canadian Journal of Forest Research* 29:1966-1978.

Chapter 2

Climate
Variability and Change in the Past and the Future

AUTHORS
Philip W. Mote, John T. Abatzoglou, and Kenneth E. Kunkel

2.1 Understanding Global and Regional Climate Change

The climate system receives energy from the Sun—mostly in the form of visible light—and balances this energy by radiation of infrared, or heat, energy back to space. Global surface temperature fluctuations are influenced by the amount of solar radiation received at the top of the atmosphere, the reflectivity or albedo, of the planet, and things that affect the efficiency of infrared energy loss to space. The solar radiation received is determined by direct solar output and the Earth's orbital fluctuations, and the albedo is largely determined by changes at the surface and by clouds and particles in the atmosphere. Things that affect the efficiency of infrared energy loss to space include both clouds and certain trace gases that absorb outgoing infrared energy and are commonly called greenhouse gases. In order of global importance to the energy balance, these greenhouse gases include water vapor, carbon dioxide (CO_2), methane (CH_4), ozone, chlorofluorocarbons (CFCs), of which CFC-12 dominates, nitrous oxide, and dozens of others. Most of these are long-lived gases, meaning that molecules emitted into the atmosphere tend to remain there for decades and their concentrations are fairly similar throughout the world; important exceptions are water vapor and ozone, which are controlled by a variety of faster processes and therefore have larger variations across the globe and change faster in time.

Human activities in the industrial era have directly and substantially increased the quantity of the long-lived greenhouse gases, and some (the CFCs among them) are entirely man-made. Observed changes in carbon dioxide account for about 63% of the radiative heating due to observed changes in long-lived greenhouse gases (Forster et al. 2007). Water vapor and ozone are also responding to human activity: tropospheric ozone has increased because of air pollution, especially nitrogen compounds, even as stratospheric ozone has decreased because of CFCs; and water vapor is closely controlled by surface temperature, so it has an important feedback that is part of the climate system response to rising long-lived greenhouse gases.

Changes in the sun's energy output and volcanic eruptions are the most important natural external forcings of climate. Changes in solar activity may be partly responsible for the cool period in the 16th–18th centuries and for the warming early in the 20th century, but observations from satellites of solar output since late 1978 demonstrate that solar changes cannot be responsible for the large increase in global temperatures during the last 34 years: solar output has not increased over that period, but has fluctuated about 0.1% with the roughly 11-year solar cycle. Since the solar cycle was in an extended minimum phase during roughly 2006-2011 (Denton and Borovsky 2012), the linear trend in solar output is actually slightly negative (see e.g. Lean and Rind 2009).

Volcanic emissions include sulfur dioxide, which turns into sulfuric acid particles that reflect sunlight. Some eruptions reach the stratosphere, but in middle and high latitudes stratospheric air is gradually sinking and the volcanic emissions are pushed into the troposphere within a month or two. The most effective volcanic eruptions that cool the Earth are tropical volcanic eruptions of sufficient force to reach the stratosphere, in the latitudes where stratospheric air is rising and hence can suspend the reflective particles.

In understanding causes of changes in global or regional climate, scientists often distinguish between processes external to the climate system and processes internal to the climate system. External processes include solar and volcanic forcings and the long-lived greenhouse gases. Internal processes include fluctuations in water vapor, surface albedo related to vegetation or snow cover, and clouds. In addition, atmospheric and oceanic circulations rearrange heat. The influence of variations in circulation patterns is more pronounced at regional to local scales than at global scales. For example, regional climate in the Northwest is strongly influenced by atmospheric circulation in the northeast Pacific ocean; to first order, atmospheric circulation merely moves heat from place to place, cancelling out in the global average, so year-to-year fluctuations in regional averages are usually much larger than the global average.

2.2 Past Changes in Northwest Climate: Means

Northwest (NW) climate is characterized by strong spatial gradients. Figure 2.1 shows the mean annual maximum temperature and precipitation. Note the strong rain shadow effects downwind of the coastal ranges and Cascades, where precipitation amounts can be reduced 10–15 fold in less than 50 km (32 mi) in many places.

Climate variability in the Northwest is affected by variations over the Pacific Ocean, especially a phenomenon known as El Niño-Southern Oscillation (ENSO). ENSO involves linked variations in the atmosphere and ocean in the equatorial region of the Pacific Ocean. Warm water north of Australia draws warm, moist air, which forms thunderstorms, so that the heaviest precipitation tends to occur with highest sea surface temperatures (SSTs). In a developing El Niño event, wind forcing or other factors may disrupt the normal distribution of SST, winds, and precipitation in such a way that the warm water and the heavy precipitation move eastward: warm SST anomalies appear along the equator as far east as the South American coast. (The name El Niño, for 'the [Christ] child,' was given centuries ago by fishermen who noticed the periodic disruption of normally productive fisheries by warm water near Christmastime). A typical El Niño event begins during northern hemisphere summer or fall, peaks around late December with warm water anomalies of 1°C (1.8°F) or more along the equator, and then fades during northern hemisphere spring. El Niño events, which occur irregularly with

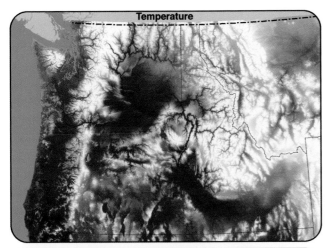

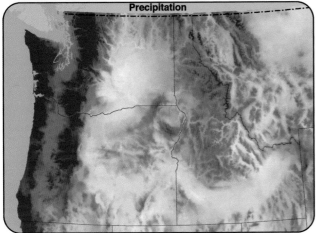

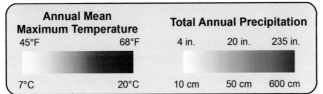

Figure 2.1 Distribution of annual mean (1981–2010) maximum temperature (top) and precipitation (bottom) from the PRISM data (Daly et al. 1994, updated; www.prism.oregonstate.edu).

a frequency of once per 2–7 years, are occasionally followed by a roughly opposite pattern called La Niña as an antonym of El Niño.

During the El Niño phase of ENSO, the wintertime jet stream in the North Pacific tends to split, with warmer air flowing into the Northwest and Alaska, and a southern branch of the jet stream directing unusually frequent and heavy storms toward southern California. Consequently winter and springs in the Northwest during El Niño events are more likely to be warmer and drier than usual (fig. 2.2; see also, e.g., Ropelewski and Halpert 1986). The warm season (not shown) shows only very weak relationships with ENSO.

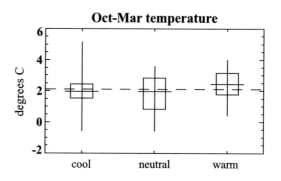

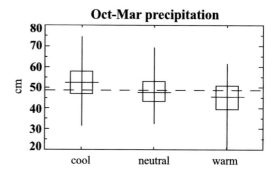

Figure 2.2 Box-and-whisker plots showing the influence of ENSO on the Northwest's cool-season climate (data are area-averaged by NOAA Climate Divisions for 1899–2000) (Mote et al. 2003). For each column, years are categorized as cool, neutral, or warm based on the Niño 3.4 index. For each climate category, the distribution of the variable is indicated as follows: range, whiskers; mean, horizontal line; top and bottom of box, 75th and 25th percentiles. The dashed line is the climatological mean.

One manifestation of ENSO in the North Pacific has been termed the Pacific Decadal Oscillation (PDO), so named because in 20th century records, variations in North Pacific SST patterns appear to have phases lasting 20–30 years (Mantua et al. 1997). However, paleo reconstructions of the PDO using tree rings (e.g., Gedalof et al. 2002) indicate a similar behavior of the PDO from the mid-18th to early 19th century, then very different behavior in the succeeding 100 years. Also, after 1998 the PDO index has shown no evidence of decadal persistence. In addition, Newman et al. (2003) show that the best statistical model of the PDO treats it not as a distinct pattern of variation independent of ENSO, but simply a slow North Pacific response to ENSO forcing.

Temperatures in the Northwest have generally been above the 20th century average for the last 30 years (fig. 2.3), with all but two years since 1998 above the 20th century average. Although the warmest year in the Northwest was 1934, most of the warmest years over the entire period of record have occurred recently, and the low-frequency variations indicate warming since the 1970s. The linear increase in temperature, over periods of record starting between 1895 and 1920 and ending in 2011, is approximately 0.7°C (1.3°F; Abatzoglou et al., in review, Kunkel et al. 2013) independent of dataset and analysis method. Trends are statistically significant and positive for every starting year before 1977. However, seasonal trends over shorter time periods can be widely varying and include a negative, albeit non-significant, trend in spring temperature for 1980–2011 (Abatzoglou et al., in review) and for the annual mean after 1985 (fig. 2.3). The occasional appearance of negative trends over short periods of record can be explained as a statistical consequence of trends that are, over short periods, small relative to variability (e.g., Easterling and Wehner 2009) and also in this case an influence of variations in

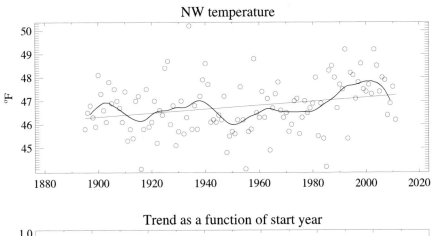

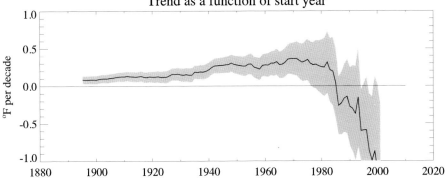

Figure 2.3 Annual mean temperature in the Northwest (Washington, Oregon, and Idaho) calculated from US Historical Climate Network data Version 2 (USHCN V2) using the Climate at a Glance utility from the National Climatic Data Center, for period of record 1895–2011. The smooth curve is computed using locally weighted regression. The bottom panel shows the slope of the linear fit to the data from starting years between 1895 and 2001, all with ending year 2011, along with the 5–95% confidence limits in the slope (shaded area).

atmospheric and oceanic circulation including ENSO conspiring to produce cooler than usual winter and spring in several recent years (Abatzoglou et al., in review).

Annual mean precipitation (fig. 2.4) has exhibited slightly (16%) higher variability since 1970, compared with the previous 75 years, a pattern observed also in streamflow in the western US (Pagano and Garen 2005). The most recent 40 years have included a number of both the wettest and driest years, including the wettest year on record, 1996, one of the driest calendar years, 1985, and the driest two "water years" (October–September), 1976–77 and 2000–01. There is no evidence to suggest that this change in precipitation variability is connected to anthropogenic climate change. The sign of linear trends has changed over time, and there is no starting year for which the trend is statistically significant either positive or negative.

Understanding the causes of these patterns of variability and change remains an active area of research. The warming trends for winter and spring can be partly attrib-

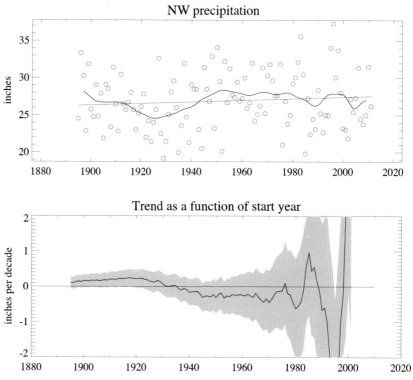

Figure 2.4 As in figure 2.3 but for precipitation.

uted to atmospheric circulation anomalies associated with ENSO and other recurrent large-scale modes of climate variability (Mote 2003; Abatzoglou and Redmond 2007). A portion of the variability in winter and spring precipitation is also associated with atmospheric circulation anomalies. Formally attributing the changes in climate to greenhouse gases and other factors, on a spatial scale this small, has not been done; Mote and Salathé (2010) noted that the average 20th century warming trend in the Northwest from climate models was very close to the observed trend of 0.8 °C (1.5 °F). Abatzoglou et al. (in review) performed statistical analysis to identify the relationships between NW seasonal climate variations and the four driving factors used by Lean and Rind (2009), viz., ENSO, volcanic, solar, and greenhouse gases; they find that seasonal trends in temperature are strongly modulated by ENSO and the Pacific North American (PNA) pattern, and that after accounting for natural factors, the remaining trends are roughly consistent with anthropogenic forcing.

2.3 Past Changes in Northwest Climate: Extremes

While the definition of mean (or average) values is straightforward, approaches to defining extremes vary considerably depending in part on application. For example, high temperature extremes could be defined by the warmest day of the year, or by a quantity that may have more relevance to impacts on human health (Gershunov et al. 2011): average minimum temperature over three consecutive days. Computing trends or long-term changes in extremes involves a tradeoff between obtaining enough events for robust statistics, and having the events be extreme enough to be consequential. It is common to achieve robust statistics in part by aggregating results over a wide area, for example the Northwest.

Bumbaco et al. (2013) examined heat waves in western Oregon and western Washington using a definition of three consecutive daytime (or nighttime) temperatures above the 99th percentile for June–September, after aggregating over sub-state spatial domains. Over the study period 1901–2009, they found no significant change in heat waves expressed as excessive daytime maximum temperatures, but a large increase since 1980 of heat waves expressed as excessively high nighttime minimum temperatures. The data had been adjusted for instrumental changes, station moves, and urban influence.

Observed changes in extreme precipitation during the past several decades are ambiguous; results depend on the period of record and the metric used. Groisman et al. (2004) examined regionally averaged trends in number of days greater than the 99th and 99.7[th] percentile of daily precipitation, over the 1908–2000 period, and trends were not statistically significant in any season. Madsen and Figdor (2007) examined station trends in the Northwest and found a statistically significant decrease in extreme precipitation in Oregon over the 1948–2006 period.

Rosenberg et al. (2010) constructed regionally averaged probability distributions from hourly station data at the Seattle, Spokane, and Portland airports, normalized by each station's long-term mean, for 1956–1980 and 1981–2005. Such analysis is necessarily restricted to the very few stations with long and fairly complete records of hourly precipitation. Results for Seattle showed increases in extreme precipitation for all definitions (annual maximum events for periods ranging from 1 hour to 10 days, and fitted 1-hour and 24-hour storms for different return periods) and ranged from about +7% for annual 1-hour storm to +37% for 50-year return period 24-hour storm. For Spokane, most definitions showed increases of 0–10%, but the largest change was -20% for 50-year 1-hour storm. For Portland, the extreme 1-hour precipitation increased across the probability distribution, whereas extreme 24-hour storms decreased slightly for the 99th percentile and increased substantially at all higher percentiles.

These analyses indicate that changes in extreme precipitation have generally been modest in the region, with some exceptions (e.g., 50-year return period for 24-hour storm in Seattle), and have been both upward and downward.

2.4 Projected Future Changes in the Northwest

Numerous modeling groups around the world have developed global climate models (GCMs) and have contributed simulations to coordinated experiments such as the Coupled Model Intercomparison Project (CMIP), which provides a framework for producing comparable simulations of global climate. The purpose of providing coordination is to help scientists and others understand and quantify the uncertainty associated with these projections. In simulating the complexities of the Earth system, many processes that are important but not completely understood (e.g., the response of cloudiness to changes in greenhouse gas forcing) are represented in different ways by different modeling groups. CMIP experiments specify a range of forcing factors that include changes in greenhouse gas concentrations that affect global and regional climate. The range of projected changes can be considered a proxy for the true uncertainty in the system; hence, the range of CMIP results provides some guidance on the range of possible outcomes. Hawkins and Sutton (2009) described the three primary contributors to uncertainties in climate projections: scenario uncertainty (i.e., concentrations of greenhouse gases and other contributors to climate change), uncertainty in the response of the climate system (usually characterized, for convenience, using the spread of results from different models), and initial condition uncertainty (usually characterized using the spread of results from different runs with the same model). The design of the CMIP experiments partly addresses these three contributors to uncertainty.

While global models were not specifically designed to simulate regional climate, the global physical consistency in GCMs along with the large number of simulations makes them a useful tool. We therefore describe below the results of two generations of CMIP experiments. Since global models' typical spatial resolution (grid boxes 100–300 km [62–186 mi] in each direction) is inadequate to represent even the largest mountain ranges in the Northwest, regional climate models (RCMs) are another way to study regional climate. Many simulations with RCMs have been performed for the Northwest at spatial scales as small as 12 km (7.5 mi), but many have only been run once, rendering estimates of uncertainty impossible. Two important exceptions are the North American Regional Climate Change Assessment Program (NARCCAP) and Regional Climate prediction.net (regCPDN).

NARCCAP is a multi-institutional program that has produced RCM simulations in a coordinated experimental approach similar to phases three and five of the CMIP (i.e., CMIP3 and CMIP5). Kunkel et al. (2013) analyzed NARCCAP results for the Northwest; at the time, there were nine simulations available using different combinations of an RCM driven by a GCM from CMIP3. Each simulation includes the periods of 1971–2000 and 2041–2070 for the SRES-A2 continued growth emissions scenario only, and is at a resolution of approximately 50 km (31 mi). Another regional modeling activity is the superensemble being generated by climateprediction.net. To date, over 200,000 one-year simulations have been generated for the period 1960–2009 using observed SSTs, and several thousand for 2029–2049 using CMIP5 SSTs. The simulations are slightly different either in how the model is formulated (i.e., parameter values are perturbed) or in the initial conditions. Volunteers contribute time on their personal computers to run the simulations. The domain is the western US, and the regional climate model, HadRM3P at 25 km (16.5 mi) resolution, is embedded in the global atmospheric model HadAM3P.

2.4.1 MEAN TEMPERATURE AND PRECIPITATION

In roughly 2005, a then-new generation of global climate model results became available from the CMIP3. These results were analyzed for the Northwest by Mote and Salathé (2010), and suggested century-scale warming (the average of years 2070–2099 minus years 1970–1999) of 3.4 °C (6.1 °F) for the continued growth peaking at mid-century (SRES-A1B) scenario of greenhouse gas emissions, and 2.5 °C (4.5 °F) for the SRES-B1 emissions scenario of substantial reductions. Projected warming varied from 1.8 to 6.1 °C (3.3 to 11 °F) across individual models and SRES scenarios, and is projected to be largest in summer. These ranges have been conditioned after considering the quality of model simulations (i.e., considering only models whose annual mean bias is less than the median of all models, from Mote and Salathé [2010], their figure 2). This consideration does not change the range in projected temperatures but does slightly reduce the upper and lower ends of the projected precipitation.

CMIP3 models project a change in annual average precipitation, averaged over the Northwest, of 3–5% with a range of -10% to +18% for 2070–2099 (Mote and Salathé 2010). Seasonally, model projections range from modest decreases to large increases in winter, spring, and fall (Mote and Salathé 2010). Projections of precipitation have larger uncertainties than those for temperature, yet one aspect of seasonal changes in precipitation is largely consistent across climate models: summer precipitation is projected to decrease by as much as 30% by the end of the century (Mote and Salathé 2010). Although NW summers are already dry, unusually dry summers have many noticeable consequences including low streamflow west of the Cascades (Bumbaco and Mote 2010) and greater extent of wildfires throughout the region (Littell et al. 2010).

We compare the newly released CMIP5 model results with CMIP3 results (fig. 2.5). The trajectories of radiative forcing are somewhat different between the two generations. By 2100, representative concentration pathway (RCP) 4.5 most closely resembles the radiative forcing of SRES-B1 (substantial reductions), whereas RCP8.5 most closely follows SRES-A1FI (very high growth) outpacing that of SRES-A2 (continued growth). Projected changes in temperature are a bit higher for the CMIP5-RCP runs than for the CMIP3-SRES runs, especially for the RCP8.5 scenario. (Note that the results of Mote and Salathé [2010] just described for CMIP3 were for later in the 21st century, so are not directly comparable to the results shown in figure 2.5). The spread in results is substantial: a factor of at least two for the annual mean and three or more for most seasons. All models project warming of at least 0.5 °C (0.9 °F) in every season. In summer, the projected warming is somewhat larger than for other seasons, especially for the CMIP5 RCP8.5 very high growth scenario, which projects changes of between 1.9 °C and 5.2 °C (3.4 °F and 9.4 °F).

For precipitation (fig. 2.6), the models have less consensus than for temperature: some models project increases and some decreases in each season. These differences originate because almost all models project increases at high latitudes and decreases in low latitudes, but vary about where in middle latitudes the zero line falls. However, a majority of models project increases in winter, spring, and fall, and a majority project decreases in summer. Annual mean changes for almost all models are small (between -5% and +14%) relative to the interannual variability (the standard deviation of the observed record is 14%), and in each season the multi-model mean changes are also small. Even in summer, when some models project decreases of 30%, the multi-model mean change is only -7%. There is a strong relationship between projected summertime changes in temperature and precipitation (not shown): the models that project the largest warming also project largest decreases in precipitation.

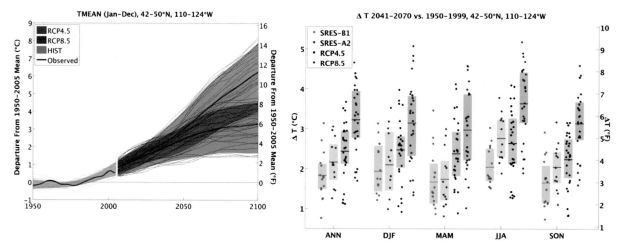

Figure 2.5 (a) Observed (1950–2011) regional mean temperature and simulated (1950–2100) regional mean temperature for selected CMIP5 global models for the emissions scenarios RCP4.5 (dashed curves, dark shading) and RCP8.5 (solid curves, light shading). (b) Changes in annual mean and seasonal temperature (2041–2070 minus 1950–1999) averaged across the Northwest, calculated from CMIP3-SRES and CMIP5-RCP simulations. Each symbol represents one simulation by one model (where more than one simulation is available, only the first is shown), and the shaded boxes indicate the interquartile range (25th to 75th percentiles). Means are indicated by thick horizontal lines in the boxes.

Figure 2.6

As in figure 2.5 (b) except for precipitation.

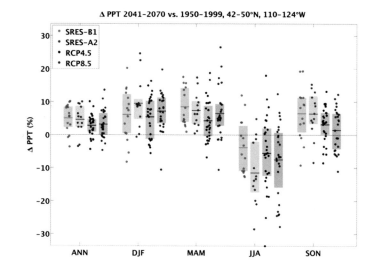

The numerical values of figures 2.5 and 2.6 are shown in table 2.1, for the CMIP5 results only (RCP4.5 and RCP8.5).

For a perspective from regional climate models, figure 2.7 compares the outputs of 15 GCMs for SRES-A2 (continued growth) and SRES-B1 (substantial reductions) for the Northwest with outputs of NARCCAP and its 4 driving GCMs. The average change in temperature of the driving GCMs (top panel) is the same as the average of the full set

Table 2.1 Summary of results shown in figures 2.5 and 2.6, for RCP4.5 and 8.5 only (labeled 4.5 and 8.5 in the table) for temperature (a) and precipitation (b).

Temp	Annual		DJF		MAM		JJA		SON	
°C	4.5	8.5	4.5	8.5	4.5	8.5	4.5	8.5	4.5	8.5
max	3.7	4.7	4.0	5.1	4.1	4.6	4.1	5.2	3.2	4.6
75th	2.9	3.9	2.8	3.8	2.9	3.9	3.3	4.4	2.8	3.7
mean	2.4	3.2	2.5	3.2	2.4	3.0	2.6	3.6	2.2	3.1
25th	2.1	2.8	2.0	2.3	1.8	2.2	2.1	3.2	1.8	2.7
min	1.1	1.7	0.9	1.3	0.5	1.0	1.3	1.9	0.8	1.6

Pcp	Annual		DJF		MAM		JJA		SON	
%	4.5	8.5	4.5	8.5	4.5	8.5	4.5	8.5	4.5	8.5
max	10.1	13.5	16.3	19.8	18.8	26.6	18	12.4	13.1	12.3
75th	4.7	6.5	10.3	11.3	8.8	9.3	2	0.7	6.7	6.5
mean	2.8	3.2	5.4	7.2	4.3	6.5	-5.6	-7.5	3.2	1.5
25th	0.9	0	-1.2	3.5	-0.4	2.8	-12.3	-15.9	0.2	-4.3
min	-4.3	-4.7	-5.6	-10.6	-6.8	-10.6	-33.6	-27.8	-8.5	-11

of GCMs, but the NARCCAP average is somewhat lower (0.3 °C [0.5 °F]). The spread in the projections is closely related to the number of ensemble members. The difference in warming projections between SRES-A2 and SRES-B1 becomes quite large by the end of the 21st century. Changes in mean annual precipitation (bottom panel) range from roughly 5% decreases to 11% increases. Multi-model mean changes are small, ranging between 0 and +3% among the different model sets for this mid-century time period.

The seasonality of change simulated by NARCCAP, as with the CMIP3 and CMIP5 global model results shown in figures 2.5 and 2.6 (Kunkel et al. 2013) is characterized by changes in temperature and precipitation that are larger in summer than other seasons, as for the GCMs. For other seasons, the spread of precipitation changes is about evenly divided between increases and decreases, but in summer (especially toward the end of the century) increases in temperature are 0.5–1 °C (1–2 °F) larger than in other seasons and a large majority of models indicate decreases in precipitation. Regionally averaged changes are similar between the GCMs and NARCCAP, indicating that while account-ing for land-atmosphere interactions at a smaller scale than can be represented at GCM grid scales may result in finer spatial patterns (fig. 2.8), it does not substantially change the regionally averaged climate response. The spatial pattern of change in NARCCAP (fig. 2.8) displays some regional texture—for instance, warming in the winter is largest in the Snake River basin, and warming in summer is smallest west of the Cascades con-sistent with the marine influence and lower rates of warming over ocean.

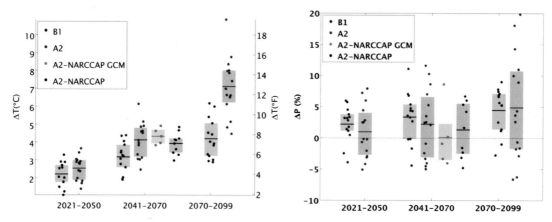

Figure 2.7 As in figures 2.5 and 2.6 but changes in annual mean temperature (a) and precipitation (b) for the time periods indicated, relative to the 1971–2000 reference period. Some of the same GCMs shown here appear also in figures 2.5 and 2.6, but with slightly different base reference periods. The 2041–2070 period also includes results from NARCCAP, both the driving GCMs (grey) and the GCM-RCM combinations (black).

Figure 2.8 Changes in temperature simulated with the NARCCAP ensemble.

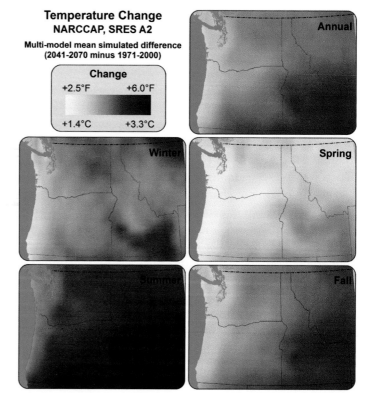

2.4.2 EXTREME TEMPERATURE AND PRECIPITATION

Climate models are unanimous that measures of heat extremes will increase and measures of cold extremes will decrease (table 2.2). For the frost-free period and number of days below cold thresholds, the changes are substantially larger than the NARCCAP standard deviations of those variables. This indicates that although all measures are consistent with an overall warming trend, the largest changes relative to the natural variability are occurring and will occur in variables measuring low temperature extremes.

Projected future changes in extreme precipitation are less ambiguous (table 2.3) than changes in total seasonal precipitation. The NARCCAP results indicate increases throughout the Northwest in the number of days above every threshold. Note that although the frequency of extremes rises in percentage with the magnitude of the extreme, the standard deviation rises faster. In other words, only modest events (>2.5 cm or 1 inch) increase by much more than one standard deviation. NARCCAP results (fig. 2.9) also indicate increases in extreme precipitation in the Northwest for 20-year return period events of 10% for the all-model average (range -4 to +22%), and 13% for 50-year events (range -5 to +28%) (Dominguez et al. 2012).

Table 2.2 The mean changes in selected temperature variables for the NARCCAP simulations (2041–2070 mean minus 1971–2000 mean, for continued growth emissions scenario SRES-A2). These were determined by first calculating the derived variable at each grid point. The spatially averaged value of the variable was then calculated for the reference and future period. Finally, the difference or ratio between the two periods was calculated from the spatially averaged values (Kunkel et al. 2013).

Variable Name	NARCCAP Mean Change	NARCCAP St. Dev. of Change
Freeze-free period	+35 days	6 days
#days Tmax > 32 °C (90 °F)	+8 days	7 days
#days Tmax > 35 °C (95 °F)	+5 days	7 days
#days Tmax > 38 °C (100 °F)	+3 days	6 days
#days Tmin < 0 °C (32 °F)	-35 days	6 days
#days Tmin < -12 °C (10 °F)	-15 days	7 days
#days Tmin < -18 °C (0 °F)	-8 days	5 days
Consecutive days > 35 °C (95 °F)	+134%	206%
Consecutive days > 38 °C (100 °F)	+163%	307%
Heating degree days	-15%	2%
Cooling degree days	+105%	98%
Growing degree days (base 10 °C [50 °F])	+51%	14%

Table 2.3 Mean changes, along with the standard deviation of selected precipitation variables from the NARCCAP simulations. As in table 2.2, values are first calculated at each grid point and then regionally averaged.

Metric of extreme precipitation	NARCCAP Mean Change	NARCCAP St. Dev. Of Change
#days with precip > 2.5 cm (1 in)	+13%	7%
#days with precip > 5.1 cm (2 in)	+15%	14%
#days with precip > 7.6 cm (3 in)	+22%	22%
#days with precip > 10.2 cm (4 in)	+29%	40%
Max run days < 0.3 cm (0.1 in)	+6 days	+3 days

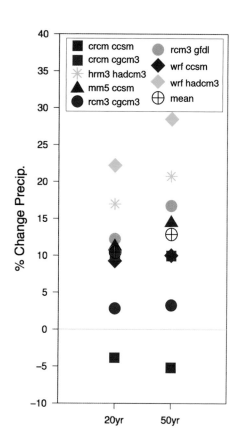

Figure 2.9 Changes in 20-year and 50-year return period precipitation events in the Northwest from NARCCAP data (model combinations indicated in legend). Adapted from Dominguez et al. (2012).

Acknowledgments

The authors acknowledge support from NOAA Climate Program Office Regional Integrated Sciences and Assessment (RISA) program for the Pacific Northwest Climate Impacts Research Consortium (CIRC) (Grant #: NA10OAR431028), and the National Institute for Food and Agriculture, award number: 2011-68002-30191. The authors also thank Francis Zwiers (Pacific Climate Impacts Consortium), Stacy Vynne (Puget Sound Partnership), and two anonymous reviewers for their comments on a previous version of this chapter.

References

Abatzoglou, J. T., and K. T. Redmond. 2007. "Asymmetry Between Trends in Spring and Autumn Temperature and Circulation Regimes over Western North America." *Geophysical Research Letters* 34: L18808. doi: 10.1029/2007GL030891.

Abatzoglou, J. T., D. E. Rupp, and P. W. Mote. "Understanding Seasonal Climate Variability and Change in the Pacific Northwest of the United States." *Journal of Climate.* In review.

Bumbaco, K. A., K. D. Dello, and N. A. Bond. 2013. "History of Pacific Northwest Heat Waves: Synoptic Patterns and Trends." *Journal of Applied Meteorology and Climatology.* doi:10.1175/JAMC-D-12-094.1.

Bumbaco, K., and P. W. Mote. 2010. "Three Recent Flavors of Drought in the Pacific Northwest." *Journal of Applied Meteorology and Climatology* 49: 2058-2068. doi:10.1175/2010JAMC 2423.1.

Daly, C., R. P. Neilson, and D. L. Phillips. 1994. "A Statistical Topographic Model for Mapping Climatological Precipitation over Mountain Terrain." *Journal of Applied Meteorology* 33: 140–158. doi: 10.1175/1520-0450(1994)033<0140:ASTMFM>2.0.CO;2.

Denton, M. H., and J. E. Borovsky. 2012. "Magnetosphere Response to High-Speed Solar Wind Streams: A Comparison of Weak and Strong Driving and the Importance of Extended Periods of Fast Solar Wind." *Journal of Geophysical Research* 117: A00L05. doi: 10.1029/2011JA017124.

Dominguez, F., E. Rivera, D. P. Lettenmaier, and C. L. Castro. 2012. "Changes in Winter Precipitation Extremes for the Western United States under a Warmer Climate as Simulated by Regional Climate Models." *Geophysical Research Letters* 39: L05803. doi: 10.1029/2011GL050762.

Easterling, D. R., and M. F. Wehner. 2009. "Is the Climate Warming or Cooling?" *Geophysical Research Letters* 36: L08706. doi: 10.1029/2009GL037810.

Forster, P., V. Ramaswamy, P. Artaxo, T. Berntsen, R. Betts, D. W. Fahey, J. Haywood, J. Lean, D. C. Lowe, G. Myhre, J. Nganga, R. Prinn, G. Raga, M. Schulz, and R. Van Dorland. 2007. "Changes in Atmospheric Constituents and in Radiative Forcing." In *Climate Change 2007: The Physical Science Basis.* Contribution of Working Group I to the Fourth Assessment Report of the Intergovernmental Panel on Climate Change, edited by S. Solomon, D. Qin, M. Manning, Z. Chen, M. Marquis, K. B. Averyt, M. Tignor, and H. L. Miller. Cambridge University Press, Cambridge, United Kingdom and New York, NY, USA.

Gedalof, Z., N. J. Mantua, and D. L. Peterson. 2002. "A Multi-Century Perspective of Variability in the Pacific Decadal Oscillation: New Insights from Tree Rings and Coral." *Geophysical Research Letters* 29 (24): 2204. doi:10.1029/2002GL015824.

Gershunov, A., Z. Johnston, H. G. Margolis, and K. Guirguis. 2011. "The California Heat Wave 2006 with Impacts on Statewide Medical Emergency: A Space-Time Analysis." *Geography Research Forum* 31: 6-31.

Groisman, P. Y., R. W. Knight, T. R. Karl, D. R. Easterling, B. Sun, and J. H. Lawrimore. 2004. "Contemporary Changes of the Hydrological Cycle over the Contiguous United States: Trends Derived from In Situ Observations." *Journal of Hydrometeorology* 5: 64-85. doi: 10.1175/1525-7541(2004)005<0064:CCOTHC>2.0.CO;2.

Hawkins, E., and R. Sutton. 2009. "The Potential to Narrow Uncertainty in Regional Climate Projections." *Bulletin of the American Meteorological Society* 90 (8): 1095-1107. doi: 10.1175/2009BAMS2607.1.

Kunkel, K. E., L. E. Stevens, S. E. Stevens, L. Sun, E. Janssen, D. Wuebbles, K. T. Redmond, and J. G. Dobson. 2013. "Regional Climate Trends and Scenarios for the U.S. National Climate Assessment. Part 6. Climate of the Northwest U.S." NOAA Technical Report NESDIS 142-6, 75 pp.

Lean, J. L., and D. H. Rind. 2009. "How Will Earth's Surface Temperature Change in Future Decades?" *Geophysical Research Letters* 36: L15708. doi:10.1029/2009GL038932.

Littell, J. S., E. E. Oneil, D. McKenzie, J. A. Hicke, J. Lutz, R. A. Norheim, and M. M. Elsner. 2010. "Forest Ecosystems, Disturbance, and Climatic Change in Washington State, USA." *Climatic Change* 102: 129-158. doi: 10.1007/s10584-010-9858-x.

Madsen, T., and E. Figdor. 2007. "When it Rains, it Pours: Global Warming and the Rising Frequency of Extreme Precipitation in the United States." Environment America Research and Policy Center, Boston, MA.

Mantua, N. J., S. R. Hare, Y. Zhang, J. M. Wallace, and R. C. Francis. 1997. "A Pacific Interdecadal Climate Oscillation with Impacts on Salmon Production." Bulletin of the American Meteorological Society 78: 1069–1079. doi: 10.1175/1520-0477(1997)078<1069:APICOW>2.0.CO;2.

Mote, P. W. 2003. "Trends in Temperature and Precipitation in the Pacific Northwest." Northwest Science 77 (4): 271-282. http://www.vetmed.wsu.edu/org_nws/NWSci%20journal%20articles/2003%20files/Issue%204/v77%20p271%20Mote.PDF.

Mote, P. W., and E. P. Salathé Jr. 2010. "Future Climate in the Pacific Northwest." Climatic Change 102: 29-50. doi: 10.1007/ s10584-010-9848-z.

Mote, P. W., E. A. Parson, A. F. Hamlet, W. S. Keeton, D. Lettenmaier, N. Mantua, E. L. Miles, D. W. Peterson, D. L. Peterson, R. Slaughter, and A. K. Snover. 2003. "Preparing for Climatic Change: The Water, Salmon, and Forests of the Pacific Northwest." Climatic Change 61, 45-88. doi: 10.1023/A:1026302914358.

Newman, M., G. P. Compo, and M. A. Alexander. 2003. "ENSO-Forced Variability of the Pacific Decadal Oscillation." Journal of Climate 16: 3853–3857. doi: 10.1175/1520-0442(2003)016<3853:EVOTPD>2.0.CO;2.

Pagano, T., and D. Garen. 2005. "A Recent Increase in Western U.S. Streamflow Variability and Persistence." Journal of Hydrometeorology 6 (2): 173-179. doi: http://dx.doi.org/10.1175/JHM410.1.

Ropelewski, C. F., and M. S. Halpert. 1986. "North American Precipitation and Temperature Patterns Associated with the El Niño/Southern Oscillation (ENSO)." Monthly Weather Review 114: 2352-2362. doi: 10.1175/1520-0493(1986)114<2352:NAPATP>2.0.CO;2.

Rosenberg, E. A., P. W. Keys, D. B. Booth, D. Hartley, J. Burkey, A. C. Steinemann, and D. P. Lettenmaier. 2010. "Precipitation Extremes and the Impacts of Climate Change on Stormwater Infrastructure in Washington State." Climatic Change 102: 319-349. doi: 10.1007/s10584-010-9847-0.

Chapter 3

Water Resources
Implications of Changes in Temperature and Precipitation

AUTHORS
Rick R. Raymondi, Jennifer E. Cuhaciyan, Patty Glick, Susan M. Capalbo,
Laurie L. Houston, Sarah L. Shafer, Oliver Grah

3.1 Introduction

Climate projections indicate that the Northwest (NW) will experience temperature increases in both cool and warm seasons and a reduction in summer precipitation with increases in fall and winter precipitation (Mote and Salathé 2010; see Chapter 2). Also, there has been an observed trend of increasing variability in cool season precipitation in the western United States since about 1973 (Hamlet and Lettenmaier 2007). Altered temperature and precipitation regimes affect snowpack (Hamlet et al. 2005), the inter-seasonal distribution of flow (Hidalgo et al. 2009), lake and stream temperatures (Mantua et al. 2010), and water quality. Changes in the seasonality and variability of temperature and precipitation have important consequences for the regional economy because of their potential impacts on irrigated agriculture, hydropower generation, floodplain infrastructure, municipal water supply, natural systems, and recreation. The effects of climate change on hydrologic systems may require adaptation initiatives and measures to reduce the potential vulnerability of natural and human systems (Walker et al. 2011).

Hydrologic responses to a changing climate are likely to display significant spatial- and temporal-variability. The magnitude and spatial distribution of future temperature and precipitation changes will be influenced by general location (e.g., east or west side of the Cascade Range as shown on fig. 3.1), while shorter-term climate patterns (e.g., Pacific Decadal Oscillation and El Niño-Southern Oscillation; see Chapter 2) are expected to periodically enhance and dampen long-term trends (Rieman and Isaak 2010; Mote et al. 2003). Hydrologic response will depend upon a watershed's dominant form of precipitation as well as other local characteristics including elevation, aspect, geology, vegetation, and changing land use (Mote et al. 2003; Safeeq et al. 2012). Safeeq et al. (2012) note the importance of watershed geology and drainage efficiency on the sensitivity of various parts of a hydrograph to climate warming effects.

Several studies have classified NW watersheds as either snowmelt dominant, rain dominant, or mix rain-snow based on the snow water equivalent (SWE) in the April 1st snowpack (Hamlet and Lettenmaier 2007; Mantua et al. 2013; Elsner et al. 2010; Hamlet et al. 2013). Figure 3.2 shows the historical distribution of these NW watersheds based on data from the 1916–2006 water years and their projected distribution as a result of climate warming (Hamlet et al. 2013).

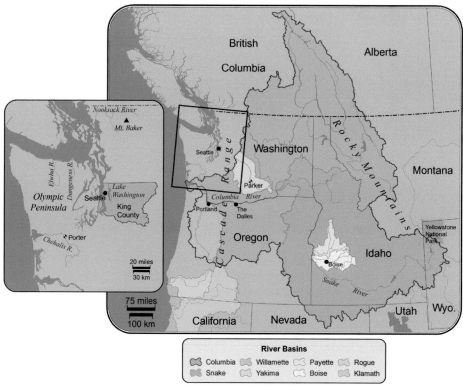

Figure 3.1 Northwest locations and river basins mentioned in this chapter.

Figure 3.2 The classification of NW watersheds into rain dominant, mixed rain-snow, and snowmelt dominant and how these watersheds are expected to change as a result of climate warming based on the SRES-A1B scenario of continued growth of greenhouse gas emissions peaking at mid-century (Hamlet et al. 2013).

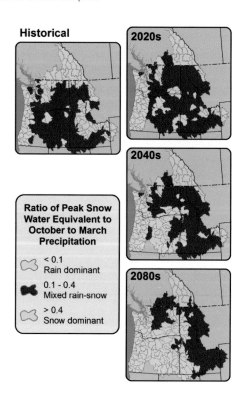

Snowmelt dominant watersheds in the Northwest are located in moderate- to high-elevation inland areas where cool season (October–March) precipitation falls as snow. In snow dominant basins, the peak runoff lags behind the peak period of precipitation, since much of the cool season precipitation occurs as snow and is stored until springtime temperatures rise above freezing (Oregon Department of Land Conservation and Development 2010). Mountain snowpack in these watersheds supply warm season (April–September) streamflows (Chang et al. 2010) that are important for migrating salmonids and are heavily relied upon by irrigators, hydropower producers, municipalities, and other users.

Rain dominant watersheds are generally in lower elevations, mostly on the west side of the Cascade Range (fig. 3.2), receive little snowfall, and produce peak flows throughout the winter months. Mixed rain-snow watersheds located in mid-range elevations (1,000–2,000 m [3,280–6,560 ft]) primarily east of the Cascade Range and in lower elevations in Idaho, receive a mix of rain and snow during the cool season (Elsner et al. 2010). These watersheds, with average mid-winter temperatures close to freezing, are particularly sensitive to the trend of increasing temperatures that shift winter precipitation toward more rain and less snow (Elsner et al. 2010; Lundquist et al. 2009). Mixed rain-snow watersheds can experience more than one peak flow event throughout the winter and are particularly susceptible to rain-on-snow events that can cause flooding in lowland areas.

Hydrographs of simulated average historical streamflows representative of the three watershed types were developed by Elsner et al. (2010) as shown in figure 3.3. The Chehalis River drains to the Pacific Ocean along the Washington coast, and the watershed is characterized as rain dominant. The Yakima River drains to the Columbia River from a characteristic mixed rain-snow watershed. Finally, the Columbia River drains from

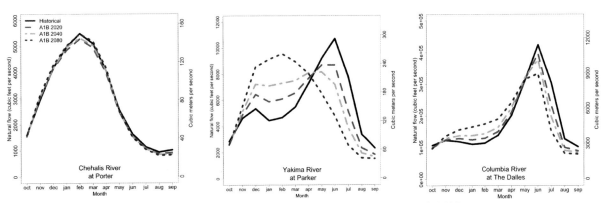

Figure 3.3 Simulated monthly streamflow hydrographs for the historical baseline (1916–2006 average, black) and the 2020s (blue), 2040s (yellow), and 2080s (red) under the SRES-A1B scenario of continued emissions growth peaking at mid-century (after Elsner et al. 2010) for three representative watershed types in the Northwest, namely rain dominant (Chehalis River at Porter, top), mixed rain-snow (Yakima River at Parker, center), and snowmelt dominant (Columbia River at The Dalles, bottom).

mountainous regions mainly in Canada, Washington, Oregon, and Idaho from a characteristic snowmelt dominant watershed overall (fig. 3.3).

Given the likelihood of increased winter air temperatures, snowmelt dominant and mixed rain-snow watersheds are projected to gradually trend towards mixed rain-snow and rain-dominant, respectively. The shift from snowmelt dominant to mixed rain-snow conditions will result in reduced peak streamflow, increased winter flow, and reduced late summer flow in these watersheds. Watersheds that shift from mixed rain-snow conditions to rain dominant will experience less snow and more rain during the winter months. Rain dominant watersheds are expected to experience minimally changed (Elsner et al. 2010) to higher winter streamflows (with relatively little change in timing) (MacArthur et al. 2012) as a result of projected increases in average winter precipitation. By the 2080s, a complete loss of snowmelt dominant basins is projected for the Northwest under the SRES-A1B emissions scenario of continued growth peaking at mid-century (fig. 3.2; Hamlet et al. 2013; Mantua et al. 2010; Nakićenović et al. 2000).

3.2 Key Impacts

3.2.1 SNOWPACK, STREAM FLOW, AND RESERVOIR OPERATIONS

A robust mountain snowpack is the most important component of the annual water supply for many watersheds in the Northwest (Graves 2009). Significant consequences of a warming climate for snowmelt dominant and mixed rain-snow watersheds are a reduction in snowpack and a substantial shift in streamflow seasonality (Barnett et al. 2005; Stewart et al. 2005; Adam et al. 2009; Leppi et al. 2011). Seasonal peak runoff timing is projected to shift, with more runoff occurring in late winter rather than during the spring and with lower summer flows (Elsner et al. 2010). Hydrologic models project that by mid-century, the peak runoff from snowmelt in NW streams will occur approximately three to four weeks earlier than the current average (US Bureau of Reclamation [USBR] 2008; Adam et al. 2009; Hamlet, Lee, et al. 2010; Elsner et al. 2010).

Water management efforts in the Northwest are likely to be affected by these hydrologic changes. Reservoir operations in regional basins (e.g., Rogue River, Oregon; Boise and Payette Rivers, Idaho; Yakima River, Washington) often have multiple objectives including irrigation delivery, hydropower production, flood control, recreation, and instream flow augmentation for fish. Projected future reductions in snowpack, shifts in streamflow seasonality, and warmer, drier summers combined with increased water demand will pose challenges for water management. These hydrologic changes require complex tradeoffs among reservoir operation objectives and have potential consequences for many important components of the regional economy, including irrigated agriculture, hydropower production, and Pacific salmon (Kunkel et al. 2013; Isaak, Muhlfeld, et al. 2012).

The design of the water management system is based upon the historical seasonal timing of snowmelt runoff and the ability of the snowpack to act as a natural reservoir by storing water during the cool season (Barnett et al. 2005; Markoff and Cullen 2008; Adam et al. 2009) and gradually releasing it in the spring and early summer. The total reservoir storage capacity in the Columbia River Basin is only about 30% of the annual flow at The Dalles, Oregon (Bonneville Power Administration 2001). The ability of water

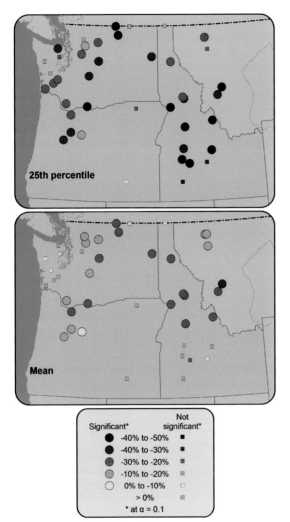

Figure 3.4 Adapted from a study by Luce and Holden (2009), these maps depict the changes in 25th percentile annual flow (top), and mean annual flow (bottom) at streamflow gauges across the Northwest for 1948–2006. Circles represent statistically significant trends (at α=0.1), whereas squares represent locations where trends were not statistically significant.

managers to capture earlier peak season runoff is limited by available storage space and the requirements for flood control operations. Reservoir managers face a difficult balance between storing as much water as possible to satisfy warm season water demands and maintaining enough space in the system to capture flood waters and minimize flood risk downstream. A shift in the timing of peak flows by several weeks to a month earlier in the year could result in an earlier release of water from reservoirs to create space for flood control (USBR 2011a) and a loss of storage supply for other objectives if the system is unable to refill.

Summers in the Northwest are relatively dry and exhibit the lowest frequency of convective storms in the conterminous United States (Kunkel et al. 2013). Higher warm season temperatures may increase evapotranspiration (Chang et al. 2010) and when combined with a reduction in summer precipitation, have the potential to further reduce stream discharge during the period of greatest water demand (Washington State Department of Ecology 2011; USBR 2011c). Recent studies of historical data highlight the trend of lower August stream discharge in Idaho and the central-Rocky Mountains

(Leppi et al. 2011) and, as illustrated in figure 3.4, a trend of the driest years becoming drier (Luce and Holden 2009). Such hydrologic impacts are likely to cause agricultural, municipal, hydropower, and instream demands during the late summer to become increasingly difficult to satisfy in any given year.

If current trends in warming and increased precipitation variability continue, extreme events (droughts, flooding, etc.) may occur with greater frequency, magnitude, and year-to-year persistence (Hamlet and Lettenmaier 2007; Pagano and Garen 2005). In other words, both extreme wet conditions and extreme dry conditions (compared to the historical record) may become more common as well as their persistence from one year to the next. Such impacts, if they continue, are particularly likely to cause problems for water managers as extended stretches of wet or dry years may overwhelm or exhaust reservoir systems.

3.2.2 WATER QUALITY

A warming climate is also likely to have important impacts on water quality. Increasing air temperatures have been shown to result in higher instream temperatures (Isaak et al. 2010; Isaak, Wollrab, et al. 2011; Bartholow 2005; Petersen and Kitchell 2001) and subsequent decreases in dissolved oxygen levels; both of which are important factors in the health and survival of endangered aquatic species. Meanwhile, higher peak flows and increased wildfire activity resulting from climate change are likely to increase sediment (Cannon et al. 2010; Goode et al. 2012) and nutrient loads to rivers and streams (Furniss et al. 2010; Chang et al. 2010) and have important consequences for water supplies and aquatic habitats. The temporal variability of these loads (sediment and phosphorus) is also expected to be altered by the changes in flow variability; as such loads typically increase during high flow events (Chang et al. 2010).

Increasing temperatures may also affect the water quality in lakes and reservoirs through earlier onset of thermal stratification and reduced mixing between layers (Meyer et al. 1999; Winder and Schindler 2004). Such conditions often result in reduced oxygen levels in bottom layers and the development of anoxic conditions in bottom sediments.

3.3 Consequences for Specific Sectors

3.3.1 IRRIGATED AGRICULTURE

Nationwide, the average value of production for an irrigated farm is more than three times the average value for a dryland farm (Schaible and Aillery 2012). Irrigated agriculture represents over 90% of the consumptive water use in the Columbia River Basin (Washington State Department of Ecology 2011) and is the predominant demand on regional reservoir systems (USBR 2011c). Current data show that 21%, 27%, and 48% of the cropland in Washington, Oregon, and Idaho, respectively, is irrigated (US Department of Agriculture Economic Research Service 2012). There are approximately 9.9 billion m³/year (8.1 million acre-feet/year) of total irrigation withdrawal and 4.4 billion m³/year (3.6 million acre-feet/year) consumptive irrigation use (45% of withdrawals) in the Columbia River Basin (excluding the part of the Columbia Basin in Canada and the area draining into the Snake River). The annual streamflow of the Columbia River at The Dalles, Oregon is 17.2 billion m³/year (139 million acre-feet/year) (Izaurralde et al. 2010).

Projected future precipitation decreases and higher temperatures during the summer months are likely to increase irrigation demand in the Northwest (USBR 2011c). According to a study by Washington State Department of Ecology (2011), the 2030 forecast demand for irrigation water across the entire Columbia River Basin (seven US states and British Columbia) is 16.8 billion m^3/year (13.6 million acre-feet/year) under average flow conditions, assuming an equivalent land base for future irrigated agriculture. Estimates range from 16.2 to 17.4 billion m^3/year (13.1 to 14.1 million acre-feet/year) during wet and dry years, respectively (20th and 80th percentile). The approximate 2.2% projected increase in irrigation demand is attributed to the combined effects of climate change and changes in crop mix driven by growth in the domestic economy and international trade.

Recent studies also indicate that a warming climate with an earlier loss of snow cover (McCabe and Wolock 2010) and a projection of at least 20 more days in the annual frost-free season in the region (Kunkel et al. 2013) would increase the length of the growing season, which could increase agricultural consumptive water use and thus water demand (USBR 2011c). Hoekema and Sridhar (2011) showed evidence of an increasing trend of springtime surface water diversions for irrigation within low- and mid-elevation reaches of the Snake River Basin that were attributed to an earlier loss in snow cover and the resulting drier early season soil moisture conditions.

Vano et al. (2010) simulated potential climate change effects on reservoir system operations and irrigated agriculture in the Yakima River Basin. Using modeled historical streamflow and current water demands and infrastructure, the simulated Yakima River Basin experienced water shortages (i.e., years in which substantial prorating of deliveries to junior water users was required) in 14% of years between 1940 and 2005. Using downscaled climate simulations from 20 climate models, Vano et al. (2010) showed that the number of years with water shortages under the SRES-A1B scenario (continued growth of greenhouse gas (GHG) emissions peaking at mid-century) is projected to increase from the historical 14% to 27% (with a range of 13–49% acres of the 20 models) in the 2020s, to 33% in the 2040s, and to 68% in the 2080s without adaptations. For the SRES-B1 scenario characterized by substantial emissions reductions, water shortages occur in 24% (7–54%) of years in the 2020s, 31% for the 2040s, and 43% for the 2080s. The scenarios also indicate an increasing frequency of historically unprecedented conditions in which senior water rights holders suffer shortfalls (Vano et al. 2010). Such water shortages could impact the amount of acreage in the region that can be irrigated and the amount of water that can be applied during the growing season.

If water shortages result in less water for irrigation, the total value of both agricultural production and agricultural land in the region may be reduced substantially, although it is difficult to predict how producers will attempt to mitigate for water shortages within a growing season. Mitigation strategies of producers might include: allowing for selective deficit irrigation of less profitable crops (Washington State Department of Ecology 2011); switching to or supplementing with groundwater for irrigation if that resource is not already fully appropriated; switching to non-irrigated crops, drought resistant crop varieties, or less intensive crop rotations; or switching to more efficient irrigation systems and intensive irrigation management techniques. These changes, combined with the impacts of warming, increasing atmospheric CO_2, precipitation variability, and other climate changes may present challenges for agronomists and farmers (Hatfield et al.

2011). The effects of increasing atmospheric CO_2 concentrations on agriculture and crop irrigation demand are discussed in greater detail in Chapter 6.

3.3.2 HYDROPOWER

Hydropower is the predominant source of electricity in the Northwest, providing about two-thirds of its electricity and 40% of all US hydropower (Northwest Power and Conservation Council 2012). According to statistics from the US Energy Information Administration (USEIA), most of this hydroelectric power is generated from facilities on the Columbia River. In 2011, Washington was the leading producer of hydroelectricity, producing 29% of the nation's net electricity generation (USEIA 2012). As such, hydropower is an extremely important factor in the NW economy.

Summer water supplies for hydropower in the region are highly dependent on snowpack. In much of the Cascade Range, snow accumulates close to the melting point, meaning that modest changes in winter temperature (e.g., 1–2 °C [1.8–3.6 °F]) can significantly increase the rate of snowmelt and cause earlier streamflow (Nolin and Daly 2006). Earlier snowmelt would reduce opportunities for hydropower generation in the late spring and summer, when rainfall is limited (Payne et al. 2004). Given that hydropower facilities have historically relied on snowmelt to provide dry season streamflows, the projected rates of accelerated snowmelt for the Cascade Range indicated by Payne et al. (2004) and Elsner et al. (2010) would substantially affect streamflow timing and hydropower generation in the Northwest.

Hamlet, Lee, et al. (2010) made use of composite temperature and precipitation simulations that are spatial (regional) and temporal (monthly) averages of climatic changes simulated by 20 general circulation models (GCMs) for three time periods (2010–2039, 2030–2059, and 2070–2099), and two emissions scenarios (SRES-A1B, continued growth peaking at mid-century; and SRES-B1, substantial reductions) to evaluate the potential impacts of climate change on hydropower production. The study projects increases in winter power production of up to 4% by 2040 compared to historical 1917–2006 levels, and about 10% by 2080, while summer power production is projected to decline by about 10%, 15%, and 20% by 2020, 2040, and 2080, respectively.

Indirect effects on hydropower production (i.e., reduced generation) related to climate change may result from adaptation for other competing water management objectives including flood control operations, instream flow augmentation, and possible renegotiation of the Columbia River Treaty (Washington State Department of Ecology 2011; Hamlet, Lee, et al. 2010). With the limited storage capacity of the Columbia River Basin and the requirement to maintain flows for endangered species, policy decisions will need to be made in order to balance the competing demands, and other sources of electricity may need to be considered.

3.3.3 FLOODPLAIN INFRASTRUCTURE

There has been an observed increase in the annual variability of cool season precipitation since about 1973 in the Northwest (Chapter 2; Hamlet and Lettenmaier 2007). Observed trends in flood risk indicate that NW basins have had a variety of responses to recent climatic variability and change. Relatively warm rain-dominant basins (>5 °C (41 °F) average in midwinter) show little systematic change. Mixed rain-snow basins show high sensitivity but no universal direction of change, with changes that range from a 30% decrease to a 30% increase in flood magnitude (Hamlet and Lettenmaier 2007). Model simulations indicate that the largest projected increases in flood magnitude and frequency are in mixed rain-snow watersheds during the winter (Mantua et al. 2010).

Urbanization of watersheds with an accompanying decrease in permeable surface area can affect the hydrologic response of basins to precipitation events. As land with permeable surface area (e.g., fields and woodlands) is converted to buildings, roads, and parking lots, it loses its ability to absorb rainfall. As a result, rainfall flows into streams at a much faster rate resulting in floodwaters that rise and peak very rapidly. Development or encroachment on floodplains and floodways may cause floodwaters to expand and rise above historical levels (Mid-Willamette Valley Council of Governments and Oregon Natural Hazards Workgroup 2005).

Projected increases in flooding related to climate change may pose even greater risks to developed areas in floodplains, urban areas, roads, stormwater systems, and other infrastructure at water crossings such as pipelines, bridges, and culverts (Climate Impacts Group 2012). Extreme precipitation events have the potential to cause localized flooding due partly to inadequate capacity of storm drain systems. Extreme events may damage or cause failure of dam spillways (Oregon Department of Land Conservation and Development 2010). Heavy rainfall can also saturate soils and increase risk of landslides, particularly in areas with unstable slopes or disturbed vegetation, potentially damaging roadways and other infrastructure (Oregon Department of Land Conservation and Development 2010). Impacts on transportation systems can impose delays on the movement of goods and the traveling public (Walker et al. 2011), and the costs of operating and maintaining transportation infrastructure (e.g., bridges and culverts) are also expected to increase (MacArthur et al. 2012). Flooding and erosion along forest road networks may damage culverts and generate increased sediment loads that can affect salmon and steelhead spawning, migration, and rearing habitat (Climate Impacts Group 2012).

3.3.4 MUNICIPAL DRINKING WATER SUPPLIES

Municipal water demands, including domestic and municipally-supplied industrial water, are likely to increase throughout the entire Columbia River Basin over the next 20 years based on population estimates and projected impacts of climate change (Washington State Department of Ecology 2011). Future hydrologic conditions are projected to include warmer stream temperatures, lower summer flows, and more frequent extreme events that may damage or stress the reliability of the current water infrastructure. NW public water suppliers facing shortages may be required to invest in capital improvements to acquire, treat, and distribute water from new sources to assure adequate availability of drinking water (Chang et al. 2010). With lower summer flows, it is projected that diversification and development of water supplies, reducing water demand, improving water-use efficiency, initiating operational changes at reservoirs, increasing water transfers between users, and increasing drought preparedness would be required (Whitely Binder et al. 2009).

NW state and local government agencies and private concerns have recently initiated planning processes in anticipation of future hydrologic conditions. The State of

Washington has developed an *Integrated Climate Change Response Strategy* that includes involving communities in water resource management approaches in highly vulnerable basins (Washington State Department of Ecology 2012). The strategy recommends expanding and accelerating implementation of municipal water efficiency improvements to reduce the amount of water used per person or household and seeking more reliable funding mechanisms to help water providers implement climate-ready plans and practices. Oregon's Integrated Water Resources Strategy advocates water conservation and reuse within municipalities to decrease water demand (Oregon Water Resources Department 2012). In 2007, the Water Utility Climate Alliance (http:// www.wucaonline.org), which includes the Portland Water Bureau and the Seattle Public Utilities, was formed to provide leadership and collaboration on climate change issues affecting drinking water utilities and to assist in integrating climate change information into local planning.

3.3.5 FRESHWATER AQUATIC ECOSYSTEMS

Rivers, lakes, and wetlands in the Northwest provide important habitat for a number of native and endangered aquatic species. The reduced resilience in some of these ecosystems, resulting from other anthropogenic pressures (urbanization, logging, agriculture, etc.) and their strong dependency on temperature and flow regimes, makes these systems particularly sensitive to the effects of climate change (Independent Science Advisory Board 2007; Rieman and Isaak 2010; Poff et al. 2002). The response of aquatic and terrestrial species to future climate changes will be complex and may be mediated by a number of other factors, including land use changes and interactions with other species (e.g., invasive species) (Chambers and Wisdom 2009). As human populations respond to climate change and make changes to the wastewater, stormwater, and water supply infrastructure, these new projects are likely to have implications for aquatic ecosystems as well.

Changes in hydrologic regimes (i.e., the timing and extent of streamflow) have been observed in recent historical data (Luce and Holden 2009). These changes are likely to result in a wide range of consequences for natural systems and are expected to alter key habitat conditions for salmon and other anadromous fish that depend on specific conditions for spawning and migration (box 3.1) (Mantua et al. 2009; Mantua et al. 2010). For example, increased winter and early-spring streamflows have the potential to scour eggs or wash away newly emerged fry from fall-spawning salmon and trout species (Isaak, Muhlfeld, et al. 2012; Mantua et al. 2010; Wenger et al. 2011). In addition, extreme low summer flows can limit the ability for some species to migrate upstream to spawn (Battin et al. 2007). The impacts of climate change on the region's salmonids will vary across the region and among different species, populations, life-stages, and site characteristics.

In addition to altered hydrologic regimes, warming stream temperatures also pose significant threats to aquatic ecosystems. Increasing trends in water temperature of lakes and streams have been observed in recent historical data (Isaak et al. 2010; Isaak, Wollrab, et al. 2011; Bartholow 2005; Petersen and Kitchell 2001). Such changes may affect the health of aquatic populations and the extent of suitable habitat for many species. Relative to the rest of the United States, NW streams dominated by snowmelt runoff appear to be temporarily less sensitive to warming due to the temperature buffering provided by snowmelt and groundwater contributions to these streams (Mohseni et

BOX 3.1
A Salmon Runs Through It

Pacific salmon (*Oncorhynchus* spp.) are an important component of many NW systems. The presence of wild salmon is an indicator of the health of the region's lakes, rivers, estuaries, and ocean. The fish play an important role in ecosystems, including providing a critical food source for a plethora of wildlife, from tiny invertebrates to bald eagles, grizzly bears, and orcas. For the people who call the Northwest home, salmon are a fundamental part of their ecological, economic, and cultural heritage. Salmon sustain the spiritual and physical well-being of the region's American Indian tribes as well as supporting recreational and commercial fishing industries that contribute millions of dollars to the regional economy each year.

The historic decline of wild salmon in the Northwest has galvanized the region and country around numerous efforts to restore and protect the populations that remain—a significant challenge that is all the more so given projected future climate change. Higher water temperatures, shifts in streamflows, and altered estuary and ocean conditions associated with projected climate change will affect the region's native salmon throughout their complex life cycles:

- Higher stream temperatures will affect habitat quality for salmon in all of their freshwater life stages (Independent Science Advisory Board 2007).

- Reduced summer streamflows will contribute to warmer temperatures and make it more difficult for migrating salmon to pass both physical and thermal obstacles (Beechie et al. 2006; Mantua et al. 2010).

- Heavier rainfall and increased flooding in the fall and winter will scour salmon nests (DeVries 1997).

- Earlier spring runoff will alter migration timing for salmon smolts in snowmelt-dominated streams (Mantua et al. 2010).

- Rising sea level, warmer ocean temperatures, and changes in freshwater flows will contribute to significant changes in estuarine habitats (Bottom et al. 2005).

- Higher average ocean temperatures and ocean acidification will alter the marine food web, reducing the survivability of salmon when conditions are unfavorable (Pearcy 1992; Orr et al. 2005).

The impacts of climate change will vary among different species and populations, and will depend on multiple and diverse factors. Indeed, the diverse habitat needs and behavior of Pacific salmon have been fundamental to their historic resilience. As different salmon species and populations within species evolved over time, they acquired diverse spawning and migratory behaviors to take advantage of variations in temperatures, streamflow, ocean conditions, and other habitat features (Mantua et al. 2010); these characteristics now shape their vulnerability to climate change. For example, steelhead (*Oncorhynchus mykiss*), "stream-type" chinook salmon (*O. tshawytscha*), sockeye salmon (*O. nerka*), and coho salmon (*O. kisutch*) are particularly sensitive to changes in stream conditions as young fish remain in freshwater habitats for a year or more after hatching before migrating to the sea. The adults then return in the spring and summer, often taking several months to migrate upstream to high-elevation headwater streams to spawn (Mantua et al. 2010). For these populations, higher stream temperatures and altered streamflows due to climate change are likely to be significant limiting factors. In contrast, young "ocean-type" chinook, pink salmon (*O. gorbuscha*), and chum salmon (*O. keta*) migrate to the sea just a few months after hatching, spend much time acclimating in estuary waters before their ocean life cycle, and the adults return to spawn in the summer and fall in the mainstream river and lower reaches of tributaries. Accordingly, changes in estuarine habitats are likely to be especially important. Understanding these complexities will be necessary to effectively address the added stressors associated with climate change in salmon restoration efforts across the Northwest.

al. 1999). However, as snowpacks decline, the future sensitivity to warming is likely to increase in these areas (Rieman and Isaak 2010).

While higher water temperatures are likely to benefit some cool- and warm-water species (both native and non-native), the consequences are likely to be adverse for cold-water species, particularly salmonids and other species that are already living under conditions near the upper range of their thermal tolerance (Richter and Kolmes 2005). Among the potential impacts, studies suggest that increasing water temperatures are likely to cause various species of salmonids to become more susceptible to disease and experience increased rates of mortality and predation (Crozier et al. 2008; Keefer et al. 2008; Keefer et al. 2009; Keefer et al. 2010; Petersen and Kitchell 2001).

The region's salmonids will respond to changes in stream temperatures and hydrology in diverse ways (Salinger and Anderson 2006). For example, in a study of 18 populations of juvenile Snake River (Idaho) spring and summer Chinook salmon, Crozier and Zabel (2006) found that populations inhabiting wider, warmer streams are likely to be more sensitive to higher summer temperatures, and those inhabiting narrower, cooler streams are more sensitive to reduced fall streamflows.

Rising stream temperatures will likely cause the suitable habitat for many species to shift further upstream. A recent study by Isaak and Rieman (2012) predicted that under a mid-range air temperature increase projection (2 °C [3.6 °F]), stream temperature gradients across the Northwest could shift 5–143 km (~3–89 miles) upstream by 2050. Culverts and other infrastructure, as well as changes to the channel structure and flow regime (lower summer streamflows resulting from earlier snowpack melt), may pose significant barriers to upstream migration and limit available habitat (Isaak, Muhlfeld, et al. 2012; Rieman et al. 2007; Mantua et al. 2010; Luce and Holden 2009; Rieman and Isaak 2010).

In general, seasonal snowpack has the most important control over streamflow in a changing climate. However, glacier melt resulting from climate change has important consequences for North Cascade rivers where glacier melt can comprise 10–30% of summer flows (Riedel and Larabee 2011). Several studies have noted the decreasing mass, extent, and volume of North Cascade glaciers due to melting, sublimation, and calving (Harper 1992; National Park Service 2012; Pelto 2006; Pelto 2010; Pelto 2011; Pelto and Brown 2012; Post et al. 1971; Riedel and Larrabee 2011). Those glaciers with a thinning accumulation zone, an emergence of new outcrops, and recession of margins, which includes 10 of 12 North Cascade glaciers with annual measurements, are not forecast to survive the current climate (Pelto 2010). Observations of accumulation area ratio (the ratio of a glacier's accumulation area to its total area) are frequently below 30%. These observations suggest a lack of consistent accumulation, a trend that may continue in the future (Pelto 2010). Those glaciers with the lowest mean elevation have experienced, and may continue to experience, the most dramatic changes in total volume. Conversely, higher elevation glaciers, like those on Mt. Baker (Washington), have the potential to approach equilibrium with the current climate conditions, however equilibrium is unlikely to occur if mean temperatures continue to increase (Pelto 2010; Pelto and Brown 2012).

Changes to glaciers as a result of climate change would have a direct effect on the magnitude and timing of streamflow and stream temperatures (Dickerson-Lange and Mitchell, in review; R. Mitchell, pers. comm.; Mantua et al. 2010; Riedel and Larabee 2011). The gross glacial melt contribution to these river systems will eventually decrease as atmospheric temperatures rise and glacial extent decreases (Dickerson-Lange and Mitchell, in review). Reduced summertime flows due to glacial ablation, and increased stream temperatures due to flow reductions, will further reduce the availability of suitable habitat for Pacific salmonids, create additional stressors, and challenge the survival and recovery of these species (Mantua et al. 2010). Implications of these climate effects are further discussed in a case study on the effect of climate change on Pacific salmon in the Nooksack River Basin (Chapter 8).

Several aquatic species are responding to higher water temperatures through changes in the timing of key life cycle events (Quinn and Adams 1996; Enquist 2012). Sockeye salmon in the Columbia River, for example, are migrating upstream to spawn an average of 10.3 days earlier in the 2000s than in the 1940s, corresponding with a 2.6 °C (4.7 °F) increase in average water temperatures (Crozier et al. 2011). Research has also shown that the 1.4 °C (2.5 °F) increase in average spring water temperature in Lake Washington (King County, Washington) has resulted in a 27-day advance in natural algal blooms (Winder and Schindler 2004). This study also noted the corresponding disruption of trophic linkages where important zooplankton species have not similarly advanced their lifecycles to take advantage of their primary algal food source.

In addition to rivers and lakes, NW wetlands provide important species habitat and a range of ecosystem services including flood storage, water quality protection, and erosion control. In general, the structure and function of NW wetlands and their associated species may be vulnerable to changes in the duration, frequency, and seasonality of precipitation and runoff; decreased groundwater recharge; and higher rates of evapotranspiration (Aldous et al. 2011; Burkett and Kusler 2000; Poff et al. 2002; Winter 2000). Reduced snowpack and altered runoff timing may contribute to the drying of many ponds and wetland habitats across the Northwest, from the Olympic Peninsula in Washington State to Yellowstone National Park in eastern Idaho and the Klamath River Basin in southern Oregon (Döll 2009; Hostetler 2009; Halofsky et al. 2011; McMenamin et al. 2008; Aldous et al. 2011). However, potential future increases in winter precipitation may lead to the expansion of some wetland systems, such as wetland prairies (Bachelet et al. 2011).

Wetlands provide key habitat for many species, including wetland prairie butterflies (e.g., great copper butterfly [*Lycaena xanthoides*], Schultz et al. 2011), amphibians (both native and invasive), and numerous birds, including migrating ducks and other wetland species (e.g., cranes, herons, shorebirds) that use the lake and wetland complexes along the NW portion of the Pacific Flyway migration route. Potential future water level decreases in these systems, coupled with increased water temperatures, may result in increased frequency of certain diseases, such as avian botulism (Rocke and Samuel 1999).

Human responses to climate change have the potential to impact aquatic ecosystems. Irrigation diversions, trade-offs between hydropower and flow augmentation (released from storage reservoirs) for endangered salmon, and changes to water supply infrastructure all have the potential to affect the survival of native and endangered species, as well as the distribution and extent of suitable habitat. As streamflow rates decline during the summer, irrigators are likely to rely more heavily on storage allocations and increased usage of groundwater supplies to fulfill their water demands. In areas where conjunctive management of ground and surface water has not been established, there is the risk that increased diversion of groundwater could further reduce streamflows in hydraulically connected aquifer-stream systems.

In Idaho, a large proportion of the storage water used for annual flow augmentation comes from willing contributions from water users with contract space in the reservoir system (USBR 2011b). With more winter rainfall, declining snowpack, and earlier spring snowmelt resulting from increasing air temperatures, drought conditions are likely to increase through the next century (Hamlet and Lettenmaier 2007). Such conditions would stress storage supplies and potentially reduce the availability of water for annual flow augmentation releases.

New storage facilities may alter temperature regimes downstream, inundate habitat, and create migration barriers. Design and construction of these facilities will need to take into account the potential for such projects to impact natural systems. Reservoir operations may mitigate temperature increases through the release of cold water from lower layers in the reservoir, however the uniform temperature regime of these bottom-draw releases may also disrupt important environmental cues for spawning and migration (Olden and Naiman 2010; Bunn and Arthington 2002; US Bureau of Reclamation and State of Washington Department of Ecology 2012).

3.3.6 RECREATION

The natural environment in the Northwest provides a variety of recreational opportunities such as fishing, hunting and wildlife viewing, swimming, boating, hiking, and skiing (Mote et al. 1999). An understanding of how climate change may impact these recreational opportunities is beginning to emerge, and it is becoming increasingly clear that water-dependent activities would be affected by extreme dry conditions, reduced snowpack, lower summer flows, impaired water quality, and exhausted reservoir storage supplies. According to Snover et al. (2007), the impacts will be variable, affecting some localities more severely than others. The impacts of climate change on recreational opportunities are also discussed in Chapter 5.

One of the more high-profile and discernible impacts from climate change is the effect on the ski industry (Irland et al. 2001). Under a warming climate, mid-elevation ski resorts throughout the region are at risk to experience precipitation falling as rain rather than snow, and snowmelt occuring earlier in the season (Nolin and Daly 2006). Reductions in snowfall and associated snowpack would result in later resort opening dates and earlier closing dates, a greater reliance on but a decreased "window" for snowmaking, an increase in costs to skiers, and significant consequences on the economic viability of ski resorts (Mote et al. 2008). This is consistent with studies by Loomis and Crespi (1999) and Mendelsohn and Markowski (1999) which conclude that the number of skiing visitor days (downhill and cross-country) would be substantially reduced under future climatic conditions. Loomis and Crespi (1999) estimate that skiing visitor days will decrease nationally by over 50% from 1990 to 2060. Shortened ski seasons will reduce visitation impacting not only resorts, but also the communities and businesses that depend on snow recreation (Nolin and Daly 2006).

One recreational activity that has received considerable attention over the years is the sport fishing industry. Hydrologic changes will reduce the ability of aquatic systems and habitats to support populations of native fish species including Pacific salmon, which are an irreplaceable asset with significant cultural and economic value. Economic assessments have placed the value of salmon in the hundreds of millions of dollars throughout the region (Helvoigt and Charlton 2009). Climate change impacts on fish, wildlife, and habitats are likely to negatively affect the estimated $2.5 billion spent annually on fish and wildlife-based recreation in Oregon (Dean Runyan Associates 2009).

3.4 Adaptation

The uncertainty and potential magnitude of the effects of climate change present great challenges to natural resource managers. Communities in the Northwest are taking steps to address adaptation. For example, Seattle Public Utilities has implemented the "Rain-Watch" program to help predict system failures during storm events. They have also implemented dynamic rule curves for some reservoirs after shortages occurred during recent dry years (USEPA 2011). The Portland Water Bureau and Seattle Public Utilities are also using climate models coupled with hydrology, population, and management models to project the potential impacts of climate change on surrounding watersheds. These efforts will help resource managers and decision-makers to make informed decisions that can reduce the negative impacts and take advantage of potential opportunities that may arise as the climate changes (Miller and Yates 2006).

State agencies are also implementing water management plans to secure water supplies for current and future uses. For example, in Washington, a new water management rule for the Dungeness River watershed and a management plan (Elwha-Dungeness Planning Unit 2005) emphasize water conservation, protection of instream flows, water reclamation and reuse, new storage studies, and other water supply strategies that benefit people and fish. Idaho's Comprehensive Aquifer Management Plans provide for strategies to conjunctively manage surface and groundwater resources that will lead to sustainable supplies and optimum use of water resources (Idaho Water Resource Board 2009).

Although there are a few examples of "adaptation in action," there are many more opportunities for management and adaptation actions that could be implemented as incentives, drivers, and climatic conditions are better understood. These include:

- *Adaptation opportunities in response to a decrease in summer streamflows.* Conservation practices and improvements in water use efficiency such as upgrading to more efficient agricultural water application systems and intensive irrigation management techniques, changing to crops that require less water, and adapting to dryland agriculture would help mitigate the effects of a reduced supply. Groundwater and surface water supply assessments, evaluations of projected drought risk, impacts, and vulnerabilities, and expanding remote sensing and streamflow monitoring capabilities would help prepare for a decrease in available supply.

- *Adaptation in response to changes in the timing of peak runoff.* The development of new storage and retention structures, modification of current water delivery systems, and aquifer recharge using early season runoff would increase the available water supply. Improved forecasting and prediction methods can be developed to assist in decision-making for water management planning and operations. Existing laws, regulations, and policies related to water allocation and management could be modified, and flood control rules for reservoir operations could be changed to allow greater flexibility and adaptation to an altered hydrologic regime.

- *Adaptation approaches to manage natural systems for resilience.* Protection of key ecosystem features, reduction of anthropogenic stresses, and restoration of critical habitat structure can improve the resilience of natural systems. Water releases could be timed to decrease temperatures during critical biological periods. On smaller streams, maintaining or restoring instream flows and improving riparian systems to increase stream shading could offset significant warming and enhance resilience. Removal of barriers to fish movement could decrease fragmentation and provide populations the flexibility to shift their distributions.

- *Adaptation for targeted species.* Restoration efforts for salmon habitat can be designed to increase species diversity or resilience and to consider how climate change is likely to alter specific recovery needs and whether restoration actions can ameliorate climate change effects (Beechie et al. 2012). Given current flood-control requirements, greater storage allocations would be needed in order to help maintain instream flows for salmonids in the Columbia River Basin that are listed as threatened or endangered under the Endangered Species Act. Such allocations would require reductions in hydropower production (Payne et al. 2004). Additional ecosystem restoration efforts targeted to natural and cultural resources are discussed in Chapter 8.

In general, adaptation efforts are likely to be more effective when partnerships between various levels of government and local organizations are developed. Improved communication and coordination within and among various local and federal water agencies throughout watersheds to incorporate all aspects of the entire water system, from headwaters to low elevations, is also critical for efficient and effective results. Explicit recognition of the increased value of water efficiency programs that address longer peak season demand patterns, stretch supplies over longer time periods, supplement conjunctive use of sources, provide for the development of emergency preparedness programs, and assess system vulnerabilities and risks are concrete examples of the output of a collaborative, integrated systems approach to adaptation strategies that would be the foundation for efficient policy. Finally, proposed adaptation strategies should be fully assessed using integrated system modeling approaches and careful planning to avoid unintended consequences.

3.5 Knowledge Gaps and Research Needs

There remain significant research and knowledge gaps in the area of climate change and water resources. Ranging from improved datasets to a more thorough knowledge of complex interactions, there are many high-priority research needs that would benefit our ability to understand and adapt to a changing climate. The most pressing of these needs include:

- *Improved monitoring networks, with greater density, for monitoring biology, stream-flow, air and stream temperatures, and snowmelt.* Data from these networks would provide important input for models and provide more accurate information for future climate assessments.

- *Improvements in tools and methods to estimate the spatial and temporal patterns of snowmelt and runoff.* For example, better information regarding whether or not peak flow has already occurred and the potential for a subsequent flow peak would increase the efficiency with which reservoirs could be managed for both irrigation storage and flood control. The complexity of such tools and the key climatic indicators likely differ between basins.

- *Methods to determine warming-induced changes to evapotranspiration from irrigated agriculture and from forested and rangeland watersheds.* This would provide important data needed to predict the water supply since changes in evapotranspiration have a large impact on the overall water budget in many basins (Barnett et al. 2005; Adam et al. 2009).

- *Localized downscaling of extreme event patterns to specific vulnerable areas.* This would allow individual communities to include such findings in their planning and program development and implementation strategies.

- *Coupling of downscaled climate and biophysical knowledge with economic knowledge on same spatial scales.* In the area of water utilization, there are gaps in physical and climatic spatial-specific knowledge that are magnified as one proceeds to estimate economic impacts. Without a spatially scaled behavioral model for these water-dependent sectors that reflects responsiveness of suppliers' behavior to changes in prices and timing of inputs, it is difficult to trace and assess the distributional costs of potential climatic changes on the products produced by these sectors.

- *Improved methods to address the impacts of reductions in hydropower in the region.* The uncertainties associated with projected changes in streamflow timing as well as the uncertainty with respect to future national and state-level energy policies and river treaties governing water usage leaves a large gap in knowledge regarding the potential impacts of reduced hydropower in the region. Better information on the prospects for utilization of the Columbia River waters over the next 50+ years is needed. Methods and frameworks are needed to quantify the technical and economic trade-offs between hydropower production, flood control, and instream flow for fish, and to better prioritize the adaptation alternatives (Hamlet et al. 2013).

- *Improved policy designs for targeting habitat restoration.* Spatial delineation of existing and potential thermal and hydrologic refugia for fish will be important for prioritizing habitat protection and restoration activities, and designing effective economic incentives (Mantua et al. 2010).

Research is also required to better address the following needs:

- *Improved understanding of nexus among energy demands, land-use changes, ecosystem services, and potential health risks.* Research on demographic changes in response to both climate change and human population increases will be needed to identify potential impacts for urban planning processes, energy demand, land use changes, and changes in public health risks.

- *Improved understanding of the consequences and tradeoffs involved in climate change, adaptation, and mitigation activities.* Since climate change is likely to exacerbate tradeoffs between energy-related demands and ecosystem needs (Mantua et al. 2010), additional research is needed to better understand and integrate human responses into impacts studies for key indicator species in the NW including salmon.

- *Linkages among institutional water rights, ecosystem protection, and effective policy alternatives.* Research is needed regarding the flexibility of water rights as well as other legal and technical issues in the region to determine efficient and creative solutions to enhance water conservation, ecosystem protection, and sustainable solutions to hydropower development and relicensing, dam decommissioning, and continued delivery of water for irrigation (Tarlock 2012).

- *Improved understanding of aquatic species and adaptation.* Research is needed to better understand how aquatic species populations are adjusting to long-term trends (Isaak, Muhlfeld, et al. 2012) and to identify the characteristics of watersheds and streams that may either enhance or offset climate change impacts (Rieman and Isaak 2010). Such work can aid in the identification and prioritization of restoration and preservation efforts, and inform policy alternatives at spatial and temporal scales that match the changes in observed and predicted aquatic species.

Acknowledgments

The authors would like to extend a special thanks to Jason Dunham and Christopher Pearl (US Geological Survey), Toni Turner (US Bureau of Reclamation), Ron Abramovich (Natural Resources Conservation Service), Philip Mote and Meghan Dalton (Oregon State University), Amy Snover (Climate Impacts Group, University of Washington), Lorna Stickel (Portland Water Bureau), Stacy Vynne (Puget Sound Partnership) and three anonymous reviewers for their review, feedback, and important contributions. S. Shafer was supported by the US Geological Survey Climate and Land Use Change Research and Development Program.

References

Adam, J. C., A. F. Hamlet, and D. P. Lettenmaier. 2009. "Implications of Global Climate Change for Snowmelt Hydrology in the 21st Century." *Hydrological Processes* 23 (7): 962-972. doi: 10.1002/hyp.7201.

Aldous, A., J. Fitzsimons, B. Richter, and L. Bach. 2011. "Droughts, Floods and Freshwater Ecosystems: Evaluating Climate Change Impacts and Developing Adaptation Strategies." *Marine and Freshwater Research* 62 (3): 223-231. doi: 10.1071/MF09285.

Bachelet, D., B. R. Johnson, S. D. Bridgham, P. V. Dunn, H. E. Anderson, and B. M. Rogers. 2011. "Climate Change Impacts on Western Pacific Northwest Prairies and Savannas." *Northwest Science* 85 (2): 411-429. doi: 10.3955/046.085.0224.

Bartholow, J. M. 2005. "Recent Water Temperature Trends in the Lower Klamath River, California." *North American Journal of Fisheries Management* 25 (1): 152-162. doi: 10.1577 /M04-007.1.

Barnett, T. P., J. C. Adam, and D. P. Lettenmaier. 2005. "Potential Impacts of a Warming Climate on Water Availability in Snow-Dominated Regions." *Nature* 438: 303-309. doi:10.1038 /nature04141.

Battin, J., M. W. Wiley, M. H. Ruckelshaus, R. N. Palmer, E. Korb, K. K. Bartz, and H. Imaki. 2007. "Projected impacts of climate change on salmon habitat restoration." *Proceedings of the National Academy of Sciences* 104 (16): 6720-6725.

Beechie, T., E. Buhle, M. Ruckelshaus, A. Fullerton, and L. Holsinger. 2006. "Hydrologic Regime and the Conservation of Salmon Life History Diversity." *Biological Conservation* 130: 560-572. doi: 10.1016/j.biocon.2006.01.019.

Beechie, T., H. Imaki, J. Greene, A. Wade, H. Wu, G. Pess, P. Roni, J. Kimball, J. Stanford, P. Kiffney, and N. Mantua. 2012. "Restoring Salmon Habitat for a Changing Climate." *River Research and Applications.* doi: 10.1002/rra.2590.

Bonneville Power Administration. 2001. "The Columbia River System Inside Story." 2nd edition. Prepared for the Federal Columbia River Power System Operation Review: Bonneville Power Administration, US Bureau of Reclamation, and US Army Corps of Engineers. http://www.bpa.gov/power/pg /columbia_river_inside_story.pdf.

Bottom, D. L., C. A. Simenstad, J. Burke, A. M. Baptista, D. A. Jay, K. K. Jones, E. Casillas, and M. H. Schiewe. 2005. "Salmon at River's End: The role of the Estuary in the Decline and Recovery of Columbia River Salmon." US Department of Commerce, NOAA Tech. Memo. NMFS-NWFSC-68.

Bunn S. E., and A. A. Arthington. 2002. "Basic Principles and Ecological Consequences of Altered Flow Regimes for Aquatic Biodiversity." *Environmental Management* 30 (4): 492-507. doi: 10.1007/s00267-002-2737-0.

Burkett, V., and J. Kusler. 2000. "Climate Change: Potential Impacts and Interactions in Wetlands of the United States." *Journal of the American Water Resources Association* 36 (2): 313-320. doi: 10.1111/j.1752-1688.2000.tb04270.x.

Cannon, S. H., J. E. Gartner, M. G. Rupert, J. A. Michael, A. H. Rea, and C. Parrett. 2010. "Predicting the Probability and Volume of Post Wildfire Debris Flows in the Intermountain Western United States." *Geological Society of America Bulletin* 122: 127–144. doi: 10.1130 /B26459.1.

Chambers, J. C., and M. J. Wisdom. 2009. "Priority Research and Management Issues for the Imperiled Great Basin of the Western United States." *Restoration Ecology* 17 (5): 707-714. doi: 10.1111/j.1526-100X.2009.00588.x.

Chang, H., J. Jones, M. Gannett, D. Tullos, H. Moradkhani, K. Vaché, H. Parandvash, V. Shandas, A. Nolin, A. Fountain, S. Johnson, I.-W. Jung, L. House-Peters, M. Steele, and B. Copeland. 2010. "Climate Change and Freshwater Resources in Oregon." In *Oregon Climate Assessment Report,* edited by K. D. Dello, and P. W. Mote, 69-149. Oregon Climate Change Research Institute, Oregon State University, Corvallis, Oregon. http://occri.net/ocar.

Climate Impacts Group. 2012. "Risk of Future Changes in Pacific Northwest Hydrologic Extremes (High, Low Flows, Stream Temperatures)." National Climate Assessment Technical Input 2012-0158, submitted by the Climate Impacts Group, University of Washington, March 1.

Crozier, L. G., A. P. Hendry, P. W. Lawson, T. P. Quinn, N. J. Mantua, J. Battin, R. G. Shaw, and R. B. Huey. 2008. "Potential Responses to Climate Change in Organisms with Complex Life Histories: Evolution and Plasticity in Pacific Salmon." *Evolutionary Applications* 1 (2): 252-270. doi: 10.1111/j.1752-4571.2008.00033.x.

Crozier, L. G., M. D. Scheuerell, and E. W. Zabel. 2011. "Using Time Series Analysis to Characterize Evolutionary and Plastic Responses to Environmental Change: A Case Study of a Shift Toward Earlier Migration Date in Sockeye Salmon." *The American Naturalist* 178 (6): 755-773. doi: 10.1086/662669.

Crozier, L., and R. W. Zabel. 2006. "Climate Impacts at Multiple Scales: Evidence for Differential Population Responses in Juvenile Chinook Salmon." *Journal of Animal Ecology* 75: 1100-1109.

Dean Runyan Associates. 2009. Fishing, Hunting, Wildlife Viewing, and Shellfishing in Oregon, 2008. Salem, OR: Oregon Department of Fish and Wildlife and Travel Oregon. http://www.dfw.state.or.us/agency/docs/Report_5_6_09--Final%20(2).pdf.

DeVries, P. E. 1997. "Riverine Salmonids Egg Burial Depths: Review of Published Data and Implications for Scour Studies." *Canadian Journal of Fisheries and Aquatic Sciences* 54: 1685-1698.

Dickerson-Lange, S. E., and R. Mitchell. "Modeling the Effects of Climate Change Projections on Streamflow in the Nooksack River Basin, Northwest Washington." *Hydrological Processes.* In review.

Döll, P. 2009. "Vulnerability to the Impact of Climate Change on Renewable Groundwater Resources: A Global-Scale Assessment." *Environmental Research Letters* 4 (3): 1-12. doi: 10.1088/1748-9326/4/3/035006.

Elsner, M. M., L. Cuo, N. Voisin, J. S. Deems, A. F. Hamlet, J. A. Vano, K. E. B. Mickelson, S. Lee and D. P. Lettenmaier. 2010. "Implications of 21st Century Climate Change for the Hydrology of Washington, State." *Climatic Change* 102: 225-260. doi: 10.1007/s10584-010-9855-0.

Elwha-Dungeness Planning Unit. 2005. "Elwha-Dungeness Watershed Plan, Water Resource Inventory Area 18 (WRIA 18) and Sequim Bay in West WRIA 17." Clallam County, Washington. http://www.clallam.net/environment/ewhadungenesswria.html.

Enquist, C .A. F. 2012. "Phenology as a Bio-Indicator of Climate Change Impacts on People and Ecosystems: Towards an Integrated National Assessment Approach." National Climate Assessment Technical Input 2011-043, submitted by USA National Phenology Network National Coordinating Office.

Furniss, M. J., B. P. Staab, S. Hazelhurst, C. F. Clifton, K. B. Roby, B. L. Ilhadrt, E. B. Larry, A. H. Todd, L. M. Reid, S. J. Hines, K. A. Bennett, C. H. Luce, P. J. Edwards. 2010. "Water, Climate Change, and Forests: Watershed Stewardship for a Changing Climate." General Technical Report PNW-GTR-812. Portland, OR: US Department of Agriculture, Forest Service, Pacific Northwest Research Station. 75 pp.

Goode, J. R., C. H. Luce, and J. M. Buffington. 2012. "Enhanced Sediment Delivery in a Changing Climate in Semi-Arid Mountain Basins: Implications for Water Resource Management and Aquatic Habitat in the Northern Rocky Mountains." *Geomorphology* 139-140: 1-15. doi: 10.1016/j.geomorph.2011.06.021.

Graves, D. 2009. "A GIS Analysis of Climate Change and Snowpack on Columbia Basin Tribal Lands." *Ecological Restoration* 27 (3): 256-257. doi: 10.3368/er.27.3.256.

Halofsky, J. E., D. L. Peterson, K. A. O'Halloran, and C. Hawkins Hoffman, eds. 2011. "Adapting to Climate Change at Olympic National Forest and Olympic National Park." General Technical Report PNW-GTR-844. Portland, OR: US Department of Agriculture, Forest Service, Pacific Northwest Research Station. 130 pp.

Hamlet, A. F., M. M. Elsner, G. S. Mauger, S.-Y. Lee, I. Tohver, and R. A. Norheim. 2013. "An Overview of the Columbia Basin Climate Change Scenarios Project: Approach, Methods, and Summary of Key Results." *Atmosphere-Ocean* 51 (4): 392-415. doi: 10.1080/07055900.2013.819555.

Hamlet, A. F., S. Lee, K. E. B. Mickelson, and M. M. Elsner. 2010. "Effects of Projected Climate Change on Energy Supply and Demand in the Pacific Northwest and Washington State." *Climatic Change* 102: 103-128. doi: 10.1007/s10584-010-9857-y.

Hamlet, A. F. and D. P. Lettenmaier. 2007: "Effects of 20th Century Warming and Climate Variability on Flood Risk in the Western U.S." *Water Resources Research* 43 (W06427): 17. doi: 10.1029/2006WR005099.

Hamlet, A. F., P. W. Mote, M. P. Clark, and D. P. Lettenmaier. 2005. "Effects of Temperature and Precipitation Variability on Snowpack Trends in the Western United States." *Journal of Climate* 18 (21): 4545-4651. doi: 10.1175/JCLI3538.1.

Harper, J. T. 1992. "The Dynamic Response of Glacier Termini to Climatic Variation During the Period 1942–1990 on Mount Baker, Washington, USA." MSc thesis, Western Washington University.

Hatfield, J. L., K. J. Boote, B. A. Kimball, L. H. Ziska, R. C. Izaurralde, D. Ort, A. M. Thomson, and D. Wolfe. 2011. "Climate Impacts on Agriculture: Implications for Crop Production." *Agronomy Journal* 103 (2): 351-370. doi:10.2134/agronj2010.0303.

Helvoigt, T. L., and D. Charlton. 2009. "The Economic Value of Rogue River Salmon." ECONorthwest. Eugene, Oregon. 26 pp. http://wildroguealliance.org/files/RogueSalmonFinalReport.pdf.

Hidalgo, H. G., T. Das, M. D. Dettinger, D. R. Cayan, D. W. Pierce, T. P. Barnett, G. Bala, A. Mirin, A. W. Wood, C. Bonfils, B. D. Santer, and T. Nozawa. 2009. "Detection and Attribution of Streamflow Timing Changes to Climate Change in the Western United States." *Journal of Climate* 22 (13): 3838–3855. doi: 10.1175/2009JCLI2470.1.

Hoekema, D. J., and V. Sridhar. 2011. "Relating Climatic Attributes and Water Resources Allocation: A Study Using Surface Water Supply and Soil Moisture Indices in the Snake River Basin, Idaho." *Water Resources Research* 47 (W07536): 17. doi: 10.1029/2010WR009697.

Hostetler, S. W. 2009. "Use of Models and Observations to Assess Trends in the 1950-2005 Water Balance and Climate of Upper Klamath Lake, Oregon." *Water Resources Research* 45 (W12409): 14. doi: 10.1029/2008WR007295.

Idaho Water Resource Board. 2009. "Eastern Snake Plain Aquifer Comprehensive Aquifer Management Plan." http://www.idwr.idaho.gov/waterboard/WaterPlanning/CAMP/ESPA/default.htm.

Independent Science Advisory Board. 2007. "Climate Change Impacts on Columbia River Basin Fish and Wildlife." ISAB Climate Change Report ISAB 2007-2. Portland, OR: Northwest Power and Conservation Council. 136 pp.

Irland, L. C., D. Adams, R. Alig, C. J. Betz, C. C. Chen, M. Hutchins, B. A. McCarl, K. Skog, and B. L. Sohnsen. 2001. "Assessing Socioeconomic Impacts of Climate Change on US Forests, Wood-Product Markets, and Forest Recreation." *BioScience.* 51 (9): 753-763. doi: 10.1641/0006-3568(2001)051[0753:ASIOCC]2.0.CO;2.

Isaak, D. J., C. H. Luce, B. E. Rieman, D. E. Nagel, E. E. Peterson, D. L. Horan, S. Parkes, and G. L. Chandler. 2010. "Effects of Climate Change and Wildfire on Stream Temperature and Salmonid Thermal Habitat in a Mountain River Network." *Ecological Applications* 20 (5): 1350-1371. doi: 10.1890/09-0822.1.

Isaak, D. J., C. C. Muhlfeld, A. S. Todd, R. Al-Chokhachy, J. Roberts, J. L. Kershner, K. D. Fausch, and S. W. Hostetler. 2012. "The Past as Prelude to the Future for Understanding 21st-Century Climate Effects on Rocky Mountain Trout." *Fisheries* 37 (12): 542-556. doi: 10.1080/03632415.2012.742808.

Isaak, D. J., and B. E. Rieman. 2012. "Stream Isotherm Shifts from Climate Change and Implications for Distributions of Ectothermic Organisms." *Global Change Biology.* 19 (3): 742-751. doi: 10.1111/gcb.12073.

Isaak, D. J., S. Wollrab, D. Horan, and G. Chandler. 2011. "Climate Change Effects on Stream and River Temperatures across the Northwest U.S. from 1980-2009 and Implications for Salmonid Fishes." *Climatic Change* 113 (2): 499-524. doi: 10.1007/s10584-011-0326-z.

Izaurralde, R. C., R. D. Sands, A. M. Thomson, and H. M. Pitcher. 2010. "Bringing Water into an Integrated Assessment Framework." Prepared by Pacific Northwest National Laboratory for the US Department of Energy under Contract DE-AC05-76RL01830. http://www.pnl.gov/main/publications/external/technical_reports/PNNL-19320.pdf.

Keefer, M. L., C. A. Peery, and M. J. Heinrich. 2008. "Temperature-Mediated en route Migration Mortality and Travel Rates of Endangered Snake River Sockeye Salmon." *Ecology of Freshwater Fish* 17: 136-145. doi: 10.1111/j.1600-0633.2007.00267.x.

Keefer, M. L., C. A. Peery, and B. High. 2009. "Behavioral Thermoregulation and Associated Mortality Trade-Offs in Migrating Adult Steelhead (Oncorhynchus mykiss): Variability among Sympatric Populations." *Canadian Journal of Fisheries and Aquatic Sciences* 66 (10): 1734-1747. doi: 10.1139/F09-131.

Keefer, M. L., G. A. Taylor, D. F. Garletts, G. A. Gauthier, T. M. Pierce, and C. C. Caudill. 2010. "Prespawn Mortality in Adult Spring Chinook Salmon Outplanted above Barrier Dams." *Ecology of Freshwater Fish* 19 (3): 361–372. doi: 10.1111/j.1600-0633.2010.00418.x.

Kunkel, K. E., L. E. Stevens, S. E. Stevens, L. Sun, E. Janssen, D. Wuebbles, K. T. Redmond, and J. G. Dobson. 2013. "Regional Climate Trends and Scenarios for the U.S. National Climate Assessment. Part 6. Climate of the Northwest U.S." NOAA Technical Report NESDIS 142-6, 75 pp. http://scenarios.globalchange.gov/report/regional-climate-trends-and-scenarios-us-national-climate-assessment-part-6-climate-northwest.

Leppi, J. C., T. H. DeLuca, S. W. Harrar, and S. W. Running. 2011. "Impacts of Climate Change on August Stream Discharge in the Central-Rocky Mountains." *Climatic Change* 112: 997-1014. doi: 10.1007/s10584-011-0235-1.

Loomis, J., and J. Crespi. 1999. "Estimated Effects of Climate Change on Selected Outdoor Recreation Activities in the United States." In *The Impact of Climate Change on the United States Economy*, edited by R. Mendelsohn and J. E. Neumann. Cambridge: Cambridge University Press.

Luce, C. H., and Z. A. Holden. 2009. "Declining Annual Streamflow Distributions in the Pacific Northwest United States, 1948-2006." *Geophysical Research Letters* 36: L16401. doi: 10.1029/2009GL039407.

Lundquist, J. D., M. Dettinger, I. Stewart, and D. Cayan. 2009. "Variability and Trends in Spring Runoff in the Western United States." In *Climate Warming in Western North America: Evidence and Environmental Effects*, edited by F. Wagner, 63-76. Salt Lake City: University of Utah Press.

MacArthur, J., P. W. Mote, J. Ideker, M. Figliozzi, and M. Lee. 2012. "Climate Change Impact Assessment for Surface Transportation in the Pacific Northwest and Alaska." Oregon Transportation Research and Education Consortium, OTREC-RR-12-01. http://www.wsdot.wa.gov/research/reports/fullreports/772.1.pdf.

Mantua, N., I. Tohver, and A. Hamlet. 2009. "Impacts of Climate Change on Key Aspects of Freshwater Salmon Habitat in Washington State." In *The Washington Climate Change Impacts Assessment: Evaluating Washington's Future in a Changing Climate*, edited by M. M. Elsner, J. Littell, L. Whitely Binder, 217-253. The Climate Impacts Group, University of Washington, Seattle, Washington.

Mantua, N., I. Tohver, and A. Hamlet. 2010. "Climate Change Impacts on Streamflow Extremes and Summertime Stream Temperature and Their Possible Consequences for Freshwater Salmon Habitat in Washington State." *Climatic Change* 102 (1): 187-223. doi: 10.1007/s10584-010-9845-2.

Markoff, M. S., and A. C. Cullen. 2008. "Impact of Climate Change on Pacific Northwest Hydropower." *Climatic Change* 87 (3): 451-469. doi: 10.1007/s10584-007-9306-8.

McCabe, G. J., and D. M. Wolock. 2010. "Long-Term Variability in Northern Hemisphere Snow Cover and Associations with Warmer Winters." *Climatic Change* 99 (1): 141-153. doi: 10.1007/s10584-009-9675-2.

McMenamin, S. K., E. A. Hadly, and C. K. Wright. 2008. "Climatic Change and Wetland Desiccation Cause Amphibian Decline in Yellowstone National Park." *Proceedings of the National Academy of Sciences* 105 (44): 16988-16993. doi: 10.1073/pnas.0809090105.

Mendelsohn, R., and M. Markowski. 1999. "The Impact of Climate Change on Outdoor Recreation." In *The Impact of Climate Change on the United States Economy*, edited by R. Mendelsohn and J. E. Neumann, 267-288. Cambridge: Cambridge University Press.

Meyer, J. L., M. J. Sale, P. J. Mulholland, and N. L. Poff. 1999. "Impacts of Climate Change on Aquatic Ecosystem Functioning and Health." *JAWRA Journal of the American Water Resources Association* 35: 1373–1386. doi: 10.1111/j.1752-1688.1999.tb04222.x.

Mid-Willamette Valley Council of Governments and Oregon Natural Hazards Workgroup. 2005. "Marion County Natural Hazards Mitigation Plan." Prepared for Marion County Emergency Management. https://scholarsbank.uoregon.edu/xmlui/handle/1794/4006.

Miller, K., and D. Yates. 2006. "Climate Change and Water Resources: A Primer for Municipal Water Providers." Denver, CO: AwwaRF and UCAR. 83 pp. http://waterinstitute.ufl.edu/WorkingGroups/downloads/WRF%20Climate%20Change%20DocumentsSHARE/Project%202973%20-%20Climate%20Change%20and%20Water%20Resources.pdf.

Mohseni, O., T. R. Erickson, and H. G. Stefan. 1999. "Sensitivity of the Stream Temperatures in the US to Air Temperatures Projected Under a Global Warming Scenario." *Water Resources Research* 35 (12): 3723-3734. doi: 10.1029/1999WR900193.

Mote, P. W., J. Casson, A. F. Hamlet, and D. C. Reading. 2008. "Sensitivity of Northwest Ski Areas to Warming." In *Proceedings of the 75th Western Snow Conference*, edited by B. McGurk, 63-67. April 16-19, 2007, Kailua-Kona, Hawaii. Soda Springs, CA: Western Snow Conference.

Mote, P. W., M. Holmberg, and N. J. Mantua. 1999. "Impacts of Climate Variability and Change on the Pacific Northwest." NOAA Office of Global Programs, and JISAO/SMA Climate Impacts Group, University of Washington, Seattle, Washington.

Mote, P. W., Parson, E. A. Parson, A. F. Hamlet, W. S. Keeton, D. P. Lettenmaier, N. Mantua, E. L. Miles, D. W. Peterson, D. L. Peterson, R. Slaughter, and A. K. Snover. 2003. "Preparing for Climatic Change: The Water, Salmon, and Forests of the Pacific Northwest." *Climatic Change* 61: 45-88. doi: 10.1023/A:1026302914358.

Mote, P. W., and E. P. Salathé. 2010. "Future Climate in the Pacific Northwest." *Climatic Change* 102 (1): 29-50. doi: 10.1007/s10584-010-9848-z.

Nakićenović, N., O. Davidson, G. Davis, A. Grübler, T. Kram, E. Lebre La Rovere, B. Metz, T. Morita, W. Pepper, H. Pitcher, A. Sankovshi, P. Shukla, R. Swart, R. Watson, and Z. Dadi. 2000. *Special Report on Emissions Scenarios: A Special Report of Working Group III of the Intergovernmental Panel on Climate Change*, Cambridge University Press, Cambridge, UK, 599 pp. http://www.grida.no/climate/ipcc/emission/index.htm.

National Park Service. 2012. "Glacier Monitoring Program." Accessed November 14. http://www.nps.gov/noca/naturescience/glacial-mass-balance1.htm.

Nolin, A. W., and C. Daly. 2006. "Mapping 'At-Risk' Snow in the Pacific Northwest, U.S.A." *Journal of Hydrometeorology* 7 (5): 1164–1171. doi: 10.1175/JHM543.1.

Northwest Power and Conservation Council. 2012. "Hydroelectricity in the Columbia River Basin." Accessed October 23. http://www.nwcouncil.org/energy/powersupply/dams/hydro.htm.

Olden, J. D., and R. J. Naiman. 2010. "Incorporating Thermal Regimes into Environmental Flows Assessments: Modifying Dam Operations to Restore Freshwater Ecosystem Integrity." *Freshwater Biology* 55: 86-107. doi:10.1111/j.1365-2427.2009.02179.x.

Oregon Department of Land Conservation and Development. 2010. "*The Oregon Climate Change Adaptation Framework.*" http://www.oregon.gov/ENERGY/GBLWRM/docs/Framework_Final_DLCD.pdf.

Oregon Water Resources Department. 2012. "Oregon's Integrated Water Resources Strategy." State of Oregon Water Resources Department, Oregon State Library Call No. WR.2In8/4:2012/final. http://www.oregon.gov/owrd/pages/law/integrated_water_supply_strategy.aspx#Oregon's_First_Integrated_Water_Resources_Strategy.

Orr, J. C., V. J. Fabry, O. Aumont, L. Bopp, S. C. Doney, R. A. Feely, A. Gnanadesikan, N. Gruber, A. Ishida, F. Joos, R. M. Key, K. Lindsay, E. Maier-Reimer, R. Matear, P. Monfray, A. Mouchet, R. G. Najjar, G.-K. Plattner, K. B. Rodgers, C. L. Sabine, J. L. Sarmiento, R. Schlitzer, R. D. Slater, I. J. Totterdell, M.-F. Weirig, Y. Yamanaka, and A. Yool. 2005. "Anthropogenic Ocean Acidification over the Twenty-First Century and its Impact on Calcifying Organisms." *Nature* 437: 681-686. doi: 10.1038/nature04095.

Pagano, T., and D. Garen. 2005. "A Recent Increase in Western U.S. Streamflow Variability and Persistence." *Journal of Hydrometeorology* 6 (2): 173-179. doi: 10.1175/JHM410.1.

Payne, J. T., A. W. Wood, A. F. Hamlet, R. N. Palmer, and D. P. Lettenmaier. 2004. "Mitigating the Effects of Climate Change on the Water Resources of the Columbia River Basin. *Climatic Change* 62 (1-3): 233-256. doi: 10.1023/B:CLIM.0000013694.18154.d6.

Pearcy, W. G. 1992. "Ocean Ecology of North Pacific Salmonids." Seattle, WA: University of Washington Press.

Pelto, M. S. 2006. "The Current Disequilibrium of North Cascade Glaciers." *Hydrological Processes* 20 (4): 769-779. doi: 10.1002/hyp.6132.

Pelto, M. S. 2010. "Forecasting Temperate Alpine Glacier Survival from Accumulation Zone Observations." *The Cryosphere* 4: 67-75. doi: 10.5194/tc-4-67-2010.

Pelto, M. S. 2011. "Skykomish River, Washington: Impact of Ongoing Glacier Retreat on Streamflow." *Hydrological Processes* 25: 3356-3363. doi: 10.1002/hyp.8218.

Pelto, M. and C. Brown. 2012. "Mass Balance Loss of Mount Baker, Washington Glaciers 1990-2010." *Hydrological Processes* 26 (17): 2601-2607. doi: 10.1002/hyp.9453.

Petersen, J. H., and J. F. Kitchell. 2001. "Climate Regimes and Water Temperature Changes in the Columbia River: Bioenergetic Implications for Predators of Juvenile Salmon." *Canadian Journal of Fisheries and Aquatic Sciences* 58 (9): 1831-1841. doi: 10.1139/f01-111.

Poff, N. L., M. M. Brinson, and J. W. Day, Jr. 2002. "Aquatic Ecosystems and Global Climate Change: Potential Impacts on Inland Freshwater and Coastal Wetland Ecosystems in the United States." Prepared for the Pew Center on Global Climate Change. http://www.pewtrusts.org/uploadedFiles/wwwpewtrustsorg/Reports/Protecting_ocean_life/env_climate_aquaticecosystems.pdf.

Post, A. P., D. Richardson, W. V. Tangborn, and F. L. Rosselot. 1971. "Inventory of Glaciers in the North Cascades, Washington." US Geological Survey Professional Paper 705-A. http://www.cr.nps.gov/history/online_books/geology/publications/pp/705-A/index.htm.

Quinn, T. P., and D. J. Adams. 1996. "Environmental Changes Affecting the Migration Timing of American Shad and Sockeye Salmon." *Ecology* 77 (4): 1151-1162. doi: 10.2307/2265584.

Richter, A., and S. A. Kolmes. 2005. "Maximum Temperature Limits for Chinook, Coho, and Chum Salmon, and Steelhead Trout in the Pacific Northwest." *Reviews in Fisheries Science* 13 (1): 23-49. doi: 10.1080/10641260590885861.

Riedel, J., and M. A. Larrabee. 2011. "North Cascades National Park Complex Glacier Mass Balance Monitoring Annual Report, Water Year 2009: North Coast and Cascades Network." Natural Resource Technical Report NPS/NCCN/NRTR—2011/483. National Park Service, Fort Collins, Colorado.

Rieman, B., and D. Isaak. 2010. "Climate change, Aquatic Ecosystems and Fishes in the Rocky Mountain West: Implications and Alternatives for Management." USDA Forest Service, Rocky Mountain Research Station, GTR-RMRS-250, Fort Collins, CO.

Rieman, B. E., D. Isaak, S. Adams, D. Horan, D. Nagel, C. Luce, and D. Myers. 2007. "Anticipated Climate Warming Effects on Bull Trout Habitats and Populations Across the Interior Columbia River Basin." *Transactions of the American Fisheries Society* 136 (6): 1552-1565. doi: 10.1577/T07-028.1.

Rocke, T. E., and M. D. Samuel. 1999. "Water and Sediment Characteristics Associated with Avian Botulism Outbreaks in Wetlands." *The Journal of Wildlife Management* 63: 1249-1260. http://forest.wisc.edu/files/pdfs/samuel/1999_rocke_samuel_water_sediment.pdf.

Safeeq, M., G. E. Grant, S. L. Lewis, and C. L. Tague. 2013. "Coupling Snowpack and Groundwater Dynamics to Interpret Historical Streamflow Trends in the Western United States." *Hydrological Processes* 27: 655-668. doi: 10.1002/hyp. 9628.

Salinger, D. H. and J. J. Anderson. 2006. "Effects of Water Temperature and Flow on Adult Salmon Migration Swim Speed and Delay." *Transactions of the American Fisheries Society* 135 (1): 188-199. doi: 10.1577/T04-181.1.

Schaible, G., and M. Aillery. 2012. "Water Conservation in Irrigated Agriculture: Trends and Challenges in the Face of Emerging Demands." USDA Economic Research Service, Economic Information Bulletin No. (EIB-99).

Schultz, C. B., E. Henry, A. Carleton, T. Hicks, R. Thomas, A. Potter, M. Collins, M. Linders, C. Fimbel, S. Black, H. E. Anderson, G. Diehl, S. Hamman, R. Gilbert, J. Foster, D. Hays, D. Wilderman, R. Davenport, E. Steel, N. Page, P. L. Lilley, J. Heron, N. Kroeker, C. Webb, and B. Reader. 2011. "Conservation of Prairie-Oak Butterflies in Oregon, Washington, and British Columbia." *Northwest Science* 85 (2): 361-388. doi: 10.3955/046.085.0221.

Snover, A. K., L. Whitely Binder, J. Lopez, E. Willmott, J. Kay, D. Howell, and J. Simmonds. 2007. "Preparing for Climate Change: A Guidebook for Local, Regional, and State Governments." ICLEI – Local Governments for Sustainability, Oakland, CA. http://www.cses.washington.edu/db/pdf/snoveretalgb574.pdf.

Stewart, I. T., D. R. Cayan, and M. D. Dettinger. 2005. "Changes toward Earlier Streamflow Timing across Western North America." *Journal of Climate* 18 (8): 1136-1155. 10.1175/JCLI3321.1.

Tarlock, A. D. 2012. "Takings, Water Rights and Climate Change." *Vermont Law Review* 36 (3): 731-757. http://lawreview.vermontlaw.edu/files/2012/06/18-Tarlock-Book-3-Vol.-36.pdf.

US Bureau of Reclamation. 2008. "The Effects of Climate Change on the Operation of Boise River Reservoirs, Initial Assessment Report." US Department of the Interior, Bureau of Reclamation, Pacific Northwest Region, Boise, ID. http://www.usbr.gov/pn/programs/srao_misc/climatestudy/boiseclimatestudy.pdf.

US Bureau of Reclamation. 2011a. "Climate and Hydrology Datasets for Use in the River Management Joint Operating Committee (RMJOC) Agencies' Longer-Term Planning Studies: Part II Reservoir Operations Assessment for Reclamation Tributary Basins." US Department of the Interior, Bureau of Reclamation, Pacific Northwest Region, Boise, ID. http://www.bpa.gov/power/pgf/ClimateChange/Part_II_Report.pdf.

US Bureau of Reclamation. 2011b. "2011 Salmon Flow Augmentation Program and Other Activities Associated with the NOAA Fisheries Service 2008 Biological Opinion and Incidental

Take Statement for Operations and Maintenance of Bureau of Reclamation Projects in the Snake River Basin above Brownlee Reservoir. Annual Progress Report." US Department of the Interior, Bureau of Reclamation, Pacific Northwest Region. http://www.usbr.gov/pn /programs/fcrps/uppersnake/2011nmfs-anrptf.pdf.

US Bureau of Reclamation. 2011c. "SECURE Water Act Section 9503(c) – Reclamation Climate Change and Water, Report to Congress." US Department of the Interior, Bureau of Reclamation, Denver, CO. http://www.usbr.gov/climate/SECURE/docs/SECUREWater Report.pdf

US Bureau of Reclamation and State of Washington Department of Ecology. 2012. "Yakima River Basin Integrated Water Resource Management Plan." Ecology Publication Number: 12-12-002. http://www.usbr.gov/pn/programs/yrbwep/reports/FPEIS/fpeis.pdf.

US Department of Agriculture Economic Research Service. 2012. "Data Products State Fact Sheets." Washington, Oregon, and Idaho. Last modified December 19. http://www.ers.usda .gov/data-products/state-fact-sheets/.

US Energy Information Administration. 2012. "Washington State Profile and Energy Estimates: Profile Overview." Accessed May 8. http://www.eia.gov/beta/state/?sid=WA.

US Environmental Protection Agency. 2011. "Climate Change Vulnerability Assessments: Four Case Studies of Water Utility Practices." EPA/600/R-10/077F. Washington, DC: National Center for Environmental Assessment. http://cfpub.epa.gov/ncea/global/recordisplay. cfm?deid=233808#Download.

Vano, J. A., M. J. Scott, N. Voisin, C. O. Stöckle, A. F. Hamlet, K. E. B. Mickelson, M. M. Elsner, and D. P. Lettenmaier. 2010. "Climate Change Impacts on Water Management and Irrigated Agriculture in the Yakima River Basin, Washington, USA." *Climatic Change* 102: 287-317. doi 10.1007/s10584-010-9856-z.

Walker, L., M. A. Figliozzi, A. R. Haire, and J. MacArthur. 2011. "Climate Action Plans and Long-Range Transportation Plans in the Pacific Northwest and Alaska, State of the Practice in Adaptation Planning." *Transportation Research Record: Journal of the Transportation Research Board* 2252: 118-126. doi: 10.3141/2252-15.

Washington State Department of Ecology. 2011. "Columbia River Basin Long-Term Water Supply and Demand Forecast." Publication No. 11-12-011. https://fortress.wa.gov/ecy/publications /publications/1112011.pdf.

Washington State Department of Ecology. 2012. "Washington State Integrated Climate Change Response Strategy." Publication No. 12-01-004. https://fortress.wa.gov/ecy/publications /publications/1201004.pdf.

Wenger, S. J., D. J. Isaak, C. H. Luce, H. M. Neville, K. D. Fausch, J. B. Dunham, D. C. Dauwalter, M. K. Young, M. M. Elsner, B. E. Rieman, A. F. Hamlet, and J. E. Williams. 2011. "Flow Regime, Temperature, and Biotic Interactions Drive Differential Declines of Trout Species under Climate Change." *Proceedings of the National Academy of Sciences* 108 (34): 14175-14180. doi: 10.1073/pnas.1103097108.

Whitely Binder, L. C. 2009. "Preparing for Climate Change in the U.S. Pacific Northwest." *Hastings West-Northwest Journal of Environmental Law and Policy* 15 (1): 183-196.

Winder, M., and D. E. Schindler. 2004. "Climate Change Uncouples Trophic Interconnections in an Aquatic Ecosystem." *Ecology* 85 (8): 2100-2106. doi: 10.1890/04-0151.

Winter, T. C. 2000. "The Vulnerability of Wetlands to Climate Change: A Hydrological Landscape Perspective." *Journal of the American Water Resources Association* 36 (2): 305-311. doi: 10.1111/j.1752-1688.2000.tb04269.x.

Chapter 4

Coasts
Complex Changes Affecting the Northwest's Diverse Shorelines

AUTHORS

W. Spencer Reeder, Peter Ruggiero, Sarah L. Shafer, Amy K. Snover, Laurie L. Houston, Patty Glick, Jan A. Newton, Susan M. Capalbo

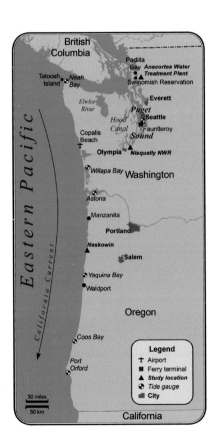

Figure 4.1 Coastal region of Washington and Oregon, including some locations mentioned in this chapter.

4.1 Introduction

The many thousands of miles of Northwest (NW) marine coastline are extremely diverse and contain important human-built and natural assets upon which our communities and ecosystems depend. Due to the variety of coastal landform types (e.g., sandy beaches, rocky shorelines, bluffs of varying slopes and composition, river deltas, and estuaries), the region's marine coastal areas stand to experience a wide range of climate impacts, in both type and severity. These impacts include increases in ocean temperature and acidity, erosion, and more severe and frequent inundation from the combined effects of rising sea levels and storms, among others.

Increases in coastal inundation and erosion are key concerns. A recent assessment determined that the coastal areas of Washington and Oregon contain over 56,656 hectares (140,000 acres) of land within 1.0-meter (3.3-feet) elevation of high tide (Strauss et al. 2012). Rising sea levels coupled with the possibility of intensifying coastal storms will increase the likelihood of more severe coastal flooding and erosion in these areas.

The Northwest is also facing the challenge of increasing ocean acidification, and is experiencing these changes earlier, and more acutely, than most other regions around the globe (NOAA OAR 2012). Changes in ocean chemistry resulting from higher global concentrations of atmospheric CO_2, combined with regional factors that amplify local acidification, are already adversely affecting important NW marine species (NOAA OAR 2012).

The combined effects of these observed and projected climate impacts represent a significant challenge to the region. The human response to the changes in our coastal systems will play a large role in determining the long-term resilience of NW coasts and the ongoing viability of the region's coastal communities, and the viability of shallow-water and estuarine ecosystems in particular (Tillmann and Siemann 2011; Huppert et al. 2009; West Coast Governors' Agreement on Ocean Health 2010; Fresh et al. 2011).

4.2 Sea Level Rise

Historical trends in sea level in the coastal marine waters of Washington and Oregon vary across the region and contain significant departures from the global mean rate of increase in sea level of approximately 3.1 mm/year (0.12 in/year), as determined by satellite altimetry for the period 1993–2012 (University of Colorado 2012; Nerem et al. 2010; National Research Council [NRC] 2012). Figure 4.2 shows: (a) time series of sea level measurements at eight NOAA tide gauge locations in Washington, Oregon, and northern California (Komar et al. 2011); and (b) derived relative sea level rates of change from various techniques for the Oregon coast. Locations in both figures display departures from the global mean. The variability among rates is due primarily to the fact that Washington and western Oregon sit above an active subduction zone, which generates forces that lead to non-uniform vertical deformation of the overlying land and are also the cause of the region's active volcanism and seismic activity (Chapman and Melbourne 2009; Harris 2005; also see section 4.2.1). Additional regional factors that cause variances in NW sea levels, when compared to the global mean, are seasonal ocean circulation and wind field effects caused by El Niño-Southern Oscillation (ENSO) events[1], as well as the gravitational effects of Alaska's extensive glaciers and deformation associated with the ongoing recovery of the region's landmass from the disappearance of the massive ice sheets (post-glacial isostatic rebound) that began to retreat approximately 19,000 years ago (NRC 2012; Yokoyama et al. 2000). Additional smaller scale factors that can appreciably affect local sea levels are described in section 4.2.1.

End-of-century sea level rise projections for Washington State released in 2008 show *relative* sea level changes ranging from a small drop of a few decimeters (result-

---

1   ENSO and other large-scale regional climatic factors are discussed in more detail in Chapter 2.

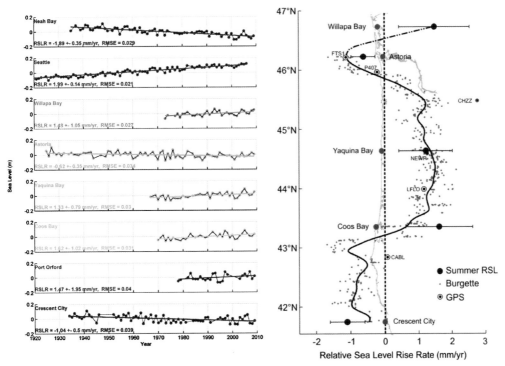

Figure 4.2.a Time series and linear trends (Komar et al. 2011) in relative sea level (RSL) as measured by NW coastal tide gauges operated by NOAA (NOAA Tides and Currents 2012).[1] The relative sea level rise (RSLR) and root mean square error (RMSE) are listed for each record. Trends in RSL range from -1.89 mm/year (-0.074 inches/year) at the Neah Bay gauge on the north coast of Washington (indicating falling relative sea level), to an increase in RSL in Seattle of +1.99 mm/year (+0.078 inches/year), and +1.33 mm/year (+0.052 inches/year) at the Yaquina Bay site. The gauges in Astoria, Oregon, and Crescent City, California, also show falling RSL with a declining linear trend of -0.62 and -1.04 mm/year (-0.024 and -0.041 inches/year), respectively. Most gauges in the NW show positive RSL trends, but less than the global mean rate of sea level increase of +3.1 mm/year (+0.12 inches/year).

Figure 4.2.b Alongshore rates of relative sea level (RSL) rise (black line) from Crescent City, California, to Willapa Bay, Washington, as determined by three methods: (1) tide-gauge records with trends based on averages of the summer only monthly-mean water levels (red circles with plus signs, error bars represent the 95% confidence interval on the trends); (2) subtracting the Burgette et al. (2009) benchmark survey estimates of uplift rates from the regional mean sea level rise rate (2.3 mm/year [0.09 inches/year]) (very small gray dots); and (3) subtracting the uplift rates estimated from global positioning system (GPS) measurements along the coast from the regional mean sea level rate (small filled black circles). After Komar et al. (2011).

---

1   Note: Naming conventions used in this figure differ from official tide gauge station names for the following stations: Toke Point (Willapa Bay), Yaquina River (Yaquina Bay), and Charleston (Coos Bay).

ing from tectonic uplift along the NW portion of the Olympic Peninsula outpacing sea level rise) to a net increase in water levels of 128 cm (50 in) in the Puget Sound (Mote et al. 2008). A 2012 assessment of West Coast sea level rise by the National Research Council (NRC 2012) suggests the upper range of the global contribution to regional sea

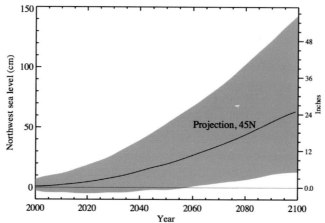

Figure 4.3 Projection for relative sea level rise at 45 °N latitude on the Northwest coast.[1] Sea level rise projections for the 21st century (in centimeters and inches) relative to the year 2000 that incorporate global and local effects of warming oceans, melting land ice, and vertical land movements along the West Coast. The shaded area shows a range of projections developed by considering uncertainties in each of those contributing factors, and also uncertainties in the global emissions of greenhouse gases. Although these projections for other latitudes in the Northwest differ by less than an inch, variation in vertical land movement within the region could add or subtract as much as 20 cm (8 in) from the projections for 2100 shown here. Additional variation in sea level could result from the local effects. Plotted with data from the NRC (2012).

---

1   Roughly the latitude of Lincoln City and Salem, Oregon.

level rise could be slightly higher than previously thought, extending the upper bound to 1.4 m (55 in) for NW ocean levels in the year 2100 (fig. 4.3). The NRC report also notes that "increases of 3-4 times the current rate [of sea level rise] would be required to realize scenarios of 1 m sea level rise by 2100" (NRC 2012).[2]

Sea level rise studies are characterized by uncertainties regarding the extent to which rates may increase over time; however, global sea levels are rising and are virtually certain to continue to do so throughout the 21st century and beyond (Meehl et al. 2007). Because the rate of sea level change is directly affected by the long term trend in global air temperature (primarily through the thermal expansion of seawater and the volumetric contribution from the melting of land-based ice), sea level rise rates are expected to accelerate in the coming decades concomitant with projected higher rates of warming (Schaeffer et al. 2012; Rahmstorf 2007, 2010; Meehl et al. 2007).

---

2   Prior regional studies used a maximum global contribution of 0.93 m (37 in; Mote et al. 2008). The recent NRC report provides a range for the global contribution of 0.5 to 1.4 m (20 to 55 in) for 2100 relative to 2000 levels.

4.2.1 EFFECTS OF TECTONIC MOTION AND OTHER LOCAL AND REGIONAL FACTORS

Because the Northwest is located in an active subduction zone, vertical land motion resulting from the forces of the subducting ocean plate can introduce significant variability in local rates of observed sea level rise (Mote et al. 2008; Komar et al. 2011). These vertical land motions can add to, or subtract from, the overall rate of regional sea level rise. On the Olympic Peninsula in Washington State, global positioning system (GPS) observations generally show a rate of vertical uplift of the same order of magnitude as sea level rise, thus creating the potential for a net decrease in local observed sea level in some locations (Mote et al. 2008). In other locations, land subsidence can create higher rates of sea level rise than that observed regionally. Other factors such as post-glacial rebound[3], local sediment loading and compaction, groundwater and hydrocarbon subsurface fluid withdrawal, and other geophysical processes can also introduce highly localized vertical deformation that further affects the observed changes in sea level at a particular location (NRC 2012).

Bromirski et al. (2011) point out that the atmospheric patterns that contribute to the Pacific Decadal Oscillation (PDO)[4] have affected upwelling along the eastern boundary of the North Pacific since the atmosphere-ocean climate system regime shift, from cold to warm phase PDO conditions, in the mid-1970s. Since roughly 1980, the predominant wind stress patterns along the US West Coast have served to regionally attenuate the otherwise rising trend in sea levels seen globally, for the most part suppressing ocean levels in the Northwest. However, recent wind stress patterns similar to pre-1970s conditions may signal a shift to the PDO cold phase that may, in turn, result in a return to

Table 4.1 Local sea level change projections (relative to the year 2000, reproduced from NRC 2012).

	2030	2050	2100
Seattle, WA	-3.7 to +22.5 cm (-1.5 to +8.9 in)	-2.5 to +47.8 cm (-1.0 to +18.8 in)	+10.0 to +143.0 cm (+3.9 to +56.3 in)
Newport, OR	-3.5 to +22.7 cm (-1.4 to +8.9 in)	-2.1 to +48.1 cm (-0.8 to +18.9 in)	+11.7 to +142.4 cm (+4.6 to +56.1 in)

3 Post glacial rebound, also known as glacial isostatic adjustment, generally results in uplift north of the 49th parallel in western North America and land subsidence of 1 mm/year (0.04 inches/year) or less in western Washington and Oregon (NRC 2012; Argus and Peltier 2010; Peltier 2004)

4 PDO and ENSO are described in more detail in Chapter 2.

Table 4.2 Year 2100 sea level rise projections (in centimeters and inches) relative to 2000 (Meehl et al. 2007; Mote et al. 2008; NRC 2012). Whereas the IPCC AR4 (Meehl et al. 2007) only provide a range, the Mote et al. (2008) and NRC (2012) studies provide central (or middle) estimates for end-of-century sea levels, with the full range of each projection shown in parentheses.[1] B1 (substantial emissions reductions), A1B (continued emissions growth peaking at mid-century), and A1FI (very high emissions growth) are IPCC Special Report on Emissions Scenarios (SRES) greenhouse gases emissions scenarios that correspond to different potential societal futures with progressively increasing levels of emissions in the latter half of the century (Nakićenović et al. 2000). Regional comparisons between Mote et al. (2008) and NRC (2012) are approximate since the NRC only assessed latitudinal variability in sea level rise and Mote et al. (2008) also considered longitudinal variability, in addition to other differences in the spatial domain covered by the estimates.

	IPCC AR4	Mote et al. (2008)	NRC (2012)
Global	B1: 18-38 cm (7.1"-15") A1B: 21-48 cm (8.3"-18.9") A1FI: 26-59 cm (10.2"-23.2")	34 cm (18-93) 13.4" (7.1-36.6)	83 cm (50-140) 32.7" (19.7-55.1)
NW Olympic Peninsula	--	4 cm (-24-88) 1.6" (-9.4-34.6)	61 cm (9-143)[2] 24" (3.5-56.3)
Puget Sound	--	34 cm (16-128) 13.4" (6.3-50.4)	62 cm (10-143)[3] 24.4" (3.9-56.3)
Central & Southern Washington Coast	--	29 cm (6-108) 11.4" (2.4-42.5)	62 cm (11-143)[4] 24.4" (4.3-56.3)
Central Oregon Coast	----	63 cm (12-142)[5] 24.8" (4.7-55.9)	

1 These central estimates are not probabilistic or statistically determined so they do not necessarily represent a "most likely" value of sea level rise.

2 Projection for Neah Bay, Washington, as estimated from fig. 5.10, NRC (2012).

3 For the latitude of Seattle, Washington (NRC 2012).

3 Projection for Aberdeen, Washington, as estimated from fig. 5.10, NRC (2012).

4 For the latitude of Newport, Oregon (NRC 2012).

5 For the latitude of Newport, Oregon (NRC 2012).

higher rates of sea level rise along the West Coast, approaching or exceeding the global rate (Bromirski et al. 2011).

Table 4.1 summarizes the net sea level change projections for Newport, Oregon, and Seattle, Washington, from the NRC (2012) report. The NRC did not incorporate the smaller scale regional heterogeneities in land deformation rates in their projections as was done in Mote et al. (2008).[5] Additional regional studies of relative sea level rise will therefore be important in assessing future risk in specific locations. See table 4.2 for a comparison of sea level rise projections from the Intergovernmental Panel on Climate Change (IPCC) Fourth Assessment Report (AR4) (Meehl et al. 2007), Mote et al. (2008), and NRC (2012); these three reports are most frequently referenced for NW sea level rise projections. Differences between the three sets of projections are primarily due to differences in assumptions concerning the contributions of Greenland and Antarctica to future global sea levels; emissions scenarios were not the main contribution to divergence among the projections.[6]

4.2.2 COMBINED IMPACTS OF SEA LEVEL RISE, COASTAL STORMS, AND ENSO EVENTS

Increases in storminess and ENSO intensity, even without substantial increases in sea levels, can substantially increase coastal flooding and erosion hazards. Both of these phenomena are complex and the specifics of how they will change under future climate conditions are uncertain.

Increasing wave heights have been observed in the northeast Pacific using instrumented NOAA buoys along the US West Coast (Allan and Komar 2000; Allan and Komar 2006; Méndez et al. 2006; Menéndez et al. 2008; Komar et al. 2009; Ruggiero et al. 2010; Seymour 2011) and from satellite altimetry (Young et al. 2011). Analyses of North Pacific extra-tropical storms have concluded that storm intensities (wind velocities and atmospheric pressures) have increased since the late 1940s (Graham and Diaz 2001; Favre and Gershunov 2006), implying that the trends of increasing wave heights perhaps began in the mid-20th century, prior to the availability of direct buoy measurements.

Studies relying solely on buoy measurements have, however, recently been called into question because of measurement hardware and analysis procedure concerns (Gemmrich et al. 2011). Subsequent analysis that accounts for the modifications of the wave measurement hardware and inhomogeneities in the records reveals trends that are smaller than those obtained from the uncorrected data. The most significant of the inhomogeneities in the buoy records occurred prior to the mid-1980s. Menéndez et al. (2008) analyzed extreme significant wave heights along the eastern North Pacific using data sets from 26 buoys over the period 1985–2007, not including the more suspect data from earlier in the buoy records. Their work revealed significant positive long-term trends

---

5   Although there is an extensive network of continuously running GPS stations throughout the western United States, NRC authors were concerned that interpolation errors between stations would be difficult to assess and characterize due to high spatial variability of vertical deformation within the region (NRC 2012, page 122).

6   The IPCC AR4 (Meehl et al. 2007) used an estimate of total ice sheet contribution to global sea level rise of 0 to 17 cm (6.7 in) by 2100. Mote et al. (2008) used a maximum value of 34 cm (13.4 in) for this term, and NRC (2012) used a total range of 50 to 67 cm (19.7 to 26.4 in) (up to 18 cm [7.1 in] from enhanced dynamics alone).

in extreme heights off the West Coast between 30–45° north latitude. Ruggiero (2013) recently showed that since the early 1980s, the increases in deep-water wave heights and periods have been more responsible for increasing the frequency of coastal erosion and flooding events along the NW outer coast than changes in sea level.

Evidence for changes in coastal storm intensity may also be revealed by examining changes in non-tidal residuals measured at tide stations (i.e., storm surge). Allan et al. (2011) analyzed the Yaquina Bay storm surge record and found no increases in surge levels and frequencies since the late 1960s.

The ongoing occurrence of periodic El Niño events, which cause higher than average regional sea levels, will compound the impacts of sea level rise, resulting in severe episodes of coastal erosion and flooding, as experienced during the El Niño winters of 1982–83 and 1997–98. Regional sea levels can be elevated as much as 30 cm (~12 in) for several months at a time during an El Niño event (Ruggiero et al. 2005). At present it is not known whether ENSO intensity and frequency will increase under a changing climate. However, in a recent modeling study, Stevenson (2012) suggested that significant changes to ENSO are not detectable by 2100 for most scenarios.

Although unequivocal evidence for climate change driven shifts in storms or ENSO characteristics in the Northwest is not yet discernible in the observational record or in model projections, future conditions that include any substantial and sustained changes in the wind environment (e.g., prevailing wind direction, magnitude, seasonality) and/or increases in precipitation intensity would have significant implications for coastal inundation and erosion risk.

4.3 Ocean Acidification

In addition to the long-recognized exposure of low-lying shorelines worldwide to sea level rise, research over the past few years has revealed a quicker-than-expected emergence of ocean acidification as a serious NW regional concern (Feely et al. 2012; Feely et al. 2010; Feely et al. 2008). The cascade of impacts related to ocean acidification, while complex, raises particular concern for regionally iconic and commercially significant marine species, including those directly affected by the observed changes in ocean chemistry (e.g., oysters) to those indirectly affected through impacts to the larger marine food web (e.g., Pacific salmon) (Ries 2009; Feely et al. 2012).

Ocean acidification is the result of a combination of factors, affected by both natural processes and human activities. Conditions in the coastal waters of the Northwest lead to some of the most highly acidified marine waters found worldwide (NOAA OAR 2012). These acidified waters appear in their most pronounced form during the spring through to the late summer months when the prevailing coastal winds seasonally shift southward, favoring upwelling of corrosive subsurface ocean waters (Feely et al. 2008; Northwest Fisheries Science Center 2012; Hickey and Banas 2003). The upwelling effect transports these subsurface waters up onto the continental shelf of the Northwest, where in some places, they reach surface waters near the coast (Feely et al. 2008; Hauri et al. 2009). These acidified waters enter sensitive estuaries in the region, such as Willapa Bay, Puget Sound, and Hood Canal, and combine with local factors to create low pH conditions (Feely et al. 2010). For example, pH values as low as 7.35 have been observed in the southern portions of Hood Canal (Feely et al. 2010).

The principal driver of acidification both globally and regionally is the increasing concentration of atmospheric CO_2, which affects the chemistry of the ocean when absorbed (Feely et al. 2008; Doney et al. 2009; NRC 2010). Atmospheric CO_2 concentrations are higher now than at any time in at least the past 650,000 years, and current estimates are that about one-quarter of the human-derived CO_2 released to the atmosphere over the last 250 years is now dissolved in the ocean (Canadell et al. 2007; Sabine et al. 2004; Sabine and Feely 2007). Once absorbed, CO_2 causes the pH and carbonate saturation state of seawater to decline, rendering ocean water corrosive to marine organisms that use carbonates (calcite and aragonite) to build shells and skeletons. These changes, commonly referred to as ocean acidification, are occurring at a rate nearly ten times faster than that of any previous period within the last 50 million years (Kump et al. 2009; Hönisch et al. 2012). The persistence of contemporary marine ecosystems is threatened (e.g., Ainsworth et al. 2011, Griffith et al. 2011), as is the persistence of shellfish aquaculture (e.g., Cheung et al. 2011, Barton et al. 2012).

Acidified waters that enter sensitive estuaries in the region can combine with inputs of nutrients and organic matter, from both natural and human sources, further reducing pH and carbonate saturation state, producing conditions that can be more corrosive than those observed off the coast (Feely et al. 2010; Cai et al. 2011; Sunda and Cai 2012). In addition, local atmospheric emissions of CO_2, nitrogen oxides, and sulfur oxides may also contribute to acidification of nearby marine waters, although further research is needed to quantify that impact (NOAA OAR 2012). Consequently, natural processes, anthropogenic additions of CO_2 and other acidifying wastes, and additions of nutrients and organic matter each play a role in intensifying ocean acidification in coastal estuaries of the Northwest. Rykaczewski and Dunne (2010) suggest that nitrate supply into the California Current System may increase in a warming climate; and, as a result, increases in acidification (and concomitant decreases in dissolved oxygen) are projected.

Due to the complexities and seriousness of the implications to commercially important and federally protected marine species, characterizing the threats of ocean acidification to the marine waters of the Northwest is a current focus of a number of research projects such as those sponsored by the National Science Foundation, NOAA, and Washington Sea Grant (also see NRC 2010). At the state level, the threat of ocean acidification to commercial shellfish production and the broader marine food web motivated Washington Governor Christine Gregoire to convene a first-in-the-nation Blue Ribbon Panel on Ocean Acidification early in 2012. The Panel's report includes more than 40 recommended actions for addressing the causes and consequences of ocean acidification in Washington State (Washington State Blue Ribbon Panel 2012; see also section 4.8).

4.4 Ocean Temperature

An increase in ocean temperature is anticipated to create shifts in the ranges and types of marine species found in coastal waters of the Northwest (Tillmann and Siemann 2011). In addition, higher temperatures may contribute to higher incidences of harmful algal blooms that have been linked to paralytic shellfish poisoning (Huppert et al. 2009; Moore et al. 2009; Moore et al. 2011).

Elevated ocean temperatures are documented for NW waters from 1900 to 2008 (Deser et al. 2010), with future increases very likely, though characterized by considerable spatial and temporal variability. Ocean heat content and average sea surface temperatures (SSTs) increased on a global-ocean scale over the periods 1993–2003 and 1979–2005, respectively (Bindoff et al. 2007; Trenberth et al. 2007).[7] Models project Washington coastal SST to increase by 1.2 °C (~2.2 °F) by the 2040s (Mote and Salathé 2010).[8] However, recent analysis by Solomon and Newman (2012) that adjusts for ENSO variability suggests the possibility of an observed weak eastern Pacific SST cooling in tropical latitudes (coupled with warming of the western Pacific) over the period 1900–2010; cooling in the eastern equatorial Pacific and ENSO related changes in wind patterns over the North Pacific might facilitate a regional moderation of warming, or perhaps even a cooling, of the northeast Pacific (also see Deser et al. 2010).

Locally, the coastal upwelling and downwelling cycle leads to strong variation in temperature annually. Hickey and Banas (2003) showed that mid-shelf seasonal SSTs off the Washington coast varied by about 6 °C (10.8 °F) and off of Oregon's coast by about 4 °C (7.2 °F), over the period from 1950 to 1984. Future changes in SST will be highly influenced by several weather-related factors, such as wind, clouds, and air temperature, as well as ocean-related factors, such as upwelling, mixing, stratification, currents, and geographic proximity to rivers and bathymetric features that cause turbulent mixing.

Moore et al. (2008) investigated the influence of climate on Puget Sound oceanographic properties at seasonal to interannual timescales using continuous profile data at 16 stations from 1993 to 2002 and records of SST and sea surface salinity (SSS) from 1951 to 2002. Variability in Puget Sound water temperature and salinity correlated well with local surface air temperatures and freshwater inflows to Puget Sound from major river basins, respectively. The study also found SST and SSS to be significantly correlated with Aleutian Low, ENSO, and PDO variations; however, these correlations were weaker when compared to those of the local environmental factors (i.e., local air temperature, freshwater inflows). Since climate change will affect both the local and regional-scale forcings of SST and SSS, there will be complexities associated with understanding the dynamics of change and projecting future conditions.

4.5 Consequences for Coastal and Marine Natural Systems

The more than 4,400 miles (~7,100 km) of tidally influenced shoreline in Washington and Oregon consist of a diversity of coastal habitats, from rocky bluffs and sandy beaches along the Pacific Ocean, to the tidal flats, marshes, mixed sediment beaches, and eelgrass beds of NW estuaries such as Puget Sound. These natural systems, along with the region's offshore marine waters, are highly exposed to climate change and associated impacts, including sea level rise, changes in storminess, and ocean acidification. Key impacts include habitat loss (from erosion and inundation), shifts in species' ranges and abundances, and altered ecological processes and changes in the marine food web. The potential consequences of these changes to the region's marine and coastal natural resources could be substantial.

---

7 The Trenberth et al. (2007) study also found progressively increasing rates of global SST warming throughout the 20th century by examining three time slices: 1850–2005, 1901–2005, and 1979–2005.

8 This is multi-model, multi-emission scenario (A2, A1B, B1) average; however, SST differences between scenarios were small.

An increasing number of studies are investigating the potential responses of coastal and marine ecosystems to future climate change (Doney et al. 2012). However, there are still uncertainties associated with both the specific nature of the projected changes and our understanding of how these ecosystems may respond to these changes. In some instances, the potential effects of climate change can be clearly conveyed. For example, we can identify, with reasonable precision, areas of low-lying terrestrial habitat that may be inundated by sea level rise under future scenarios (Sallenger et al. 2003). However, the sensitivities and adaptive capacity of many coastal and marine species and ecosystems are not yet well-understood.

Climate change may create new stressors, or amplify existing stressors, on species and ecosystems that could interact synergistically and vary through time (Doney et al. 2012). In the following three subsections, we outline key concerns for coastal and marine ecosystems with regards to habitat loss, changes in species' ranges and abundances, and effects on ecological processes and the marine food web.

4.5.1 HABITAT LOSS

Climate change is expected to have significant physical impacts along the coast and estuarine shorelines of the Northwest, ranging from increased erosion and inundation of low-lying areas to incremental loss of coastal wetlands. Sea level rise, in particular, is considered to be one of the most certain and direct threats to the region's coastal systems resulting in progressive habitat loss in some areas. While the specific amount of relative sea level rise throughout the region can vary with the rate of change in coastal land elevation (Mote et al. 2008; see section 4.2.1), it is very likely that, with continuing global sea level rise, much of the NW coast will experience increased erosion and inundation (Glick et al. 2007). This includes coastal wetlands, tidal flats, and beaches; systems that are often highly susceptible to loss and alteration, particularly in low-lying areas, in locations with erodible sediments, or in areas where upland migration of a coastal habitat is hindered by natural bluffs or human-built structures such as dikes (Glick et al. 2007).

Ruggiero et al. (in press) examined physical shoreline changes along the Oregon and Washington outer coast. They noted significant beach erosion in north-central Oregon along the 150 km (~93 mi) section between the towns of Waldport and Manzanita, a region where relative sea levels have been rising (fig. 4.2). Specifically, they documented erosion rates of approximately -0.5 m/year (-1.6 ft/year) with local beach retreat rates as high as -4.4 m/year (-14.4 ft/year) over the period from 1967 to 2002. In contrast, they found that beaches along the southern Oregon coast (where land uplift rates exceed the local rate of sea level rise) have been relatively stable. The study also pointed out that major El Niño events (see section 4.2.2), which elevate water levels and wave heights (and change wave approach angles), can alter the NW coastline by redistributing beach sand alongshore. This redistribution creates "hot-spot" erosion sites and the potential for associated habitat losses near headlands, inlets, bays, and estuaries. The authors noted the current dearth of sources of sand for Oregon's beaches (compared to the number of sources available thousands of years ago at a time of lowered sea levels) and highlighted the fact that many of the state's beaches are presently deficient in sand volume, and as a result, do not provide sufficient buffer protection to backshore areas during winter storms and are susceptible to increased erosion hazards as sea levels continue to rise.

Physical changes to coastal wetlands, tidal flats, and beaches may have significant ecological implications for the fish and wildlife species they support. Beach erosion, as noted above, can be exacerbated by sea level rise and potential changes in storminess. In some locations, these changes will lead to increased exposure of upland areas to extreme tides and storm surges and may affect beach and upland habitat, such as haul-out sites used for resting, breeding, and rearing of pups by NW pinnipeds (e.g., harbor seals [*Phoca vitulina*]). Projections for sea level rise impacts on the coastal habitats of Puget Sound and parts of the outer Oregon and Washington coasts (Glick et al. 2007) suggest that nearshore habitats in the region are likely to face a dramatic shift in their composition, even under the relatively moderate IPCC AR4 scenario of 39 cm (~15 in) global sea level rise (Meehl et al. 2007). While there is considerable variability among different sites, much of the region's coastal freshwater marsh and swamp habitats are projected to convert to salt marsh or transitional marsh due to increases in saltwater inundation (Glick et al. 2007). These changes would include a reduction in the extent of tidal flats and estuarine and outer coast beaches (Glick et al. 2007), affecting associated species such as shorebirds and forage fish (Drut and Buchanan 2000; Krueger et al. 2010).

Nearshore ecosystems play a critical role in the life cycle of anadromous fish (e.g., salmon), many of which use coastal marshes and riparian areas for feeding and refuge as they transition between their freshwater and ocean life stages (Independent Scientific Advisory Board 2007; Bottom et al. 2005; Williams and Thom 2001). At particular risk are juvenile chum (*Onchoryncus keta*) and Chinook (*Onchorynchus tshawytcha*) salmon, which are considered to be the most estuarine-dependent species. For example, Hood (2005) estimated that rearing capacity in marshes for threatened juvenile Chinook salmon would decline by 211,000 and 530,000 fish, respectively, for 0.45- and 0.80-meter (17.7- and 31.5-inch) sea level rise scenarios. Sea level rise also may alter the salinity of surface and groundwater in coastal ecosystems. Many coastal plant and animal species are adapted to a certain level of salinity and prolonged salinity changes may result in habitat loss for some species (Burkett and Davidson 2012). Changes in salinity may also facilitate invasion by non-native species better adapted to salinity variations, such as the invasive New Zealand mud snail (*Potamopyrgus antipodarum*), which has been found in the Columbia River estuary (Hoy et al. 2012).

Coastal habitats may be able to accommodate, to some extent, moderate changes in sea level by migrating inland. Shaughnessy et al. (2012) estimated the effects of sea level rise on the availability of eelgrass (*Zostera marina*) for foraging black brant geese (*Branta bernicula* ssp. *nigricans*) in Willapa Bay and in the Padilla Bay complex (consisting of Padilla, Fidalgo, and Samish bays) in Washington. Under three future sea level rise scenarios of 2.8, 6.3, and 12.7 mm/year (0.11, 0.25, and 0.50 in/year), eelgrass habitat moved inland; but, the area of eelgrass habitat accessible to foraging black brant was projected to remain relatively constant in the Padilla Bay complex and expand in Willapa Bay over the next 100 years (Shaughnessy et al. 2012). However, in many other areas along the NW coast, the opportunity for inland migration has been considerably reduced by the development of dikes, seawalls, and other forms of armoring structures. Coastal armoring, while generally effective at protecting coastal property, may limit natural beach replenishment by cutting off backshore sediment sources.

For the region's river deltas, natural deposition of river sediments may enable at least some habitats to keep pace with sea level rise. However, modifications that inhibit the natural flow of sediments, such as dams and levees, are limiting this sedimentation (Redman et al. 2005) and thus a river's ability to keep pace with higher sea levels in the future. Site-specific studies are necessary to determine how changes in sedimentation rates associated with upstream activities might affect the localized impacts of sea level rise. The removal of two upriver dams in the Elwha River basin of the Olympic Peninsula of Washington State offers an excellent opportunity to monitor how restored sediment flow to a river delta might enhance the adaptive capacity of coastal systems in the region to sea level rise (Warrick et al. 2011).

Coastal dunes are often the "first line of defense" in terms of protecting coastal ecosystems and the backshore from storm damage. Dunes comprise approximately 45% of the outer Oregon and Washington coasts (Cooper 1958) and were historically managed to maximize coastal protection through the planting of European beach grass (*Ammophila arenaria*) and later American beach grass (*Ammophila breviligulata*). The switch in dominance from native species to exotic dune species resulted in a complete state change in coastal dune systems (Seabloom and Wiedemann 1994) with the creation of stable foredunes, reaching 15–20 meters (49–66 feet) in height, allowing for the interception of sand and decreased sand supply to the backshore. Foredunes dominated by *A. bre-viligulata* are lower and wider than foredunes dominated by *A. arenaria* due to the inferior ability of *A. breviligulata* to accumulate sand (Seabloom and Wiedemann 1994; Hacker et al. 2012; Zarnetske et al. 2012). Seabloom et al. (2013) modeled the exposure to storm-wave induced dune overtopping posed by the *A. breviligulata* invasion and the influence of projected multi-decadal changes in sea level and storm intensity. In their models, storm intensity was the largest driver of overtopping extent; however, the invasion by *A. breviligulata* tripled the area made vulnerable to overtopping and posed a fourfold larger exposure than sea level rise alone, over multi-decadal time scales.

4.5.2 CHANGES IN SPECIES' RANGES AND ABUNDANCES

Climate change is expected to have a significant impact on the geographical ranges, abundances, and diversity of marine species, including those that inhabit the waters off the Pacific Coast (Hollowed et al. 2001; Tillmann and Sieman 2011). Changes in pelagic (open ocean) fish species ranges and production associated with Pacific Ocean temperature variability during cyclical events, such as ENSO, PDO, and North Pacific Gyre Oscillation (NPGO), are an important indicator for potential species responses to climate change in the future (Cheung et al. 2009; Menge et al. 2010). For example, during ENSO and/or warm phase PDO, higher ocean temperatures and changes in wind patterns can change the timing and distribution of Pacific mackerel and hake, which are drawn to the region's coastal waters by warmer SSTs (Pearcy 1992; Peterson and Schwing 2003; Worm et al. 2005).

Longer-term trends also show a strong relationship between ocean temperatures and landings of anchovies and sardines in the eastern Pacific Ocean (Chavez et al. 2003). During periods when the Pacific Ocean has been warmer than average, sardines become more prevalent; and, during cold-water regimes, the relative abundance of anchovies rises. Considerable uncertainty remains as to whether climate change influences these relationships; however, they do illustrate important interconnections between marine species and climatic conditions. Moreover, Overland and Wang (2007) suggest that the anthropogenic influence on SSTs in the North Pacific Ocean may be as large as those of natural climate variability within the next 30–50 years, which could significantly alter marine species distributions and abundance. Indeed, several studies have detected the relative importance of climate variability versus long-term climate change in influencing patterns of change among certain species. For example, in a study of seabirds and climate in the California Current System, Sydeman et al. (2009) found long-term climate change to be the predominant factor in changes in the timing of breeding, productivity, and abundance of several seabird species, such as Cassin's auklet (*Ptychoramphus aleuticus*) (Becker et al. 2007). Since the mid-1980s, species generally associated with colder water (shearwaters and auklets) have become less abundant in the southern California Current System as SSTs have increased in the region. Research by Wolf et al. (2010) for California suggests that projected higher SSTs and changes in the intensity and timing of peak upwelling for 2080–2099 would contribute to an 11–45% decline in the population growth rate of the Farallon Island Cassin's auklet population by the end of the century.

The distribution and abundance of NW marine mammal species is also projected to change in the future. Davidson et al. (2012) identified the NW coastal region as a current area of relatively high extinction risk for marine mammals in a study that included historical SST anomalies and human impacts (e.g., fishing). Hazen et al. (2012) used climate simulations to examine habitat changes over the next century for fifteen North Pacific marine predator species, including three marine mammals. Blue whale (*Balaenoptera musculus*) and California sea lion (*Zalophus californianus*) habitats were projected to decrease over this time period while northern elephant seal (*Mirounga angustirostris*) habitat was projected to increase (Hazen et al. 2012). The future distribution and abundance of NW marine mammal species also may be altered by the potential effects of climate change on important habitat outside of the NW region. For example, the timing of gray whale (*Eschrichtius robustus*) migration along the NW coast may be affected by potential future changes in ocean temperatures and sea ice occurrence at summer feeding grounds in the Arctic (Moore and Huntington 2008; Robinson et al. 2009). A number of studies indicate that gray whales have responded to recent observed climate-related changes, such as sea-ice decline (Moore and Huntington 2008; Grebmeier et al. 2006).

4.5.3 ALTERED ECOLOGICAL PROCESSES AND CHANGES IN THE MARINE FOOD WEB

Climate change is likely to alter key ecological processes in both the open ocean and estuarine systems of the Northwest (Doney et al. 2012). Multiple segments of the marine food web may be altered by climate change effects on marine systems, such as potential changes in the timing and strength of coastal ocean upwelling (Barth et al. 2007), gradual and abrupt changes in the distribution of sea surface temperatures (Payne et al. 2012), ocean acidification (Hofmann et al. 2010), the salinity of estuaries (Ruggiero et al. 2010), and the occurrence of anoxic zones (Chan et al. 2008). These processes are intimately tied to the abundance, productivity, range, and distribution of both zooplankton and phytoplankton, which form the foundation of the marine food web. Climate change factors that play the most prominent role in affecting ecological processes include changes in SST, vertical stratification of the water column, depth of the mixing layer, wind patterns, freshwater input, eddy formation, pH, and calcium carbonate saturation states.

As discussed in section 4.3, of particular concern in the Northwest is ocean acidification. In the California Current System, for example, preliminary research suggests that ocean acidification could alter the composition of open ocean phytoplankton, with diatoms potentially gaining at the expense of calcifying phytoplankton (Hauri et al. 2009). In addition, research near Tatoosh Island, Washington, has already identified complex interactions between species under lower pH conditions (Wootton et al. 2008). This study suggests that declining ocean pH may have contributed to a decline in the abundance and mean size of the California mussel (*Mytilus californianus*), the dominant predator in the system, as well as the blue mussel (*Mytilus trossulus*) and goose barnacle (*Pollicipes polymerus*).[9] In contrast, the abundance of acorn barnacles (*Balanus glandula*, *Semibalanus cariosus*) and fleshy algae (*Halosaccion glandiforme*) has increased, likely due to decreased competition and predation from affected calcareous species. There is compelling evidence that ocean acidification associated with upwelling along the Oregon Coast was a major factor in recent die-off of oyster larvae at a regional hatchery, which validates laboratory-based acidification experiments and suggests that natural shellfish populations also may be vulnerable to increasing CO_2 (Barton et al. 2012).

A study by Kaplan et al. (2010) simulated ocean acidification impacts on shelled benthos and plankton, using an Atlantis ecosystem model for the US West Coast. Their model resulted in a 20–80% decline in the abundance of commercially important groundfish such as English sole (*Pleuronectes vetulus*), arrowtooth flounder (*Atheresthes stomias*), and yellowtail rockfish (*Sebastes flavidus*), owing to the loss of shelled prey items from their diet.

Bivalves exhibit a high sensitivity to pH and carbonate saturation state (Green et al. 2004; Gazeau et al. 2007; Talmage and Gobler 2009; Hettinger et al. 2012; Barton et al. 2012) particularly during larval and juvenile stages. Gazeau et al. (2007) projected decreases in mussel (*Mytilus edulis*) and Pacific oyster (*Crassostrea gigas*) calcification rates of 25% and 10% respectively by 2100 (see also Ries et al. 2009). Olympia oyster (*Ostrea lurida*) larvae reared under low pH conditions displayed juvenile shell growth rates up to 41% slower a week after settlement, compared with growth rates under control conditions (Hettinger et al. 2012). Slower shell growth rates persisted for over seven weeks after the oysters were returned to control conditions that replicated present-day CO_2 levels in seawater. These results suggest the existence of carry-over effects of acidification from larval to adult stages (Hettinger et al. 2012).

While some marine animal species, such as shelled invertebrates, typically respond negatively to ocean acidification conditions, certain marine aquatic plants, such as some seagrass species, appear to benefit from CO_2 enrichment (e.g., Hendriks et al. 2010). Much is still unknown, however, about the effects of ocean acidification on many

---

9 The primary cause of the rapid decline in pH observed at Tatoosh Island by Wootton et al. (2008) has been assessed by others (for example, see Brown 2012) and those studies indicate that local factors, such as variances in regional river discharge, may better explain the bulk of the transient declines in pH, rather than a larger scale acidification mechanism.

organisms and how changes in ocean pH and carbonate saturation state may interact with other environmental factors (e.g., temperature, dissolved oxygen, nitrogen) and human impacts (e.g., pollution, fisheries, habitat modification and loss) (Harley et al. 2006; Whitney et al. 2007; Doney et al. 2012). For example, studies suggest that the toxicity of certain phytoplankton associated with harmful algal blooms (HABs), including the dinoflagellate *Karlodinium* and two species of the diatom *Pseudo-nitzschia*, may increase under ocean acidification (Fu et al. 2010; Reusink et al. 2012; Sun et al. 2011; Tatters et al. 2012). Climate change may also contribute to greater risks from the dinoflagellate *Alexandrium catenella* in Puget Sound, along with associated accumulation of paralytic shellfish toxins (Moore et al. 2011). Specific conditions that appear to favor HABs of *A. catenella* include a combination of warmer air and water temperatures, low streamflow, low winds, and small tidal height variability. Under the SRES-A1B scenario of continued emissions growth peaking at mid-century, models project the window of opportunity for *A. catenella* in Puget Sound to increase by an average of 13 days by the end of the century (Moore et al. 2011). Furthermore, the onset of favorable conditions is projected to begin up to two months earlier and persist for up to one month later than it does currently (Moore et al. 2011).

4.6 Consequences for Coastal Communities and the Built Environment

NW coastal communities will be affected by climate change through changes in both the terrestrial and marine environments, with potential issues of concern including erosion, temporary flooding, and permanent inundation from sea level rise, coastal storms, and river flooding; local flooding and landslides due to high-intensity precipitation events; water supply and water quality impacts; direct heat effects; and ecological changes. These changes will affect coastal transportation and navigation, engineered coastal structures (seawalls, riprap, jetties, etc.), flood and erosion control infrastructure, water supply and waste and storm water systems, public health and safety, and the coastal recreation, travel, and hospitality sectors more broadly. Details of these general impact pathways and associated consequences have been reviewed elsewhere (e.g., Oak Ridge National Laboratory 2012).[10]

The varying characteristics of coastlines throughout the Northwest (see sections 4.1 and 4.5) lead to sub-regional differences in the degree to which coastal infrastructure is exposed to climate change impacts. Quantifying the potential extent of climate change impacts on coastal communities and infrastructure at the regional scale is complicated by (1) local variations in projected drivers of community impacts (e.g., sea level rise, landslide and erosion risk, evolving floodplains), (2) fine-scale coastal topography, (3) limited site-specific elevation data for quantifying the exposure of critical infrastructure to sea level rise and other hazards, and (4) compounding effects of multiple climate impacts (e.g., sea level rise, coastal flooding, landslides). To date, most large-scale analyses of consequences of climate change for the built environment and human communities of

---

10 Note: non-coastal specific impact pathways, such as climate change impacts on urban water supplies (e.g., Vano et al. 2010), are not addressed in this chapter.

the NW coast have focused on transportation-related impacts (WSDOT 2011; MacArthur et al. 2012). A handful of individual communities have begun assessing local impacts and vulnerabilities across multiple sectors and various hazards, with some implementing adaptive actions as described in section 4.8.

4.6.1 COASTAL TRANSPORTATION INFRASTRUCTURE

Approximately 4,500 km (2,800 mi) of roads in the coastal counties of Washington and Oregon are in the 100-year flood plain (Douglass et al. 2005); many important roadways in coastal counties run along rivers or creeks and may experience increasing damage from river flooding, debris flows, bridge scouring, and/or landslides (MacArthur et al. 2012; WSDOT 2011).

The Washington State Department of Transportation (WSDOT) assessed the climate change vulnerability of state-owned transportation infrastructure (i.e., state highways, roads, bridges, tunnels, railroads, ferry terminals, airports, maintenance facilities) by considering the implications of multiple climate drivers and impacts, including sea level rise and changes in temperature, precipitation, flooding, landslides, and wildfire (WSDOT 2011). WSDOT's qualitative analysis combined information about climate change impacts with agency staff's knowledge of local roadway characteristics and current vulnerabilities, weighted by an assessment of the asset's importance ("criticality") to local and regional connectivity. Under a scenario of 2 feet (0.6 meters) of sea level rise (consistent with NRC [2012] end-of-century projections for Washington State, see table 4.2) a few low-lying Puget Sound roadways and highways along the outer coast could see significant long-term inundation (fig. 4.4). However, most major state highways within the Puget Sound region are situated high enough to avoid *permanent* inundation under this scenario. More likely impacts include temporary closures and reduced vehicle capacity due to highly localized and intermittent flooding resulting from storm surge and culvert backups. In some locations, such impacts already occur during high tides, or during average tides combined with heavy rain events. Under higher sea level rise scenarios, additional roadway segments in Washington become vulnerable (e.g., sections of State Routes 3 and 101). Impacts would be exacerbated in those areas where the risk of landslides and river flooding is projected to increase (fig. 4.4; WSDOT 2011).

Changing sediment transport regimes, due to both changing river flows and receding glaciers, which are projected to alter the shape and depth of river channels, also increase the risk of flooding damage to state highways. Although impacts on Oregon's roads and highways have not been assessed in similar detail, a regional-scale study identified highways near the mouth of the Columbia River and near Astoria, Oregon as most at risk, after Puget Sound highways in Washington (MacArthur et al. 2012).

Other state-owned coastal transportation modes are thought to be largely robust to projected changes, with a few exceptions. The Copalis Beach airport in Washington (fig. 4.1), which already closes at high tide, is expected to close more frequently, if not permanently, as sea level rises. Washington's ferry terminals are expected to be able to accommodate 2 feet (0.6 meters) of sea level rise with minor impacts, with the exception of West Seattle's Fauntleroy terminal (fig. 4.1), which WSDOT determined to have a slightly higher risk of adverse impact. At four feet (1.2 meters) of sea level rise, the Bainbridge Island, Edmonds and Keystone terminals become highly vulnerable as well (WSDOT

Figure 4.4 Vulnerability of Washington State transportation assets (state-owned roads, bridges, tunnels, ferry terminals and maintenance facilities, and airports) in Washington's coastal counties to the climate change impacts associated with 2 feet (0.6 meters) of sea level rise and a range of temperature and precipitation changes. Washington State Department of Transportation (WSDOT) agency staff qualitatively evaluated (1) the likelihood of asset failure due to the combined impacts of sea level rise (erosion, inundation, storm surge, and flooding), changes in mean and extreme temperature and precipitation, and changes in snowpack, streamflow, river flooding, landslides, and wildfire—taking into account local infrastructure characteristics, and (2) the "criticality" of each asset to the regional and local transportation system (i.e., consequences of failure). Adapted from WSDOT (2011) Exhibit B-4.15. Note: For Planning Purposes Only. Not suitable for site-specific use.

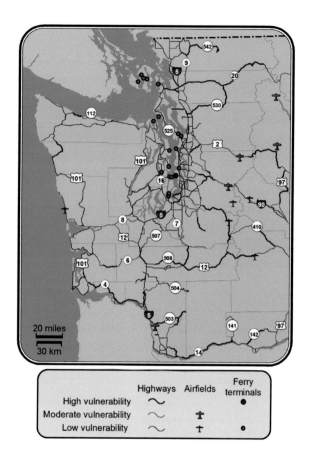

2011). The assessment also noted that sea level rise might lead to fewer ferry terminal closures as a result of extreme low water levels, a potential benefit of higher sea levels.

Climate impacts on secondary transportation routes can be extremely important to local communities, even if effects on the region's overall interconnectivity are small. Key access routes to the Swinomish Indian Reservation on Fidalgo Island in northern Puget Sound (see fig. 4.1), for example, are located in low-lying areas at risk of inundation. The Swinomish Indian Tribal Community estimates that a 4-foot (1.2-meter) tidal surge could cut off access to their reservation entirely, isolating residents from the mainland (Swinomish Indian Tribal Community 2009).

Additional potential climate change impacts on coastal transportation infrastructure, beyond the risks posed by sea level rise, are just beginning to be examined in detail in the Northwest. Examples include: direct impacts of increases in temperature on pavement longevity, rail track deformities, and rail speed restrictions[11] (currently being examined for Sound Transit, e.g., A. Shatzkin, Sound Transit, pers. comm.); increased landslide risk for coastal highways and rail lines (e.g., MacArthur et al. 2012); future reliability of coastal tsunami evacuation pathways; and bridge clearance issues caused by higher river flows and/or sea level rise (MacArthur et al. 2012; T. Morgenstern, City of

---

11 For example, the Portland, Oregon, transit agency, TriMet, mandates reducing train speeds by 10 mph in areas with speed limits at or above 35 mph when temperatures exceed 32 °C (90 °F) (TriMet 2010).

Seattle, pers. comm.), although most of Washington State's newer bridges are thought to be robust up to levels of 4 feet (1.2 meters) of sea level rise (WSDOT 2011).

4.6.2 COASTAL COMMUNITIES

A few local governments in the Northwest are evaluating—and in some cases preparing for (see section 4.8)—climate-related coastal risks and vulnerabilities. Whereas some efforts focus on a single aspect of climate change (e.g., sea level rise) and a single issue area (e.g., wastewater treatment facilities), others are assessing the combined implications of multiple risks, climate and otherwise, for overall community values and priorities. We provide some illustrative examples of both; these and others have been compiled elsewhere (Bierbaum et al. 2013) and new examples continue to emerge. The City of Seattle has assessed, and continues to evaluate, the risks posed by sea level rise and storm surge, examining public utility infrastructure, including maintenance holes, water mains, and drainage outfalls and pump stations with proximity to the shoreline (P. Fleming and J. Rufo-Hill, Seattle Public Utilities, pers. comm.). Initial results include a partial inventory of vulnerable assets and maps indicating future coastal inundation

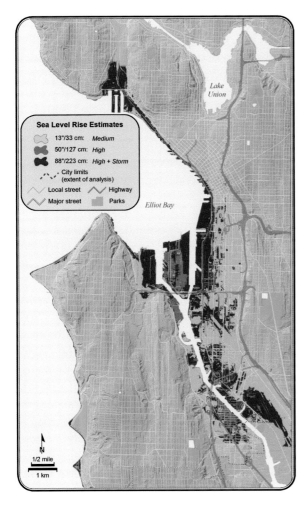

Figure 4.5 Rising sea levels and changing inundation risks in the City of Seattle. Areas of Seattle projected by Seattle Public Utilities to be below sea level during high tide (mean higher high water) and therefore at risk of inundation are shaded in blue under three levels of sea level rise (Mote et al. 2008) assuming no adaptation (P. Fleming and J. Rufo-Hill, Seattle Public Utilities, pers. comm.). High (50 in [127 cm]) and medium (13 in [33 cm]) levels are within the range projected for the Northwest by 2100; the highest level incorporates the compounding effect of storm surge. Unconnected inland areas shown to be below sea level may not be inundated, but could experience localized flooding due to areas of standing water caused by a rise in the water table and drainage pipes backed up with sea water. (Adapted figure courtesy of Seattle Public Utilities).

(fig. 4.5); identified assets are being manually inspected to confirm vulnerability and to develop adaptation options. The City of Olympia has similarly used high-resolution land elevation data to assess areas of future exposure to inundation in the downtown core under various sea level rise scenarios (box 4.1).

King County and the City of Anacortes, as well as other local governments across the nation, are considering sea level rise and precipitation driven impacts in their risk assessments and design of storm, wastewater, and drinking water treatment infrastructure (King County 2009; City of Anacortes 2012; Solecki and Rosensweig 2012).

The King County Wastewater Treatment Division, in response to the County's climate action plan released in 2007, assessed wastewater infrastructure (e.g., treatment plants, pump stations) at 40 separate locations for vulnerability to coastal climate impacts. They examined facility elevations, historical tide levels and storm surge, and projected future sea level rise to create a "vulnerable facilities inventory" that identified the five most vulnerable facilities for which more detailed site analyses, and ultimately design modifications, were made (King County 2009).

Anacortes, located approximately 80 miles (~130 km) north of Seattle (fig 4.1), has altered design criteria to account for the projected increased risk of flooding on the adjacent Skagit River, and the accompanying dramatic increased sediment loading of the drinking water source waters. The City's new $65 million water treatment plant (under construction in 2013) includes elevated structures, water tight construction with minimal structural penetrations below the (current) 100-year flood elevation, relocation of electric control equipment above the (current) 100-year flood level, and, for the first time, active rather than gravity-based sediment removal processes. Future analyses will examine the degree to which the plant's source water intake is likely to be contaminated with saltwater, due to its current proximity to the salt wedge and the combined future pressures of sea level rise and lower summer streamflows (City of Anacortes 2012; Zemtseff 2012).

In one of the most comprehensive assessments yet conducted for a small coastal community, the Swinomish Indian Tribal Community examined a wide range of climate vulnerabilities and corresponding adaptation strategies (see also Chapter 8). Aggregating climate impacts into three primary risk zones (i.e., sea level rise inundation, tidal surge inundation, and wildfire), the tribe created an inventory of potentially affected assets and resources, mapped impact areas, and provided a detailed accounting of the major risks facing their community and the local ecosystems upon which they depend. Specific issues considered include: vulnerability of vital transportation linkages, risks to agricultural and economic development lands, resilience of cultural sites and practices, tribal member health, and potential economic consequences (Swinomish Indian Tribal Community 2009). Approximately 15% of Swinomish tribal land is at risk of inundation from rising sea level, potentially threatening major investments and enterprises in the Tribe's primary economic development lands, in addition to potential impacts on low-lying agricultural land, culturally important shellfish beds, fishing docks, and commercial and private residential development. Upland areas containing extensive forest resources and developed property worth over $518 million may be at risk from potentially destructive wildfire. Within the tribe's low-lying inundation prone areas are approximately 160 residential structures with a total estimated value of over $83 million and a number of commercial structures with a total estimated value of almost $19 million (Swinomish Indian Tribal Community 2009). Moving forward, a primary question is how to reconcile the

BOX 4.1

Coping with sea level rise risks today and tomorrow in Olympia, Washington

Washington State's capital, located at the southern tip of Puget Sound with most of its downtown built on low-lying fill, has long been recognized as vulnerable to sea level rise (e.g., Craig 1993). Past and current sea level rise mapping show that, with 15 cm (~6 in) of sea level rise, an extreme high tide could flood vital public infrastructure, high-density development, and the City's historic district. City planners discovered that areas projected to be at significant risk to flooding and inundation included some that were not just adjacent to the shoreline (fig. 4.6). Based on both sea level rise scenarios and existing experience with high tide and high river flow events, climate change is projected to affect downtown Olympia via (1) marine waters entering stormwater outfalls and flowing up and discharging into downtown streets from inland storm drains during high tides; (2) overloading of the stormwater system (including piped streams) during high-intensity precipitation events coincident with a high tide, causing storm drain back-up and discharge; and (3) marine waters overtopping the bank resulting in saltwater inundation (City of Olympia 2012).

Technical work in 2009–2010 provided sophisticated hydraulic simulation and landform analysis to improve the City's understanding of how tidal elevations and precipitation events could interact and affect downtown infrastructure systems and buildings. More recently, Olympia completed an engineering analysis of potential sea wall designs and responses to an increase in sea level of 127 cm (50 in) (Coast and Harbor Engineering 2011). The City is in the process of incorporating sea level rise issues into its Comprehensive Plan and Shoreline Master Program revisions (City of Olympia 2012). Because City policy directs department staff to investigate how to protect the downtown from sea level rise, various adaptation options (both engineering approaches and regulatory measures) are being examined. The investigations to date have resulted in a new recognition of the current vulnerability of Olympia's downtown properties, emergency transportation corridors, and essential public services (including stormwater and wastewater systems) and led Olympia to enact temporary emergency measures (e.g., sealing specific storm drains), and begin small projects to reduce current risks (e.g., consolidating stormwater outfalls and raising shorelines), while planning for the more significant investments necessary to lower longer term risks (A. Haub, City of Olympia, pers. comm.).

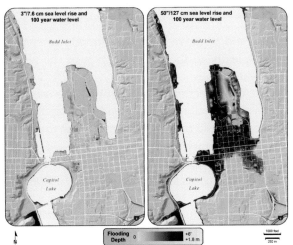

Figure 4.6 Projected flooding of downtown Olympia with a 100-year water level (0.01 average annual exceedance probability for storm tides, wave effects on mean water level at the shoreline, and precipitation run-off) and 7.6 cm (3 in) of sea level rise (left) and a 100-year water level and 127 cm (50 in) of sea level rise (right). Redrawn from Coast and Harbor Engineering (2011), courtesy City of Olympia.

plans to continue development of the Tribe's primary economic development zone with the vulnerability of those lands to inundation by projected sea level rise. The tribe has begun, within the context of the master planning for their economic development zone, to evaluate new flexible approaches to waterfront development that explicitly integrate sea level rise considerations (e.g., designs that accommodate progressively higher levels of inundation over time) while also balancing the economic, social, and environmental goals of the Swinomish Indian Tribal Community (Knight et al. 2013).

4.7 Economic Consequences of Coastal Impacts

Coastal impacts from climate change have the potential to substantially affect the economies of coastal communities and a number of regionally important sectors. These sensitivities stem primarily from the region's extensive seaport and municipal coastal infrastructure, the limited options for alternative transportation corridors in many locations along NW coasts, and the local and regional importance of the marine-based fishing industry.

Marine and coastal resources provide communities in the Northwest with numerous economic benefits including: natural harbors and deep-water ports for commerce, trade, and transportation; shorelines that attract residents and tourists; and wetlands and estuaries that are critical for the productivity of fisheries and marine biodiversity. Coastal ecosystems also contain economic value through their ability to cycle and move nutrients, store carbon, detoxify wastes, and mitigate damages from floods and coastal storms. Scavia et al. (2002) provide an overview of climate change impacts on US coastal and marine ecosystems that can serve as a foundation for economic assessments. However, translating these impacts into monetary units is challenging and research has been limited to isolated case studies. Such information, however, is needed for robust risk assessment, policy design, and adaptation planning.

In the following section we use recent landings and revenue data to illustrate the potential significance of climate impacts on the Northwest's marine fishing industries. Robust economic evaluations of the impacts of climate change on other coastal relevant sectors have yet to be conducted.

4.7.1 MARINE FISHERIES

Climate change will have both positive and negative economic impacts on commercial and recreational fisheries, adding complexity to the determination of the net overall economic impact to the region. Different species and population patterns will vary in their responses to climate change, as noted in Section 4.5. In addition, the robustness of commercial fishing in the Northwest is dependent upon the physical characteristics and conditions throughout the marine waters of western North America, with the economics driven to a large extent by the markets into which these products are sold. In general, cool-water species are expected to decline in abundance while warm-water species become more abundant in response to a warming ocean (Scavia et al. 2002; Roessig et al. 2004; Harley et al. 2006; Brander 2007).

Changes in distribution, abundance, and productivity of marine populations due to climate-related changes in ocean conditions will impact the level and composition of landings and the value of landings in the Oregon and Washington commercial fisheries. Currently, the key commercial species for Oregon and Washington are: crab, clams, oysters, salmon, albacore tuna, sablefish, shrimp, hake, halibut, flatfish, and Pacific sardine.

Figure 4.7 presents total landings statistics for all fish species in Oregon and Washington from 1980 to the present. This graph illustrates the general upward trend in landings and value of landings as well as the yearly fluctuations which are dependent on a combination of harvest rules (based on stock assessments and allocations) and oceanographic variations such as temporary warming or cooling events. Total revenue from these species averaged around $275 million per year between 2000 and 2009 but rose sharply from 2009 to 2011 (National Marine Fisheries Service 2010, updated). In 2009, the region's seafood industry is estimated to have generated $8.4 billion in sales, $2 billion in income, and 71 million jobs. These impacts reflect the overall impact at the harvesting, processing, and retailing levels (National Marine Fisheries Service 2010).

Ocean acidification is projected to adversely affect NW coastal estuaries that are the source of highly valued shellfish fisheries (Barton et al. 2012). Figure 4.8 illustrates the importance of shellfish to the overall fishing industry in the region. For example, shellfish landings represent 49% and 72% of the total landing values of Oregon and Washington commercial fisheries, respectively, over the period 2000–2009 (fig. 4.8); and, in Washington State alone, shellfish growers in 2010 produced more than $150 million in product (Pacific Shellfish Institute 2013). Shellfish aquaculture is an important source of jobs in the Northwest with revenues directly benefiting state and local economies. The loss or decline of shellfish aquaculture could have significant social and economic effects (National Marine Fisheries Service 2010). However, some adaptation may be possible; commercial oyster growers in the region have successfully altered seed production techniques by leveraging water chemistry monitoring resources to minimize the exposure of new oyster seed to particularly acidified waters (Scigliano 2012; Washington State Blue Ribbon Panel 2012, chapter 6), although the long-term viability of this strategy is unknown. Additionally, the negative impact of ocean acidification on shelled benthos (prey for groundfish) will very likely have negative effects on commercially important groundfish in the region (Kaplan et al. 2010).

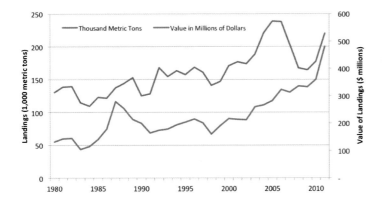

Figure 4.7 Landing statistics for Oregon and Washington from 1980 to the present. Data obtained from National Marine Fisheries Service (2012).

Figure 4.8 Commercial shellfish landings revenue for Oregon and Washington as a percent of total commercial landings revenue for each state (2000–2009). Data obtained from National Marine Fisheries Service (2010).

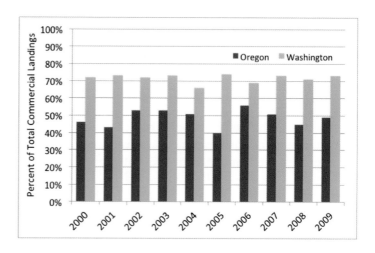

4.7.2 OTHER ECONOMIC IMPACTS

Impacts on coastal systems are considered among the most costly consequences of a warming climate (Burkett and Davidson 2012), due in part to the combined impacts of sea level rise, increased ocean acidification, increased probability of extreme weather events, as well as growing populations in many coastal communities (Crossett et al. 2004). However, quantifying the economic impacts of a changing climate is complicated by the uncertainty in the physical, biological, and socio-economic factors. Despite these uncertainties, several studies have attempted to quantify at least some of the economic outcomes of climate change for coastal areas. Burkett and Davidson (2012) examined the effects of climate change on coastal economies in US and concluded that adapting to a changing climate will be a challenge for these economies, which are highly dependent on marine resources.

Seo (2007) also notes that it has been difficult to quantify impacts on natural systems and thus current estimates are speculative. Yohe (1989) and Yohe et al. (1996) developed a cost-benefit approach that examined properties at risk for flooding and compared the value of those properties to the cost of building protective structures such as seawalls. Heberger et al. (2011) provided planning-level estimates of economic vulnerability by examining the replacement value of properties vulnerable to damaging floods in the future, assuming no adaptation. Although highly variable, the potential negative economic consequences of damage and degradation to infrastructure and ecosystems (described throughout this chapter) resulting from projected changes in climate are substantial and pose further threats to public health, safety, and the economic vitality in many NW coastal areas.

4.8 Adaptation

Adaptation to sea level rise and other climate change impacts in coastal systems has received considerable attention in the region over the past decade. Both Oregon and Washington have developed state-based climate change adaptation strategies to address impacts across multiple sectors (Oregon Coastal Management Program 2009;

Washington Department of Ecology 2008, 2012). The Washington Department of Ecology has developed guidance for addressing sea level rise in Shoreline Master Programs, which are the locally developed land-use policies and regulations designed to manage shoreline use under Washington's Shoreline Management Act. *The 2012/2013 Action Agenda for Puget Sound* (Puget Sound Partnership 2012) acknowledges the threat of climate change and suggests near-term actions to address the challenges. Washington and Oregon joined California in the West Coast Governors' Agreement (WCGA) on Ocean Health, which is working to develop a framework to assist state and local governments in planning for climate change impacts to coastal areas and communities throughout the region (WCGA 2010).

In 2012, a panel of scientists; state, local, federal, and tribal policy makers; shellfish industry representatives; and conservation community representatives came together at the request of Washington State Governor Christine Gregoire to recommend actions for addressing the causes and consequences of ocean acidification (Washington State Blue Ribbon Panel 2012). In a report informed by a scientific summary documenting the current state of knowledge and outlining specific research and monitoring needs for ocean acidification in Washington State (NOAA OAR 2012), the Panel recommended 42 actions, including 18 "key early actions," grouped into 6 major categories: reduction of carbon dioxide emissions, control of land-based pollutants, adaptation and remediation of the impacts, monitoring and investigation, stakeholder and public engagement and education, and government action. Examples of "key early actions" in these categories include (Washington State Blue Ribbon Panel 2012):

- Work with international, national, and regional partners to advocate for a comprehensive strategy to reduce carbon dioxide emissions;
- Strengthen local source control programs to achieve needed reductions in nutrients and organic carbon that enhance the ocean acidification problem;
- Investigate and develop commercial-scale water treatment methods or hatchery designs to protect larvae from corrosive seawater;
- Identify, protect, and manage refuges for organisms vulnerable to ocean acidification and other stressors;
- Establish an expanded and sustained ocean acidification monitoring network to measure trends in local acidification conditions and related biological responses;
- Establish the ability to make short-term forecasts of corrosive conditions for application to shellfish hatcheries, growing areas, and other areas of concern; and
- Provide a forum for agricultural, business, and other stakeholders to engage with coastal resource users and managers in developing and implementing solutions.

International collaboration on the scientific research and policy response to the coastal impacts of climate change was initiated in 2008 by Washington State and British Columbia (British Columbia and Washington State MOU 2008). This collaboration has led to the sharing of information on sea level rise and related research and an expansion of the *Green Shores* program, first developed in British Columbia and currently being piloted in Washington State. The *Green Shores* program fosters "softer" coastal engineering alternatives that mimic natural shoreline features instead of traditional engineered shoreline armoring techniques, such as concrete bulkheads or riprap. As part of this collaboration, Washington and British Columbia initiated a "king tides" photo initiative to document extreme high winter tides and build awareness around the potential impacts of future sea level rise (Washington Department of Ecology 2013).

Numerous adaptation efforts are emerging at the site- or community-level, for both natural and human systems. In addition, there are examples of actions that are primarily motivated by other factors—habitat restoration or community protection, for example—that also deliver important adaptive benefit, as the following case studies illustrate.

4.8.1 NISQUALLY DELTA CASE STUDY: RESTORING SALMON AND WILDLIFE HABITAT IN PUGET SOUND

In the Nisqually River Delta in Washington, estuary restoration on a large scale to assist salmon and wildlife recovery provides an example of adaptation to climate change and sea level rise. After a century of isolation behind dikes, a large portion of the Nisqually National Wildlife Refuge was reconnected in 2009 with tidal flow by removal of a major dike and restoration of 308 hectares (761 acres; see fig. 4.9). These restoration efforts, with the assistance of Ducks Unlimited, the Nisqually Indian Tribe, and others, have reconnected more than 33.8 km (21 mi) of historical tidal channels and floodplains with Puget Sound (US Fish and Wildlife Service 2010). A new exterior dike was constructed to protect freshwater wetland habitat for migratory birds from tidal inundation and future sea level rise. More than 57 hectares (141 acres) of tidal wetlands were also restored by the Nisqually Tribe. Combined with expansion of the authorized Refuge boundary, ongoing acquisition efforts to expand the Refuge will further enhance the ability of the Nisqually River Delta to provide diverse estuary and freshwater habitats despite rising sea level, increasing river floods, and loss of estuarine habitat elsewhere in Puget Sound. This project is considered a major step in increasing estuary habitat and recovering the greater Puget Sound estuary.

4.8.2 NESKOWIN, OREGON, CASE STUDY: ORGANIZING TO COPE WITH AN ERODING COASTLINE

Erosion and flooding have been particularly acute along portions of the Neskowin littoral cell (a section of coast characterized by sediment sources, transport pathways, and sinks), in southern Tillamook County, Oregon, since the late 1990s. The Neskowin Coastal Hazards Committee (NCHC 2013) is a local community group, formed in response to these coastal hazards in order to support the protection of the Neskowin beach and community and explore ways to plan for and adapt to the potential future changes in the Neskowin coastal area.

Despite uncertainty over the future frequency and magnitude of flood and erosion hazards, the seriousness of existing risks has motivated Neskowin to conduct a community-wide risk and vulnerability assessment and to plan for hazard reduction, including the examination of the costs and benefits of various decisions within the context of a range of climate change scenarios (figs. 4.10 and 4.11). Baron (2011) and Ruggiero et al.

Figure 4.9 Adapting the Nisqually River Delta to Sea Level Rise. Photo Credits: Backhoe (a), Jesse Barham/US Fish and Wildlife Service http://www.flickr.com/photos/usfwspacific/5791362738 /in/set-72157626745822317); Aerial (b), Jean Takekawa/US Fish and Wildlife Service (http://www .flickr.com/photos/usfwspacific/5790804083/in/set-72157626745822317)

(2011) developed coastal vulnerability assessments for all of Tillamook County by exploring a range of possible climate futures using a suite of simple coastal change models (fig. 4.10). Exposure analyses were performed by superimposing relevant infrastructure asset information, such as locations of structures and roads, on the hazard zones (fig. 4.11) while also considering climate change uncertainty. The ultimate aim of this effort is to provide coastal planners with the tools and information to allow for science-based decisions that will increase the adaptive capacity of coastal communities in Tillamook County as they prepare for future climate change.

Concerned about ongoing dramatic erosion threatening the community beach and private property, the NCHC commissioned an engineering study to examine the costs and benefits of beach and community protection via elevating and maintaining riprap revetments, constructing extensive coastal barriers, engaging in costly and perpetual beach nourishment programs, or migrating infrastructure inland. Although the cheapest

Figure 4.10 Recent coastal flooding (Allan et al. 2009), erosion, and failures of coastal protection structures in the community of Neskowin, Oregon. Both photographs were taken by Jonathan Allan of the Oregon Department of Geology and Mineral Industries (DOGAMI) in 2008 at approximately the location of the red star (right hand panel), within a shaded "coastal change hazard zone" (Ruggiero et al. 2011; Baron et al. 2010). These zones have been incorporated into Tillamook County, Oregon's Coastal Change Adaptation Plan (Rhose 2011). Hazard zones were developed for both the annual and 100-year storm events for the time periods of 2009, 2030, 2050, and 2100. Coastal change hazard zones were derived from 1,800 scenarios using an array of climate change projections and accounting for coastal morphological variability.

(a)

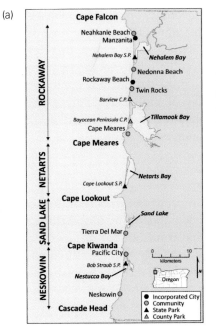

Figure 4.11 Tillamook County infrastructure coastal change hazard exposure analysis (Baron 2011; Ruggiero et al. 2011). Panel (b) shows the number of exposed structures; and panel (c) the length of affected roadway, both organized by littoral cell, shown in panel (a). Storm exposure analyses were performed for both the annual and 100-year storm events for the time periods of 2009, 2030, 2050, and 2100, assuming local rates of sea level rise within the range recently projected for the Central Oregon Coast (see table 4.2; NRC 2012). Results are shown for confidence intervals of 98%, 50%, and 2%. The number of structures in the Neskowin littoral cell exposed to the annual storm event increases from 161 in 2009 to 421 in 2100 for the 2% confidence interval. The length of roadway impacted by the 100-year storm more than doubles by 2100 (from 5 to 11 km [3.1 to 6.8 mi]).

(b)

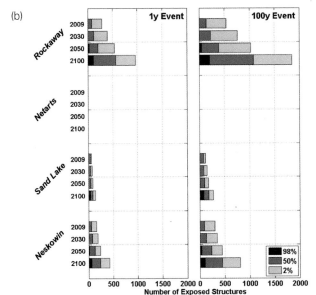

(c)

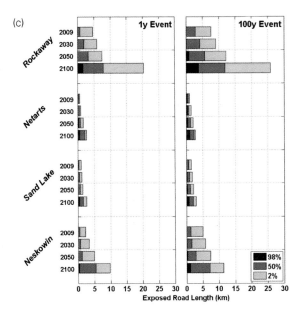

alternative was to raise the height of the 7,000-foot (2,134-meter) long revetment protecting the shoreline, its estimated cost of $7 million in construction costs alone has raised interest in additional options, including: managed retreat, protection of critical infrastructure (e.g., transportation access, water and sewage systems), new land use and construction permitting requirements, and the establishment of a geologic hazard abatement district (M. Labhart, Tillamook County Commissioner, pers. comm.).

The NCHC has proposed the establishment of a "geologic hazard abatement district," as well as new land use recommendations and ordinances for adoption by the County. The Neskowin Coastal Erosion Adaptation Plan, proposed as a new component of the County's Comprehensive Plan, identifies hazard overlay zones and establishes permitting, construction, and reconstruction rules for these zones. Proposed ordinances would prohibit new "slab-on-grade" foundations in the hazard zone; require that new structures be moveable, either vertically or horizontally on the lot (for example, either stem wall or pile foundations); limit creation of parcels to those that include a building site located outside the hazard risk zone; and require a 50-year annual erosion rate, plus a 20-foot (6.1-meter) buffer distance, to be utilized for construction on any bluff-backed building sites (M. Labhart, Tillamook County Commissioner, pers. comm.).

4.9 Knowledge Gaps and Research Needs

There are a number of priority research topics that will improve our understanding of how coastal marine ecosystems and human systems will be impacted by future climate change. Further study is needed to more fully understand contributions of terrestrial factors (see section 4.2.1) to relative sea level rise, current and projected trends in ocean acidification along our coasts, climate impacts on commercial and recreational fisheries, and whether and how coastal winds and cyclical events such as ENSO and PDO may change in the future.

To better estimate local sea level change, more rigorous long and medium wavelength vertical deformation field analysis of coastal landforms is needed, particularly for the region north of the Mendocino triple junction where subduction dominates and vertical deformation rates are of the same order of magnitude as that of sea level rise. There is also little information on sediment dynamics and how they would contribute to consequences of sea level rise and other coastal hazards (Dalton et al. 2012).

Further study and improved projections of the rate of change of offshore (open ocean and deep water) pH and carbonate saturation states along the West Coast are needed, including further analysis of regionally specific lag times between surface absorption of anthropogenic CO_2 and its appearance as upwelled waters along the NW coast. Some studies note the possibility for synergies between acidification and other stressors, such as temperature or nutrient changes. A better understanding of these interactions is necessary to determine how NW marine species and communities may respond to future acidification in the context of other coincident stressors, knowledge that is currently not available or very limited for local species.

Better information is required on the resilience of marine and coastal species and the likely shifts in migration patterns (or other adaptive responses) that may result from projected changes in ocean conditions, including those associated with extreme events. The adaptability of many commercial and recreational fisheries, and the coastal economies reliant on them, may depend on the ability to anticipate the general magnitude and direction of these changes. Additional research is needed regarding how the loss of certain existing species may be offset by possible enhancements in other local species, or the in-migration of non-native species better suited to future ocean conditions. Improved spatial information related to the climate sensitivities of specific NW marine and coastal habitats is needed to assess the potential future economic impacts to the fishing industry and coastal communities.

Major uncertainty still exists in terms of how coastal winds, and hence upwelling, will change with climate change and regarding a possible link between warmer ocean temperatures and hypoxic events. Lack of time series data and adequate spatial coverage make risk assessments challenging (Dalton et al. 2012). Considerable uncertainty also remains in how changes to ENSO and PDO might impact the geographical ranges, abundances, and diversity of marine species.

Additional research topics that should be considered for future funding include the development of more robust and comprehensive NW regional climate change economic impact assessments and benefit-cost studies of leading coastal adaptation strategies most applicable to the region.

Acknowledgements

The authors would like to extend a special thanks to Nathan Mantua (NOAA NMFS, Southwest Fisheries Science Center), Lara Whitely Binder (University of Washington, Climate Impacts Group), Sharon Melin (Alaska Fisheries Science Center/NOAA), and Rob Suryan, Philip Mote, and Meghan Dalton (Oregon State University) for their review, feedback, and important contributions. The authors would also like to thank Stacy Vynne (Puget Sound Partnership) and four anonymous reviewers of this chapter. Their comments spurred us on to make a number of tangible improvements. S. Shafer was supported by the US Geological Survey Climate and Land Use Change Research and Development Program.

References

Ainsworth, C. H., J. F. Samhouri, D. S. Busch, W. W. L. Chueng, J. Dunne, and T. A. Okey. 2011. "Potential Impacts of Climate Change on Northeast Pacific Marine Fisheries and Food Webs." *ICES Journal of Marine Science* 68 (6): 1217-1229. doi: 10.1093/icesjms/fsr043.

Allan, J. C., and P. D. Komar. 2000. "Are Ocean Wave Heights Increasing in the Eastern North Pacific?" *EOS, Transactions of the American Geophysical* Union 81 (47): 561-567. doi: 10.1029/EO081i047p00561-01.

Allan J. C., and P. D. Komar. 2006. "Climate Controls on US West Coast Erosion Processes." *Journal of Coastal Research* 22 (3): 511-529. doi: 10.2112/03-0108.1.

Allan, J. C., P. D. Komar, and P. Ruggiero. 2011. "Storm Surge Magnitudes and Frequency on the Central Oregon Coast." In *Solutions to Coastal Disasters 2011*, 53-64, ASCE Conference Proceedings, Anchorage, AK. doi: 10.1061/41185(417)6.

Allan, J. C., R. C. Witter, P. Ruggiero, and A. D. Hawkes. 2009. "Coastal Geomorphology, Hazards, and Management Issues along the Pacific Northwest Coast of Oregon and Washington." In *Volcanoes to Vineyards: Geologic Field Trips through the Dynamic Landscape of the Pacific Northwest*: Geological Society of America Field Guide 15, edited by J. E. O'Connor, R. J. Dorsey and I. P. Madin, pp. 495-519, The Geological Society of America.

Argus, D. F., and W. R. Peltier. 2010. "Constraining Models of Postglacial Rebound Using Space Geodesy: A Detailed Assessment of Model ICE-5G (VM2) and Its Relatives." *Geophysical Journal International* 191 (2): 697-723. doi: 10.1111/j.1365-246X.2010.04562.x.

Baron, H. M. 2011. "Coastal Hazards and Community Exposure in a Changing Climate: The Development of Probabilistic Coastal Change Hazard Zones." MS thesis, Oregon State University. http://hdl.handle.net/1957/21811.

Baron, H. M., N. J. Wood, P. Ruggiero, J. C. Allan, and P. Corcoran. 2010. "Assessing Societal Vulnerability of U.S. Pacific Northwest Communities to Storm-Induced Coastal Change." In

Shifting Shorelines: Adapting to the Future, The 22nd International Conference of the Coastal Society, June 13-16, 2010, Wilmington, NC. http://aquaticcommons.org/3882/.

Barth, J. A., B. A. Menge, J. Lubchenco, F. Chan, J. M. Bane, A. R. Kirincich, M. A. McManus, K. J. Nielsen, S. D. Pierce, and L. Washburn. 2007. "Delayed Upwelling Alters Nearshore Coastal Ocean Ecosystems in the Northern California Current." *Proceedings of the National Academy of Sciences* 104 (10): 3719-3724. doi: 10.1073/pnas.0700462104.

Barton, A., B. Hales, G. G. Waldbuster, C. Langdon, and R. Feely. 2012. "The Pacific Oyster, *Crassostrea gigas*, Shows Negative Correlation to Naturally Elevated Carbon Dioxide Levels: Implications for Near-Term Ocean Acidification Effects." *Limnology and Oceanography* 57 (3): 698-710. doi: 10.4319/lo.2012.57.3.0698.

Becker, B. H., M. Z. Peery, and S. R. Beissinger. 2007. "Ocean Climate and Prey Availability Affect the Trophic Level and Reproductive Success of the Marbled Murrelet, an Endangered Seabird." *Marine Ecology Progress Series* 329: 267-279. doi: 10.3354/meps329267.

Bierbaum, R., J. B. Smith, A. Lee, M. Blair, L. Carter, F. Stuart Chapin, III, P. Fleming, S. Ruffo, M. Stults, S. McNeely, E. Wasley, and L. Verduzco. 2013. "A Comprehensive Review of Climate Adaptation in the United States: More Than Before, But Less Than Needed." *Mitigation and Adaptation Strategies for Global Change* 18 (3): 361-406. doi: 10.1007/s11027-012-9423-1.

Bindoff, N. L., J. Willebrand, V. Artale, A. Cazenave, J. Gregory, S. Gulev, K. Hanawa, C. Le Quéré, S. Levitus, Y. Nojiri, C. K. Shum, L. D. Talley, and A. Unnikrishnan. 2007. "Observations: Oceanic Climate Change and Sea Level." In *Climate Change 2007: The Physical Science Basis. Contribution of Working Group I to the Fourth Assessment Report of the Intergovernmental Panel on Climate Change*, edited by S. Solomon, D. Qin, M. Manning, Z. Chen, M. Marquis, K. B. Averyt, M. Tignor, and H. L. Miller, Cambridge University Press, Cambridge, United Kingdom and New York, NY, USA.

Bottom, D. L., C. A. Simenstad, J. Burke, A. M. Baptista, D. A. Jay, K. K. Jones, E. Sasillas, and M. H. Schiewe. 2005. "Salmon at the River's End: The Role of the Estuary in the Decline and Recovery of Columbia River Salmon." US Department of Commerce, NOAA Tech. Memo. NMFS-NWFSC-68.

Brander, K. M. 2007. "Global Fish Production and Climate Change." *Proceedings of the National Academy of Sciences* 104 (50): 19709–19714. doi: 10.1073/pnas.0702059104.

British Columbia and Washington State Memorandum of Understanding on Coastal Climate Change Adaptation (BC and WA MOU), June 20, 2008. http://www.ecy.wa.gov/climatechange/docs/BC/CoastalClimateAdaptationMOU.pdf.

Bromirski, P. D., A. J. Miller, R. E. Flick, and G. A. Auad. 2011. "Dynamical Suppression of Sea Level Rise along the Pacific Coast of North America: Indications for Imminent Acceleration." *Journal of Geophysical Research* 116: C07005. doi: 10.1029/2010JC006759.

Brown, C. A. 2012. "Factors Influencing Trends in pH in the Wootton et al. (2008) Dataset." Report from the US Environment Protection Agency, Pacific Coastal Ecology Branch. EPA/600/R/12/676. http://www.epa.gov/region10/pdf/water/303d/washington/wootton_review_brown2012.pdf.

Burgette, R. J., R. J. Weldon, and D. A. Schmidt. 2009. "Interseismic Uplift Rates for Western Oregon and Along-Strike Variation in Locking on the Cascadia Subduction Zone." *Journal of Geophysical Research* 114: B01408. doi: 10.1029/2008JB005679.

Burkett, V. R., and M. A. Davidson, eds. 2012. "Coastal Impacts, Adaptation and Vulnerability: A Technical Input to the 2013 National Climate Assessment." Cooperative Report to the 2013 National Climate Assessment, 150 pp. http://www.coastalstates.org/wp-content/uploads/2011/03/Coastal-Impacts-Adaptation-Vulnerabilities-Oct-2012.pdf.

Cai, W.-J., X. Hu, W.-J. Huang, M. C. Murrell, J. C. Lehrter, S. E. Lohrenz, W.-C. Chou, W. Zhai, J. T. Hollibaugh, Y. Wang, P. Zhao, X. Guo, K. Gundersen, M. Dai, and G.-C. Gong. 2011. "Acidification of Subsurface Coastal Waters Enhanced by Eutrophication." *Nature Geoscience* 4: 766–770, doi: 10.1038/ngeo1297.

Canadell, J. G., C. Le Quéré, M. R. Raupach, C. B. Field, E. T. Buitenhuis, P. Ciais, T. J. Conway, N. P. Gillett, R. A. Houghton, and G. Marland. 2007. "Contributions to Accelerating Atmospheric CO_2 Growth from Economic Activity, Carbon Intensity, and Efficiency of Natural Sinks." *Proceedings of the National Academy of Sciences* 104 (47): 18866-18870. doi: 10.1073/pnas.0702737104.

Chan, F., J. A. Barth, J. Lubchenco, A. Kirincich, H. Weeks, W. T. Peterson, and B. A. Menge. 2008. "Emergence of Anoxia in the California Current Large Marine Ecosystem." *Science* 319 (5865): 920. doi: 10.1126/science.1149016.

Chapman, J., and T. Melbourne. 2009. "Future Cascadia Megathrust Rupture Delineated by Episodic Tremor and Slip." *Geophysical Research Letters* 36: L22301. doi: 10.1029/2009GL040465.

Chavez, F. P., J. Ryan, S. E. Lluch-Cota, and M. C. Ñiquen. 2003. "From Anchovies to Sardines and Back: Multidecadal Change in the Pacific Ocean." *Science* 299 (5604): 217-221. doi: 10.1126/science.1075880.

Cheung, W. W. L., J. Dunne, J. L. Sarmiento and D. Pauly. 2011. "Integrating Ecophysiology and Plankton Dynamics into Projected Maximum Fisheries Catch Potential under Climate Change in the Northeast Atlantic." *ICES Journal of Marine Science: Journal du Conseil*, 68 (6): 1008-1018, doi: http://dx.doi.org/10.1093/icesjms/fsr012.

Cheung, W. W. L., V. W. Y. Lam, J. L. Sarmiento, K. Kearney, R. Watson, and D. Pauly. 2009. "Projecting Global Marine Biodiversity Impacts Under Climate Change Scenarios." *Fish and Fisheries* 10 (3): 235-251. doi: 10.1111/j.1467-2979.2008.00315.x.

City of Anacortes. 2012. "City of Anacortes, Water Treatment Plant, Climate Change Impact Mitigation." Presentation to Washington State Senate Environment Committee by City of Anacortes Public Works, Committee Working Session, November 30.

City of Olympia. 2012. "Sea Level Rise 2012 Community Update." Transition Olympia-Climate Change and City of Olympia Public Works Department Staff. Last modified January 27, 2012. http://olympiawa.gov/community/sustainability/climate-change.

Coast and Harbor Engineering. 2011. City of Olympia Engineered Response to Sea Level Rise. Prepared for the City of Olympia Public Works Department, Planning and Engineering. Available at: http://olympiawa.gov/community/sustainability/~/media/Files/PublicWorks/Sustainability/Sea%20Level%20Rise%20Response%20Technical%20Report.ashx.

Cooper, W. S. 1958. "Coastal Sand Dunes of Oregon and Washington." *Geological Society of America* Memoir 72. Denver, Colorado, USA.

Craig, D. 1993. *Preliminary Assessment of Sea Level Rise in Olympia, Washington: Technical and Policy Implications*. City of Olympia Public Works Department Policy and Program Development Division, Olympia, Washington.

Crossett, K. M., T. J. Culliton, P. C. Wiley, and T. R. Goodspeed. 2004. "Population Trends Along the Coastal United States: 1980-2008." NOAA Coastal Trends Report Series. http://oceanservice.noaa.gov/programs/mb/pdfs/coastal_pop_trends_complete.pdf.

Dalton, M., P. Mote, J. A. Hicke, D. Lettenmaier, J. Littell, J. Newton, P. Ruggiero, and S. Shafer. 2012. "A Workshop in Risk-Based Framing of Climate Impacts in the Northwest: Implementing the National Climate Assessment Risk-Based Approach" Technical Input Report to the Third National Climate Assessment. http://downloads.usgcrp.gov/NCA/Activities/northwestncariskframingworkshop.pdf.

Davidson, A. D., A. G. Boyer, H. Kim, S. Pompa-Mansilla, M. J. Hamilton, D. P. Costa, G. Ceballos, and J. H. Brown. 2012. "Drivers and Hotspots of Extinction Risk in Marine Mammals." *Proceedings of the National Academy of Sciences* 109 (9): 3395-3400. doi: 10.1073/pnas.1121469109.

Deser, C., A. Phillips, and M. Alexander. 2010. "Twentieth Century Tropical Sea Surface Temperature Trends Revisited." *Geophysical Research Letters* 37 (10): L10701. doi: 10.1029/2010GL043321.

Doney, S. C., V. J. Fabry, R. A. Feely, and J. A. Kleypas. 2009. "Ocean Acidification: The Other CO_2 Problem." *Annual Review of Marine Science* 1: 169-192. doi: 10.1146/annurev.marine.010908.163834.

Doney, S. C., M. Ruckelshaus, J. E. Duffy, J. P. Barry, F. Chan, C. A. English, H. M. Galindo, J. M. Grebmeier, A. B. Hollowed, N. Knowlton, J. Polovina, N. N. Rabalais, W. J. Sydeman, and L. D. Talley. 2012. "Climate Change Impacts on Marine Ecosystems." *Annual Review of Marine Science* 4: 11-37. doi: 10.1146/annurev-marine-041911-111611.

Douglass, S. L., J. Lindstrom, J. M. Richards, and J. Shaw. 2005. "An Estimate of the Extent of U.S. Coastal Highways." Presentation to Transportation Research Board, Hydraulics, Hydrology, and Water Quality Committee Meeting, Washington DC. www.southalabama.edu/usacterec/chighwayestimate.pdf.

Drut, M., and J. B. Buchanan. 2000. "U.S. National Shorebird Management Plan: Northern Pacific Coast Working Group Regional Management Plan." US Fish and Wildlife Service, Portland, OR.

Favre, A., and A. Gershunov. 2006. "Extra-Tropical Cyclonic/Anticyclonic Activity in North-Eastern Pacific and Air Temperature Extremes in Western North America." *Climate Dynamics* 26 (6): 617-629. doi: 10.1007/s00382-005-0101-9.

Feely, R. A., S. Alin, J. Newton, C. Sabine, M. Warner, A. Devol, C. Krembs, and C. Maloy. 2010. "The Combined Effects of Ocean Acidification, Mixing, and Respiration on pH and Carbonate Saturation in an Urbanized Estuary." *Estuarine, Coastal and Shelf Science* 88 (4): 442-449. doi: 10.1016/j.ecss.2010.05.004.

Feely, R. A., T. Klinger, J. A. Newton, and M. Chadsey. 2012. "Scientific Summary of Ocean Acidification in Washington State Marine Waters." NOAA OAR Special Report, Publication No. 12-01-016.

Feely, R. A., C. L. Sabine, J. M. Hernandez-Ayon, D. Ianson, and B. Hales. 2008. "Evidence for Upwelling of Corrosive 'Acidified' Water onto the Continental Shelf." *Science* 320 (5882): 1490–1492. doi: 10.1126/science.1155676.

Fresh, K., M. Dethier, C. Simenstad, M. Logsdon, H. Shipman, C. Tanner, T. Leschine, T. Mumford, G. Gelfenbaum, R. Shuman, and J. Newton. 2011. "Implications of Observed Anthropogenic Changes to the Nearshore Ecosystems in Puget Sound." Technical Report 2011-03 prepared for the Puget Sound Nearshore Ecosystem Restoration Project. http://www.pugetsoundnearshore.org/technical_papers/implications_of_observed_ns_change.pdf.

Fu, F. X., A. R. Place, N. S. Garcia, and D. A. Hutchins. 2010. "CO_2 and Phosphate Availability Control the Toxicity of the Harmful Bloom Dinoflagellate *Karlodinium veneficum*." *Atlantic Microbial Biology* 59: 55-65. doi: 0.3354/ame01396.

Gazeau, F., C. Quiblier, J. M. Jansen, J. P. Gattuso, J. J. Middelburg, and C. H. R. Heip. 2007. "Impact of Elevated CO_2 on Shellfish Calcification." *Geophysical Research Letters* 34 (7): L07603. doi: 10.1029/2006GL028554.

Gemmrich, J., B. Thomas, and R. Bouchard. 2011. "Observational Changes and Trends in Northeast Pacific Wave Records." *Geophysical Research Letters* 38 (22): L22601. doi: 10.1029/2011GL049518.

Glick, P., J. Clough, and B. Nunley. 2007. "Sea-Level Rise and Coastal Habitats in the Pacific Northwest." National Wildlife Federation, Seattle, WA.

Graham, N. E., and H. F. Diaz. 2001. "Evidence for Intensification of North Pacific Winter Cyclones Since 1948." *Bulletin of American Meteorological Society* 82 (9): 1869-1893. doi: 10.1175/1520-0477(2001)082<1869:EFIONP>2.3.CO;2.

Grebmeier, J. M., J. E. Overland, S. E. Moore, E. V. Farley, E. C. Carmack, L. W. Cooper, K. E. Frey, J. H. Helle, F. A. McLaughlin, and S. L. McNutt. 2006. "A Major Ecosystem Shift in the Northern Bering Sea." *Science* 311 (5766): 1461-1464. doi: 10.1126/science.1121365.

Green, M. A., M. E. Jones, C. L. Boudreau, R. L. Moore, and B. A. Westman. 2004. "Dissolution Mortality of Juvenile Bivalves in Coastal Marine Deposits." *Limnology and Oceanography* 49 (3): 727-734. doi: 10.4319/lo.2004.49.3.0727.

Griffith, G. P., E. A. Fulton, and A. J. Richardson. 2011. "Effects of Fishing and Acidification Related Benthic Mortality on the Southeast Australian Marine Ecosystem." *Global Change Biology* 17 (10): 3058-3074, doi: 10.1111/j.1365-2486.2011.02453.x.

Hacker, S., P. Zarnetske, E. Seabloom, P. Ruggiero, J. Mull, S. Gerrity, and C. Jones. 2012. "Subtle Differences in Two Non-Native Congeneric Beach Grasses Significantly Affect their Colonization, Spread, and Impact." *Oikos* 121 (1): 138-148. doi: 10.1111/j.1600-0706.2011.18887.x.

Harley, C. D. G., A. R. Hughes, K. M. Hultgren, B. G. Miner, C. J. B. Sorte, C. S. Thornber, L. F. Rodriguez, L. Tomanek, and S. L. Williams. 2006. "The Impacts of Climate Change in Coastal Marine Systems." *Ecology Letters* 9 (2): 228-241. doi: 10.1111/j.1461-0248.2005.00871.x.

Harris, S. L. 2005. *Fire Mountains of the West: The Cascade and Mono Lake Volcanoes (3rd Edition)*. Mountain Press Publishing. ISBN-10: 087842511X.

Hauri, C., N. Gruber, G.-K. Plattner, S. Alin, R. A. Feely, B. Hales, and P. A. Wheeler. 2009. "Ocean Acidification in the California Current System." *Oceanography* 22 (4): 60-71. doi: 10.5670/oceanog.2009.97.

Hazen, E. L., S. Jorgensen, R. R. Rykaczewski, S. J. Bograd, D. G. Foley, I. D. Jonsen, S. A. Shaffer, J. P. Dunne, D. P. Costa, L. B. Crowder, and B. A. Block. 2012. "Predicted Habitat Shifts of Pacific Top Predators in a Changing Climate." *Nature Climate Change* 3: 234-238. doi: 10.1038/nclimate1686.

Heberger, M., H. Cooley, P. Herrera, P. H. Gleick and E. Moore. 2011. "Potential Impacts of Increased Coastal Flooding in California due to Sea-Level Rise." *Climatic Change* 109, (1S): 229-249. doi: 10.1007/s10584-011-0308-1.

Hendriks, I. E., C. M. Duarte, and M. Alvarez. 2010. "Vulnerability of Marine Biodiversity to Ocean Acidification: A Meta-Analysis." *Estuarine, Coastal and Shelf Science* 86 (2): 157-164. doi: 10.1016/j.ecss.2009.11.022.

Hettinger, A., E. Sanford, T. M. Hill, A. D. Russell, K. N. S. Sato, J. Hoey, M. Forsch, H. N. Page, and B. Gaylord. 2012. "Persistent Carry-Over Effects of Planktonic Exposure to Ocean Acidification in the Olympia Oyster." *Ecology* 93 (12): 2758-2768. doi: 10.1890/12-0567.1.

Hickey, B., and N. Banas. 2003. "Oceanography of the U.S. Pacific Northwest Coastal Ocean and Estuaries with Application to Coastal Ecology." *Estuaries* 26 (4): 1010-1031. doi: 10.1007/BF02803360.

Hofmann, G. E., J. P. Barry, P. J. Edmunds, R. D. Gates, D. A. Hutchins, T. Klinger, and M. A. Sewell. 2010. "The Effect of Ocean Acidification on Calcifying Organisms in Marine Ecosystems: An Organism-to-Ecosystem Perspective." *Annual Review of Ecology Evolution, and Systematics* 41: 127-147. doi: 10.1146/annurev.ecolsys.110308.120227.

Hollowed, A. B., S. R. Hare, and W. S. Wooster. 2001. "Pacific Basin Climate Variability and Patterns of Northeast Pacific Marine Fish Production." *Progress in Oceanography* 49: 257-282. doi: 10.1016/S0079-6611(01)00026-X.

Hönisch, B., A. Ridgwell, D. N. Schmidt, E. Thomas, S. J. Gibbs, A. Sluijs, R. Zeebe, L. Kump, R. C. Martindale, S. E. Greene, W. Kiessling, J. Ries, J. C. Zachos, D. L. Royer, S. Barker, T. M. Marchitto, R. Moyer, C. Pelejero, P. Ziveri, G. L. Foster, and B. Williams. 2012. "The Geological Record of Ocean Acidification." *Science* 335 (6072): 1058-1063. doi: http://dx.doi .org/10.1126/science.1208277.

Hood, W. G. 2005. "Sea Level Rise in the Skagit Delta." *Skagit River Tidings*. Skagit Watershed Council, Mount Vernon, WA.

Hoy, M., B. L. Boese, L. Taylor, D. Reusser, and R. Rodriguez. 2012. "Salinity Adaptation of the Invasive New Zealand Mud Snail (*Potamopyrgus antipodarum*) in the Columbia River Estuary (Pacific Northwest, USA): Physiological and Molecular Studies." *Aquatic Ecology* 46 (2): 249-260. doi: 10.1007/s10452-012-9396-x.

Huppert, D. D., A. Moore, and K. Dyson. 2009. "Impacts of Climate Change on the Coasts of Washington State." In *Washington Climate Change Impacts Assessment: Evaluating Washington's Future in a Changing Climate*, edited by M. M. Elsner, J. Littell, and L. Whitely Binder. Climate Impacts Group, University of Washington, Seattle, WA. www.cses.washington.edu/db/pdf /wacciaexecsummary638.pdf.

Independent Scientific Advisory Board. 2007. "Climate Change Impacts on Columbia River Basin Fish and Wildlife." Northwest Power and Conservation Council, Portland, OR.

Kairis, P. A., and J. M. Rybczyk. 2010. "Sea Level Rise and Eelgrass (*Zostera marina*) Production: A Spatially Explicit Relative Elevation Model for Padilla Bay, WA." *Ecological Modeling* 221 (7): 1005–1016. doi: 10.1016/j.ecolmodel.2009.01.025.

Kaplan, I. C., P. S. Levin, M. Burden, and E. A. Fulton. 2010. "Fishing Catch Shares in the Face of Global Change: A Framework for Integrating Cumulative Impacts and Single Species Management." *Canadian Journal of Fisheries and Aquatic Sciences* 67 (12): 1968–1982. doi: 10.1139/F10-118.

King County. 2009. "Vulnerability of Major Wastewater Facilities to Flooding from Sea-Level Rise." Presentation by King County Wastewater Treatment Division at the NOAA/WA Department of Ecology Coastal Training Program in Padilla Bay, WA, March 4. http://www .coastaltraining-wa.org.

Knight, E., S. Keithly, and S. Moddemeyer. 2013. "Facing the Future: Waterfront Development Challenges in a Changing Climate." Presentation at the National Working Waterfronts & Waterways Symposium, March 27. http://wsg.washington.edu/mas/pdfs/nwwws/D5/D5 _Knight.pdf.

Komar, P. D., J. C. Allan, and P. Ruggiero. 2009. "Ocean Wave Climates: Trends and Variations due to Earth's Changing Climate." Handbook of Coastal and Ocean Engineering, edited by Y.C. Kim, World Scientific Publishing Co., 971-975.

Komar, P. D., J. C. Allan, and P. Ruggiero. 2011. "Sea Level Variations along the U.S. Pacific Northwest Coast: Tectonic and Climate Controls." *Journal of Coastal Research* 27 (5): 808-823. doi: 10.2112/JCOASTRES-D-10-00116.1.

Krueger, K. L., K. B. Pierce, Jr., T. Quinn, and D. E. Penttila. 2010. "Anticipated Effects of Sea Level Rise in Puget Sound on Two Beach-Spawning Fishes." In *Puget Sound Shorelines and the Impacts of Armoring*, edited by H. Shipman, M. N. Dethier, G. Gelfenbaum, K. L. Fresh, and R. S. Dinicola, 171-178. Proceedings of a State of the Science Workshop, May 2009: USGS Scientific Investigations Report 2010-5254.

Kump, L. R., T. J. Bralower, and A. Ridgwell. 2009. "Ocean Acidification in Deep Time." *Oceanography* 22 (4): 94-107. doi: 10.5670/oceanog.2009.100.

MacArthur, J., P. W. Mote, J. Ideker, M. Figliozzi, and M. Lee. 2012. "Climate Change Impact Assessment for Surface Transportation in the Pacific Northwest and Alaska." Oregon Transportation Research and Education Consortium, OTREC-RR-12-01. http://www.wsdot.wa.gov/research/reports/fullreports/772.1.pdf.

Meehl, G. A., T. F. Stocker, W. D. Collins, P. Friedlingstein, A. T. Gaye, J. M. Gregory, A. Kitoh, R. Knutti, J. M. Murphy, A. Noda, S. C. B. Raper, I. G. Watterson, A. J. Weaver, and Z.-C. Zhao. 2007. "Global Climate Projections." In *Climate Change 2007: The Physical Science Basis. Contribution of Working Group I to the Fourth Assessment Report of the Intergovernmental Panel on Climate Change*, edited by S. Solomon, D. Qin, M. Manning, Z. Chen, M. Marquis, K. B. Averyt, M. Tignor, and H. L. Miller. Cambridge University Press, Cambridge, United Kingdom and New York, NY, USA.

Méndez, F. J., M. Menéndez, A. Luceno, and I. J. Losada, 2006. "Estimation of the Long-Term Variability of Extreme Significant Wave Height Using a Time-Dependent Peak Over Threshold (POT) Model." *Journal of Geophysical Research: Oceans* 111: C07024, doi: 10.1029/2005JC003344.

Menéndez, M., F. J. Méndez, I. Losada, and N. E. Graham. 2008. "Variability of Extreme Wave Heights in the Northeast Pacific Ocean Based on Buoy Measurements." *Geophysical Research Letters* 35: L22607. doi: 10.1029/2008GL035394.

Menge, B. A., F. Chan, K. Nielsen, and T. Freidenburg. 2010. "Community Consequences of Climate Change in the Oregon Coastal Rocky Intertidal Ecosystem." Proceedings from the 2010 AGU Ocean Sciences Meeting, February 22-26. 2010.

Moore, S. E., and H. P. Huntington. 2008. "Arctic Marine Mammals and Climate Change: Impacts and Resilience." *Ecological Applications* 18: S157-S165. doi: 10.1890/06-0571.1.

Moore, S. K., N. J. Mantua, B. M. Hickey, and V. L. Trainer. 2009. "Recent Trends in Paralytic Shellfish Toxins in Puget Sound, Relationships to Climate, and Capacity for Prediction of Toxic Events." *Harmful Algae* 8 (3): 463-477. doi:10.1016/j.hal.2008.10.003.

Moore, S. K., N. J. Mantua, J. P. Kellogg, J. A. Newton. 2008. "Local and Large-Scale Climate Forcing of Puget Sound Oceanographic Properties on Seasonal to Interdecadal Timescales." *Limnology and Oceanography* 53 (5): 1746–1758. doi: 10.4319/lo.2008.53.5.1746.

Moore, S. K., N. J. Mantua, and E. P. Salathé. 2011. "Past Trends and Future Scenarios for Environmental Conditions Favoring the Accumulation of Paralytic Shellfish Toxins in Puget Sound Shellfish." *Harmful Algae* 10 (5): 521-529. doi: 10.1016/j.hal.2011.04.004.

Mote, P., A. Petersen, S. Reeder, H. Shipman, L. Whitely Binder. 2008. "Sea Level Rise in the Coastal Waters of Washington State." Joint Institute for the Study of the Atmosphere and Ocean publication #1474, University of Washington, Seattle, Washington. http://cses.washington.edu/db/pdf/moteetalslr579.pdf.

Mote, P. W., and E. P. Salathé. 2010. "Future Climate in the Pacific Northwest." *Climatic Change* 102: 29-50. doi: 10.1007/s10584-010-9848-z.

Nakićenović, N., O. Davidson, G. Davis, A. Grübler, T. Kram, E. Lebre La Rovere, B. Metz, T. Morita, W. Pepper, H. Pitcher, A. Sankovshi, P. Shukla, R. Swart, R. Watson, and Z. Dadi. 2000. *Special Report on Emissions Scenarios: A Special Report of Working Group III of the Intergovernmental Panel on Climate Change*, Cambridge University Press, Cambridge, UK, 599 pp. http://www.grida.no/climate/ipcc/emission/index.htm.

National Marine Fisheries Service. 2010. "Fisheries Economics of the United States, 2009." US Deptartment of Commerce, NOAA Tech. Memo. NMFS-F/SPO-118, 172p. https://www.st .nmfs.noaa.gov/st5/publication/index.html.

National Marine Fisheries Service. 2012. "Annual Commercial Landings Statistics." Accessed December 2012. http://www.st.nmfs.noaa.gov/commercial-fisheries/commercial-landings /annual-landings/index.

National Research Council (NRC). 2010. *Ocean Acidification: A National Strategy to Meet the Challenges of a Changing Ocean*. Committee on the Development of an Integrated Science Strategy for Ocean Acidification Monitoring, Research, and Impacts Assessment. National Research Council. The National Academies Press, Washington, DC. http://www.nap.edu/catalog /12904.html.

National Research Council (NRC). 2012. *Sea-Level Rise for the Coasts of California, Oregon, and Washington: Past, Present, and Future*. Committee on Sea Level Rise in California, Oregon, and Washington, Board on Earth Sciences and Resources, and Ocean Studies Board Division on Earth and Life Sciences, National Research Council. The National Academies Press, Washington, DC. http://www.nap.edu/catalog.php?record_id=13389.

Nerem, R. S., D. Chambers, C. Choe, and G. T. Mitchum. 2010. "Estimating Mean Sea Level Change from the TOPEX and Jason Altimeter Missions." *Marine Geodesy* 33 (1) Supplement 1: 435. doi: 10.1080/01490419.2010.491031.

Neskowin Coastal Hazards Committee (NCHC). 2013. "Neskowin Community Association." Accessed January. http://www.neskowincommunity.org/neskowincoastalhazards.html.

Nicholls, R. J., P. P. Wong, V. Burkett, J. Codignotto, J. Hay, R. McLean, S. Ragoonaden, and C. Woodroffe. 2007. "Coastal Systems and Low-Lying Areas." In *Climate Change 2007: Impacts, Adaptations and Vulnerability. Contribution of Working Group II to the Fourth Assessment Report of the Intergovernmental Panel on Climate Change*, edited by M. L. Parry, O. F. Canziani, J. P. Palutikof, P. J. van der Linden, and C. E. Hanson, 315-356. Cambridge University Press, Cambridge, UK.

NOAA OAR Special Report. 2012. *Scientific Summary of Ocean Acidification in Washington State Marine Waters*, edited by R. A. Feely, T. Klinger, J. Newton, and M. Chadsey. Washington Shellfish Initiative – Blue Ribbon Panel on Ocean Acidification. https://fortress.wa.gov/ecy /publications/SummaryPages/1201016.html.

NOAA Tides and Currents. 2012. "Center for Operation Oceanographic Products and Services." Accessed 2012. http://tidesandcurrents.noaa.gov/index.shtml.

Northwest Fisheries Science Center, NOAA Fisheries Service. "Coastal Upwelling." Last modified July 31, 2012. http://www.nwfsc.noaa.gov/research/divisions/fed/oeip/db-coastal -upwelling-index.cfm.

Oak Ridge National Laboratory. 2012. "Climate Change and Infrastructure, Urban Systems, and Vulnerabilities." Technical Report to the US Department of Energy in Support of the National Climate Assessment, February 29, 2012. http://www.esd.ornl.gov/eess/Infrastructure.pdf.

Oregon Coastal Management Program. 2009. "Climate Ready Communities: A Strategy for Adapting to Impacts of Climate Change on the Oregon Coast." Oregon Department of Land Conservation and Development, Salem, OR.

Overland, J., and M. Wang. 2007. "Future Climate of the North Pacific Ocean." *Eos Transactions, American Geophysical Union* 88 (16): 178-182. doi: 10.1029/2007EO160003.

Pacific Shellfish Institute. 2013. Bobbi Hudson, personal communication, unpublished data collected via NOAA grant NA10OAR4170057 to the Pacific Shellfish Institute, http://www .pacshell.org.

Payne, M. C., C. A. Brown, D. A. Reusser, and H. Lee, II. 2012. "Ecoregional Analysis of Near-shore Sea-Surface Temperature in the North Pacific." *PLoS ONE* 7 (1): e30105. doi: 10.1371/journal.pone.0030105.

Pearcy, W. G. 1992. "Ocean Ecology of North Pacific Salmonids." University of Washington Press, Seattle, WA.

Peltier, W. R. 2004. "Global Glacial Isostasy and the Surface of the Ice-Age Earth: The ICE-5G (VM2) Model and GRACE." *Annual Review of Earth and Planetary Sciences* Vol. 32: 111-149. doi: 10.1146/annurev.earth.32.082503.144359.

Pendleton, L., P. King, C. Mohn, D. G. Webster, R. Vaughn, and P. N. Adams. 2011. "Estimating the Potential Economic Impacts of Climate Change on Southern California Beaches." *Climatic Change* 109 (1S): 277-298. doi: 10.1007/s10584-011-0309-0.

Peterson, W. T., and F. B. Schwing. 2003. "A New Climate Regime in Northeast Pacific Eco-systems." *Geophysical Research Letters* 30 (17): 1896-1899. doi: 10.1029/2003GL017528.

Puget Sound Partnership. 2012. *The 2012/2013 Action Agenda for Puget Sound.* Olympia, WA. http://www.psp.wa.gov/downloads/AA2011/083012_final/Action%20Agenda%20Book%202_Aug%2029%202012.pdf.

Rahmstorf, S. 2007. "A Semi-Empirical Approach to Projecting Future Sea-Level Rise." *Science* 315 (5810): 368-370. doi: 10.1126/science.1135456.

Rahmstorf, S. 2010. "A New View on Sea Level Rise." *Nature Reports Climate Change* (1004): 44-45. doi: 10.1038/climate.2010.29.

Redman, S., D. Myers, and D. Averill. 2005. "Regional Nearshore and Marine Aspects of Salmon Recovery in Puget Sound." Puget Sound Action Team, Olympia, WA. http://www.ecy.wa.gov/programs/sea/shellfishcommittee/mtg_may08/Salmon_Recovery_Report.pdf.

Reusink, J., D. S. Busch, G. Waldbusser, T. Klinger, and C. S. Friedman. 2012. "Responses of Species Assemblages to Ocean Acidification," In NOAA OAR Special Report: *Scientific Summary of Ocean Acidification in Washington State Marine Waters*, edited by R. A. Feely, T. Klinger, J. A. Newton, and M. Chadsey, 57-80.

Rhose, M. 2011. "Adapting to Coastal Erosion Hazards in Tillamook County: Framework Plan." Oregon Department of Land Conservation and Development, Coastal Management Program. http://www.neskowincommunity.org/Adaptation%20Plan%20Final%20Draft%2010%20Jun%202011%20(2).pdf.

Ries, J. B., A. L. Cohen, and D. C. McCorkle. 2009. "Marine Calcifiers Exhibit Mixed Responses to CO_2-Induced Ocean Acidification." *Geology* 37 (12): 1131-1134. doi: 10.1130/G30210A.1.

Robinson, R. A., H. Q. P. Crick, J. A. Learmonth, I. M. D. Maclean, C. D. Thomas, F. Bairlein, M. C. Forchhammer, C. M. Francis, J. A. Gill, B. J. Godley, J. Harwood, G. C. Hays, B. Huntley, A. M. Hutson, G. J. Pierce, M. M. Rehfisch, D. W. Sims, M. Begoña Santos, T. H. Sparks, D. A. Stroud, and M. E. Visser. 2009. "Travelling through a Warming World: Climate Change and Migratory Species." *Endangered Species Research* 7: 87-99. doi: 10.3354/esr00095.

Roessig, J. M., C. M. Woodley, J. J. Cech, and L. J. Hansen. 2004. "Effects of Global Climate Change on Marine and Estuarine Fishes and Fisheries." *Reviews in Fish Biology and Fisheries* 14 (2): 251-275. doi: 10.1007/s11160-004-6749-0.

Ruggiero, P. 2013. "Is the Intensifying Wave Climate of the U.S. Pacific Northwest Increasing Flooding and Erosion Risk Faster than Sea Level Rise?" *Journal of Waterway, Port, Coastal, and Ocean Engineering* 139 (2): 88-97. doi: 10.1061/(ASCE)WW.1943-5460.0000172.

Ruggiero, P., H. Baron, E. Harris, J. Allan, P. Komar, and P. Corcoran. 2011. "Incorporating Uncertainty Associated with Climate Change into Coastal Vulnerability Assessments." In *Solutions to Coastal Disasters 2011*, 53-64, ASCE Conference Proceedings, Anchorage, AK. doi: 10.1061/41185(417)6.

Ruggiero, P., G. M. Kaminsky, G. Gelfenbaum, and B. Voigt. 2005. "Seasonal to Interannual Morphodynamics along a High-Energy Dissipative Littoral Cell." *Journal of Coastal Research* 21 (3): 553-578. doi: 10.2112/03-0029.1.

Ruggiero, P., P. D. Komar, and J. C. Allan. 2010. "Increasing Wave Heights and Extreme-Value Projections: The Wave Climate of the U.S. Pacific Northwest." *Coastal Engineering* 57 (5): 539-552. doi: 10.1016/j.coastaleng.2009.12.005.

Ruggiero, P., M. A. Kratzmann, E. G. Himmelstoss, D. Reid, J. Allan, and G. Kaminsky. In press. *National Assessment of Shoreline Change: Historical Shoreline Change along the Pacific Northwest Coast.* US Geological Survey Open-File Report 2012–1007, 55 p.

Rykaczewski, R. R., and J. P. Dunne. 2010. "Enhanced Nutrient Supply to the California Current Ecosystem with Global Warming and Increased Stratification in an Earth System Model." *Geophysical Research Letters* **37**: L21606. doi: 10.1029/2010GL045019.

Sabine, C. L., and R. A. Feely. 2007. "The Oceanic Sink for Carbon Dioxide." In *Greenhouse Gas Sinks*, edited by D. Reay, N. Hewitt, J. Grace, K. Smith, 31-49. CABI Publishing, Oxfordshire, UK.

Sabine, C. L., R. A. Feely, N. Gruber, R. M. Key, K. Lee, J. L. Bullister, R. Wanninkhof, C. S. Wong, D. W. R. Wallace, B. Tilbrook, F. J. Millero, T.-H. Peng, A. Kozyr, T. Ono, and A. F. Rios. 2004. "The Oceanic Sink for Anthropogenic CO_2." *Science* 305 (5682): 367-371. doi: 10.1126/science .1097403.

Sallenger, A. H., Jr., W. B. Krabill, R. N. Swift, J. Brock, J. List, Mark Hansen, R. A. Holman, S. Manizade, J. Sontag, A. Meredith, K. Morgan, J. K. Yunkel, E. B. Frederick and H. Stockdon. 2003. "Evaluation of Airborne Topographic Lidar for Quantifying Beach Changes." *Journal of Coastal Research* 19: 125-133. http://www.jstor.org/stable/4299152.

Scavia, D., J. C. Field, D. F. Boesch, R. W. Buddemeier, V. Burkett, D. R. Cayan, M. Fogarty, M. A. Harwell, R. W. Howarth, C. Mason, D. J. Reed, T. C. Royer, A. H. Sallenger, and J. G. Titus. 2002. "Climate Change Impacts on U. S. Coastal and Marine Ecosystems." *Estuaries* 25 (2): 149-164. http://www.jstor.org/stable/1353306.

Schaeffer, M., W. Hare, S. Rahmstorf, and M. Vermeer. 2012. "Long-Term Sea-Level Rise Implied by 1.5 °C and 2 °C Warming Levels." *Nature Climate Change* 2: 867-870.. doi: 10.1038/ NCLIMATE1584.

Scigliano, E. 2012. "Sweetening the Waters: The Feasibility and Efficacy of Measures to Protect Washington's Marine Resources from Ocean Acidification." White paper commissioned by the Global Ocean Health Program in support of the Washington State Blue Ribbon Panel on Ocean Acidification. http://www.ecy.wa.gov/water/marine/oa/2012report_app9 .pdf.

Seabloom, E., P. Ruggiero, S. Hacker, J. Mull, and P. Zarnetske. 2013. "Invasive Grasses, Climate Change, and Flood Risk in Coastal Ecosystems." *Global Change Biology* 19 (3): 824-832. doi: 10.1111/gcb.12078.

Seabloom, E. W., and A. M. Wiedemann. 1994. "Distribution and Effects of Ammophila Brevil-igulata Fern. (American beachgrass) on the Foredunes of the Washington Coast." *Journal of Coastal Research* 10 (1): 178-188. http://www.jstor.org/stable/4298202.

Seo, S. N. 2007. "Climate Change Impacts on Non-Market Activities." In *Encyclopedia of Earth*, edited by C. J. Cleveland. Environmental Information Coalition, National Council for Science and the Environment, Washington, DC. http://www.eoearth.org/article/Climate _change_impacts_on_non-market_activities.

Seymour, R. J. 2011. "Evidence for Changes to the Northeast Pacific Wave Climate." *Journal of Coastal Research* 27 (1): 194-201. doi: 10.2112/JCOASTRES-D-09-00149.1.

Shaughnessy, F. J., W. Gilkerson, J. M. Black, D. H. Ward, and M. Petrie. 2012. "Predicted Eelgrass Response to Sea Level Rise and its Availability to Foraging Black Brant in Pacific Coast Estuaries." *Ecological Applications* 22 (6): 1743-1761. doi: 10.1890/11-1083.1.

Solecki, W., and C. Rosenzweig eds. 2012. "U.S. Cities and Climate Change: Urban, Infrastructure, and Vulnerability Issues." Technical Report submitted in support of the National Climate Assessment.

Solomon, A., and M. Newman. 2012. "Reconciling Disparate Twentieth-Century Indo-Pacific Ocean Temperature Trends in the Instrumental Record." *Nature Climate Change* 2: 691-699. doi: 10.1038/nclimate1591.

Stevenson, S. L. 2012. "Significant Changes to ENSO Strength and Impacts in the Twenty-First Century: Results from CMIP5." *Geophysical Research Letters* 39: L17703. doi: 10.1029/2012GL052759.

Strauss, B. H., R. Ziemlinski, J. L. Weiss, and J. T. Overpeck. 2012. "Tidally Adjusted Estimates of Topographic Vulnerability to Sea Level Rise and Flooding for the Contiguous United States." *Environmental Research Letters* 7 (1): 014033. doi: 10.1088/1748-9326/7/1/014033.

Sun, J., D. A. Hutchins, Y. Y. Feng, E. L. Seubert, D. A. Caron, and F. X. Fu. 2011. "Effects of Changing pCO_2 and Phosphate Availability on Domoic Acid Production and Physiology of the Marine Harmful Bloom Diatom *Pseudo-nitzschia multiseries*." *Limnology and Oceanography* 56 (3): 829-840. doi: 10.4319/lo.2011.56.3.082.

Sunda, W. G., and W.-J. Cai. 2012. "Eutrophication Induced CO_2-Acidification of Subsurface Coastal Waters: Interactive Effects of Temperature, Salinity, and Atmospheric PCO_2." *Environmental Science & Technology* 46 (19): 10651–10659. doi: 10.1021/es300626f.

Swinomish Indian Tribal Community. 2009. "Swinomish Climate Change Initiative Impact Assessment Technical Report." Swinomish Indian Tribal Community, Office of Planning and Community Development, La Conner, WA. http://www.swinomish.org/climate_change /Docs/SITC_CC_ImpactAssessmentTechnicalReport_complete.pdf.

Sydeman, W. J., K. L. Mills, J. A. Santora, S. A. Thompson, D. F. Bertram, K. H. Morgan, B. K. Wells, J. M. Hipfner, and S. G. Wolf. 2009. "Seabirds and Climate in the California Current: A Synthesis of Change." California Cooperative Oceanic Fisheries Investigations Reports 50: 82-104.

Talmage, S. C., and C. J. Gobler. 2009. "The Effects of Elevated Carbon Dioxide Concentrations on the Metamorphosis, Size and Survival of Larval Hard Clams (*Mercenaria mercenaria*), Bay Scallops (*Argopecten irradians*), and Eastern Oysters (*Crassostrea virginica*)." *Limnology and Oceanography* 54 (6): 2072-80. doi: 10.4319/lo.2009.54.6.2072.

Tatters, A. O., F. X. Fu, and D. A. Hutchins. 2012. "High CO_2 and Silicate Limitation Synergistically Increase the Toxicity of *Pseudo-nitzschia fraudulenta*." *PloS ONE* 7 (2): e32116. doi: 10.1371/journalpone.0032116.

Tillmann, P., and D. Siemann. 2011. "Climate Change Effects and Adaptation Approaches in Marine and Coastal Ecosystems of the North Pacific Landscape Conservation Cooperative Region." National Wildlife Federation.

Trenberth, K. E., P. D. Jones, P. Ambenje, R. Bojariu, D. Easterling, A. Klein Tank, D. Park, F. Rahimzadeh, J. A. Renwick, M. Rusticucci, B. Soden, and P. Zhai. 2007. "Observations: Surface and Atmospheric Climate Change." In *Climate Change 2007: The Physical Science Basis. Contribution of Working Group I to the Fourth Assessment Report of the Intergovernmental Panel on Climate Change*, edited by S. Solomon, D. Qin, M. Manning, Z. Chen, M. Marquis, K. B. Averyt, M. Tignor, and H. L. Miller, 235-336. Cambridge University Press, Cambridge, United Kingdom and New York, NY, USA.

TriMet. 2010. "Portland-Milwaukie Light Rail Project Navigational Results Report." October 2010. http://trimet.org/pdfs/pm/FEIS/Navigational__Final_Oct2010.pdf

University of Colorado: CU Sea Level Research Group. 2012. "Global Mean Sea Level Time Series (seasonal signals removed)." Accessed December 3. http://sealevel.colorado.edu/.

US Fish and Wildlife Service. 2010. "Rising to the Urgent Challenge: Strategic Plan for Responding to Accelerating Climate Change." http://www.fws.gov/home/climatechange/pdf/CCStrategicPlan.pdf.

Vano, J. A., N. Voisin, L. Cuo, A. F. Hamlet, M. M. Elsner, R. N. Palmer, A. Polebitski, and D. P. Lettenmaier. 2010. "Climate Change Impacts on Water Management in the Puget Sound Region, Washington State, USA." *Climatic Change* 102 (1-2): 261-286. doi: 10.1007/s10584-010-9846-1.

Warrick, J. A., A. E. Draut, M. L. McHenry, I. M. Miller, C. S. Magirl, M. M. Beirne, A. W. Stevens, and J. B. Logan. 2011. "Geomorphology of the Elwha River and its Delta." In *Coastal Habitats of the Elwha River, Washington: Biological and Physical Patterns and Processes Prior to Dam Removal*, edited by J. J. Duda, J. A. Warrick, and C. S. Magirl. Scientific Investigations Report 2011-5120. U.S. Geological Survey.

Washington Department of Ecology. 2008. "Leading the Way: Preparing for the Impacts of Climate Change in Washington: Recommendations of the Preparation and Adaptation Working Groups." Olympia, WA.

Washington Department of Ecology. 2012. "Preparing for a Changing Climate: Washington State's Integrated Climate Response Strategy." Olympia, WA

Washington Department of Ecology. 2013. "King Tides in Washington State." Accessed March 2013. http://www.ecy.wa.gov/climatechange/ipa_hightide.htm.

Washington State Blue Ribbon Panel on Ocean Acidification. 2012. *Ocean Acidification: From Knowledge to Action, Washington State's Strategic Response*, edited by H. Adelsman and L. Whitely Binder. Washington Department of Ecology, Olympia, Washington. Publication No. 12-01-015.

Washington State Department of Transportation (WSDOT), 2011. "Climate Impacts Vulnerability Assessment." Report to the Federal Highway Administration. February. http://www.wsdot.wa.gov/NR/rdonlyres/B290651B-24FD-40EC-BEC3-EE5097ED0618/0/WSDOTClimateImpactsVulnerabilityAssessmentforFHWAFinal.pdf

West Coast Governors' Agreement (WCGA) on Ocean Health. 2010. "Climate Change Action Coordination Team Work Plan." http://www.westcoastoceans.org/media/Climate_Final_Work_Plan.pdf.

Whitney, F., H. Freeland, and M. Robert. 2007. "Persistently Declining Oxygen Levels in the Interior Waters of the Eastern Subarctic Pacific." *Progress in Oceanography* 75 (2): 179–199. doi: 10.1016/j.pocean.2007.08.007.

Williams, G. D., and R. M. Thom. 2001. "Marine and Estuarine Shoreline Modification Issues." Battelle Marine Sciences Laboratory and Pacific Northwest National Laboratory, Sequim, WA.

Wolf, S. G., M. A. Snyder, W. J. Sydeman, D. F. Doaks, and D. A. Croll. 2010. "Predicting Population Consequences of Ocean Climate Change for an Ecosystem Sentinel, the Seabird Cassin's Auklet." *Global Change Biology* 16 (7): 1923-1935. doi: 10.1111/j.1365-2486.2010.02194.x.

Wootton, J. T., C. A. Pfister, and J. D. Forester. 2008. "Dynamic Patterns and Ecological Impacts of Declining Ocean pH in a High-Resolution Multi-Year Dataset." *Proceedings of the National Academy of Sciences* 105 (48): 18848–18853. doi: 10.1073/pnas.0810079105.

Worm, B., M. Sandow, A. Oschlies, H. K. Lotze, and R. A. Myers. 2005. "Global Patterns of Predator Diversity in the Open Oceans." *Science* 309 (5739): 1365-1369. doi: 10.1126/science.1113399.

Yohe, G. 1989. "The Cost of Not Holding Back the Sea: Phase 1 Economic Vulnerability." In *The Potential Effects of Global Climate Change on the United States*. Report to Congress. Appendix B: Sea Level Rise. US Environmental Protection Agency, EPA 230-05-89-052, Washington, DC.

Yohe, G., J. Neumann, P. Marshall, H. Ameden. 1996. "The Economic Cost of Greenhouse-Induced Sea-Level Rise for Developed Property in the United States." *Climatic Change* 32 (4): 387-410. doi: 10.1023/A:1005338413531.

Yokoyama, Y., K. Lambeck, P. De Deckker, P. Johnston, and L. Fifield. 2000. "Timing of the Last Glacial Maximum from Observed Sea-Level Minima." *Nature* 406: 713-716. doi: 10.1038/35021035.

Young, I. R., S. Zeiger, and A. V. Babanin. 2011. "Global Trends in Wind Speed and Wave Height." *Science* 332 (6028): 451-455. doi: 10.1126/science.1197219.

Zarnetske, P. L., S. D. Hacker, E. W. Seabloom, P. Ruggiero, J. R. Killian, T. B. Maddux, and D. Cox. 2012. "Biophysical Feedback Mediates Effects of Invasive Grasses on Coastal Dune Shape." *Ecology* 93 (6): 1439-1450. doi: 10.1890/11-1112.1.

Zemtseff, K. 2012. "Anacortes Building a $56M Water Plant." *Seattle Daily Journal of Commerce*. January 12. Accessed February 27, 2013.

Chapter 5

Forest Ecosystems
Vegetation, Disturbance, and Economics

AUTHORS
Jeremy S. Littell, Jeffrey A. Hicke, Sarah L. Shafer, Susan M. Capalbo,
Laurie L. Houston, Patty Glick

5.1 Introduction

Forests cover about 47% of the Northwest (NW–Washington, Oregon, and Idaho) (Smith et al. 2009, fig. 5.1, table 5.1). The impacts of current and future climate change on NW forest ecosystems are a product of the sensitivities of ecosystem processes to climate and the degree to which humans depend on and interact with those systems. Forest

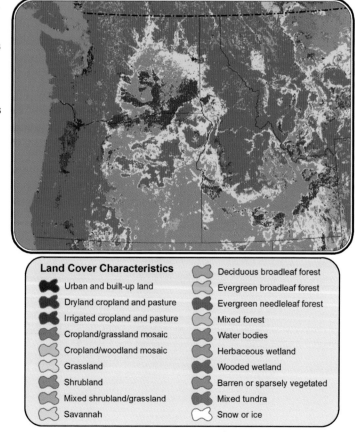

Figure 5.1 Land cover characteristics and vegetation types of the Northwest. Forests cover about 52% of Washington, 49% of Oregon, and 41% of Idaho. Data: National Center for Earth Resources Observation and Science, US Geological Survey, 2002.

Land Cover Characteristics

- Urban and built-up land
- Dryland cropland and pasture
- Irrigated cropland and pasture
- Cropland/grassland mosaic
- Cropland/woodland mosaic
- Grassland
- Shrubland
- Mixed shrubland/grassland
- Savannah
- Deciduous broadleaf forest
- Evergreen broadleaf forest
- Evergreen needleleaf forest
- Mixed forest
- Water bodies
- Herbaceous wetland
- Wooded wetland
- Barren or sparsely vegetated
- Mixed tundra
- Snow or ice

ecosystem structure and function, particularly in relatively unmanaged forests where timber harvest and other land use have smaller effects, is sensitive to climate change because climate has a strong influence on ecosystem processes. Climate can affect forest structure directly through its control of plant physiology and life history (establishment, individual growth, productivity, and mortality) or indirectly through its control of disturbance (fire, insects, disease). As climate changes, many forest processes will be affected, altering ecosystem services such as timber production and recreation. These changes have socioeconomic implications (e.g., for timber economies) and will require changes to current management of forests. Climate and management will interact to determine the forests of the future, and the scientific basis for adaptation to climate change in forests thus depends significantly on how forests will be affected.

Climate change impacts on NW forests were recently summarized in assessments of climate impacts in Washington (Littell, Oneil, et al. 2009; Littell et al. 2010) and Oregon (Shafer et al. 2010), as well as in a review of ecophysiological and other responses of NW forests to climate change (Chmura et al. 2011). Recent NW and western US regional studies have also reported climate effects on wildfire (Littell, McKenzie, et al. 2009; Littell et al. 2010; Littell and Gwozdz 2011; Rogers et al. 2011), insects such as the mountain pine beetle (*Dendroctonus ponderosae*) (Hicke et al. 2006; Bentz et al. 2010) and spruce beetle (*D. rufipennis*) (Bentz et al. 2010), diseases (Kliejunas et al. 2009; Sturrock et al. 2011), and vegetation (Littell et al. 2010; Rogers et al. 2011). We draw on this and other literature to describe the ways climate change may affect NW forests, with some attention to other biomes including high-elevation systems, grasslands, and shrublands, and what those impacts may mean for the region's forest economy and ecosystem services. In this chapter, we discuss the primary mechanisms by which climate change will affect the region's forests through the direct effects of climate on vegetation (establishment, growth/productivity, and distribution of plant species) and on disturbances (fire, insect outbreaks, and disease). We also describe the vulnerability of forest ecosystem services to climate change and identify key gaps in knowledge.

5.2 Direct Climate Sensitivities: Changes in Distribution, Abundance, and Function of Plant Communities and Species

Forest sensitivities to climate and expected outcomes vary with forest type and the factors that limit ecological processes. Temperature and precipitation are closely related to plant function because of their interacting effects on water supply (soil moisture) and demand (relative humidity). Energy and water interact to affect plant establishment, growth, and mortality and can be integrated as potential and actual evapotranspiration (PET and AET, respectively; Stephenson 1990; Churkina and Running 1998; Churkina et al. 1999; Nemani et al. 2003). When PET is greater than AET, for example, vegetation productivity is considered "water-limited," and when AET is greater than PET, it is considered "energy-limited" (Churkina et al. 1999; Littell et al. 2010). Water balance deficit (PET–AET or AET/PET, e.g., Stephenson 1990, Churkina et al. 1999) is a good correlate of the distribution of biome vegetation (e.g., forest, grassland, shrubland; Stephenson 1990). Historical and future summer water balance deficits are shown for the western United States in figure 5.2.

Figure 5.2 Historical (top, 1916–2006 average) June–August total water balance deficit (PET–AET) and future (bottom, 2030–2059 average) change in deficit as calculated using Variable Infiltration Capacity hydrologic model runs forced with projections from an ensemble of 10 global climate models (as in Littell, Elsner, et al. 2011). Except for the higher elevations of the Olympic Mountains, Cascade Range, and northern Rocky Mountains, summer deficit in the 2040s is projected to increase in most of the Northwest due to increased temperature and decreased precipitation. These changes are expected to change the geography of climatic suitability for current species and vegetation and alter disturbance regimes.

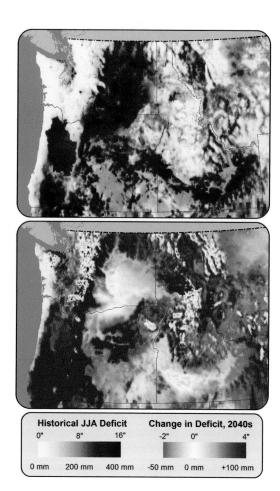

Water is a major limiting factor in the forests of the Northwest. Water limitation in this region occurs seasonally even in the western Cascade Range because the timing of supplies of water and energy in this region is asynchronous: more than 75% of the precipitation arrives outside the growing season (Waring and Franklin 1979; Stephenson 1990). In energy- (temperature-) limited vegetation, either there is sufficient water availability that thermal energy is the primary limiting factor (as in cool, moist temperate climates) or there is a chronic thermal limitation on plants (as in cold, dry climates). Projected increases in temperature (annual, spring, and summer) and changes in precipitation (increases in cool season, decreases in summer) for the Northwest (Mote and Salathé 2010; Chapter 2) will reduce regional April 1 snowpack and July 1 soil moisture, and increase summer (June–August) water balance deficit (Elsner et al. 2010) for many of the forests in the Northwest by the 2040s. Most lower-elevation forests that currently experience chronic or seasonal water limitation will therefore experience more severe and/or longer duration water limitation under projected future climate change than under historical climate (Littell et al. 2010, fig. 5.2). The near-term consequences for water-limited forests can be expected to manifest as decreases in successful seedling regeneration and tree growth, and increases in mortality, vulnerability to insects due to host tree stress, and area burned (Littell et al. 2010).

Forests that were historically energy-limited (primarily thermal limitation) will, in most cases, become less energy-limited and climate change might be expected to be favorable for existing forests. However, the impacts of climate change will depend on the degree of seasonal water limitation. For example, tree growth in Douglas-fir at mid elevations of the Cascade Range could increase or decrease, but if summer precipitation decreases, the water demand associated with the increased temperature is likely to outpace the increased supply of energy (e.g., Case and Peterson 2005; Littell et al. 2008). The near-term consequences for energy-limited forests will likely manifest as increases in seedling establishment and tree growth, but also increases in the frequency of disturbance, and so net outcomes for landscapes will depend on the interaction between direct and indirect pathways (see section 5.3).

There is substantial evidence that climate change will directly affect the abundance, geographical distributions, and function of NW plant species through climate effects on plant processes such as growth, phenology, and mortality (see Chmura et al. 2011 for a review of functional mechanisms pertaining to the Northwest). The paleoenvironmental record demonstrates that plant species have responded individualistically to past climate changes (Davis and Shaw 2001). Changes in the distribution and abundance of plant species have been observed over the past century in nearby regions, for example, in changes in subalpine tree populations (e.g., lodgepole pine [*Pinus contorta* var. *murrayana*] in the central Sierra Nevada; Millar et al. 2004). Climate changes affect plant phenology, such as plant flowering dates (e.g., common purple lilacs [*Syringa vulgaris* f. *purpurea*] and honeysuckle [*Lonicera* spp.] in the western United States; Cayan et al. 2001). Changes in phenology in turn alter the timing and availability of plant resources used by other species (e.g., pollinators). Interannual and interdecadal climate variability has been observed to affect the growth of trees in the Northwest, and the effects depend on the species and climatically limiting factors across their habitats (e.g., Peterson and Peterson 2001, Littell et al. 2008). There is evidence that, for some species, plant responses to climate change may be mediated by the physiological response of plants to changes in atmospheric CO_2 concentrations, and these responses may vary within species, geographically across a species' range (Chmura et al. 2011) and with other limiting factors such as nutrient and water availability. Observed relationships between climate and plant response, taken together, form the basis of future projections of species and ecosystem responses to climate change.

In general, model simulations indicate large potential changes in the climatic suitability for some plant species and habitats in the Northwest (e.g., McKenzie et al. 2003, Rehfeldt et al. 2006), such as the simulated loss of subalpine habitat (Millar et al. 2006; Rogers et al. 2011; fig. 5.3). Both statistical and mechanistic models have been used to estimate changes in forests in response to climate change. Statistical models of tree species-climate relationships show that each tree species has unique climatic tolerances (McKenzie et al. 2003; Rehfeldt et al. 2006; Rehfeldt et al. 2008; McKenney et al. 2011) and therefore is likely to respond individualistically to changes in temperature and precipitation. These relationships have been used to project potential future distributions of favorable climates for species in western North America (McKenney et al. 2007, 2011; Rehfeldt et al. 2006, 2008) and in Washington (e.g., Littell et al. 2010 after Rehfeldt et al. 2006). Climate is projected to become unfavorable for Douglas-fir (*Pseudotsuga menziesii*) over 32% of its current range in Washington by the 2060s using climate simulations from the HadCM3 and CGCM2 global climate models (GCMs) under a scenario that assumes a 1%/year increase in greenhouse gas emissions (Littell et al. 2010; data: Rehfeldt 2006). For three NW pine species susceptible to mountain pine beetle (ponderosa pine [*Pinus ponderosa*], lodgepole pine [*P. contorta*], whitebark pine [*P. albicaulis*]), 15% of their current range in Washington is projected to remain suitable for all three species by the 2060s, whereas 85% of their current range is projected to be outside the climatically suitable range for one or more of the three species (Littell et al. 2010; data: Rehfeldt 2006). McKenney et al. (2011) summarized species responses across western North America using future climate simulations from three GCMs (CGCM3.1, CSIRO-MK3.5, CCSM3.0) produced under the SRES-A2 scenario of continued growth of greenhouse gas emissions. The authors concluded that the change in number of tree species having suitable climate in the Northwest is often either near 0 (range of responses from gain of 10 to loss of 5 species) or net loss (range of 6 and 20 species), although some projections have subregional responses of greater net loss (range of 21 to 38 species). Coops and Waring (2011) used a mechanistic model (3PG) driven by future climate from a single GCM (CGCM3) to estimate the response of 15 tree species in the western United States to potential climate change. They assessed the area of the historical (1950–1975) and current (1976–2006) ranges that will be stressed under the greenhouse gases emissions scenarios SRES-A2 (continued growth) and SRES-B1 (substantial reductions) (Nakićenović et al. 2000), concluding that important NW species were climatically stressed over significant fractions of their historical range. For example, the modeled climate for Douglas-fir indicated it was stressed over 19% of its historical area in the Northwest, particularly at lower elevations of the eastern Cascade Range and the western slopes of the Rocky Mountains. Similarly, western hemlock and ponderosa pine were stressed over 12% and 27%, respectively, of their historical area. Coops and Waring (2011) also projected responses of the 15 species in the 2080s for the SRES-A2 scenario (continued growth) only, and concluded that five species, including Douglas-fir, will have relatively little loss of total current climatically suitable range (<20% loss), whereas 10 species (including ponderosa pine) will have as much as a 70% decline in current climatically suitable area. Their results for future changes in the 2080s were reported in aggregate for 10 US states and Canadian provinces, so species-level results are not available for the Northwest past the current period. However, they present results suggesting that some species (such as Douglas-fir) may balance the decline in suitability at lower elevations with increases in suitability at higher elevations, although this response is highly species- and location-dependent. Statistical models have also been used to assess changes in biotic communities (multiple plant species). Rehfeldt et al. (2006) estimated large changes in some communities in the Northwest in response to future warming. For example, desert scrub increases in the Northwest whereas grasslands and forests decrease.

Most of these statistically based approaches use very basic relationships between species or vegetation distributions and climate to infer plausible future distributions (but see Coops and Waring 2011 as an example of a more mechanistic approach). There are limitations of such projections; particularly those based solely on statistical comparison between species ranges and climate variables. Very few consider the ecophysiological basis for the observed patterns or complex ecological interactions, particularly disturbance, that also determine species' actual distributions. Almost all use climate

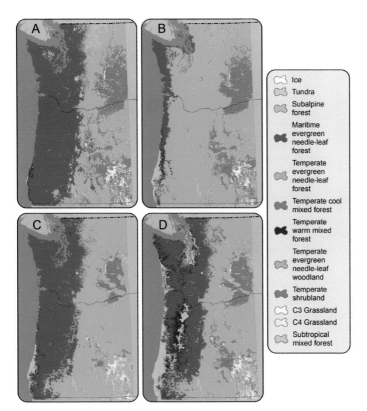

Figure 5.3 Historical (1971–2000) vegetation and three projections of future (2070–2099, SRES-A2 scenario of continued growth of greenhouse gas emissions) vegetation for parts of Washington and Oregon simulated by MC1, a dynamic global vegetation model (after Rogers et al. 2011). (A) Simulated historical vegetation; (B) HadCM3; (C) CSIRO Mk3; (D) MIROC 3.2 medres. Note the simulated future contraction of subalpine forest and expansion of temperate forest, particularly in B and D. Data: http://bitly.com/JU39Zy, 2010, http://databasin.org/.

simulations from one to a handful of climate models without justifying the choice of models or discussing the climatic context of the results compared to other models that might have been used. There are, therefore, significant uncertainties associated with model simulations of future vegetation change (McMahon et al. 2011; Littell, McKenzie, et al. 2011) that limit the use of such results for inferring the actual changes that will occur in species' distributions.

Mechanistic models also have been used to simulate potential future changes in biomes as well as individual species, and in contrast to statistical models, explicitly represent mechanisms such as photosynthesis and evapotranspiration to model ecosystems. Using the dynamic global vegetation model (DGVM) MC1, Bachelet et al. (2011) estimated relatively little change in regional biomes over the 21st century under a moderate warming, wetter climate (CSIRO Mk3, SRES-B1 [substantial reductions in emissions]). However, a nearly complete conversion of maritime conifer forests to temperate conifer forest (figs. 5.4, 5.5) and subtropical mixed forest in western Oregon and Washington was reported under a warmer, drier climate (HadCM3, SRES-A2 [continued growth in emissions]) (Rogers et al. 2011; fig. 5.3). Lenihan et al. (2008) used climate simulations from three GCMs under emissions scenarios of continued growth (SRES-A2) and substantial reductions (SRES-B1) to drive MC1, and responses in the Northwest (Washington, Oregon, Idaho, western Montana) indicate vegetation shifts under most scenarios, although the magnitude and diversity of these shifts vary considerably with the projected changes in precipitation and temperature.

Figure 5.4 Maritime evergreen coniferous forests (left) are characteristic of wetter, cooler habitats of the Oregon Coast Range, west slopes of the north and central Cascade Range, and some parts of the interior Northwest. Drier temperate evergreen coniferous forests (right) are characteristic of interior forests of most of the Northwest and the southern central Cascade Range. Under two of three future climate simulations evaluated by Rogers et al. (2011, fig. 5.3), most of the area currently in maritime forests will have climates associated with temperate forests by the late 21st century. Photos: J. Littell

Figure 5.5 Subalpine forests in the drier eastern Cascade Range (left) and wetter western Cascade Range (right) are projected to decrease in area substantially as temperatures increase in most of the Northwest (Rogers et al. 2011). Particularly in eastern Cascade Range forests, the effects of disturbance from fire, insects, and disease may accelerate vegetation changes. Photos: J. Littell

In addition to changes in species or community distributions, climate change will affect the productivity of trees. Tree growth response to climate change is likely to be dependent on local-to-subregional characteristics that increase or decrease the sensitivity of species along the climatic gradients of their ranges (e.g., Peterson and Peterson 2001, Littell et al. 2008, Chen et al. 2010). Douglas-fir is expected to decrease in growth in much of the drier part of its range (Littell et al. 2008; Chen et al. 2010), but may be currently increasing in growth in some locations in the Northwest (Littell et al. 2008, 2010). Observed growth-climate relationships indicate that, given projections of warmer, possibly drier summers in the Northwest (Mote and Salathé 2010; Chapter 2), tree growth will increase where trees are energy-limited (e.g., higher elevations) and decrease where trees are water-limited (e.g., drier areas) (Case and Peterson 2005; Holman and Peterson 2006; Littell et al. 2008). Changes in growth at mid-elevations probably depend on changes in summer precipitation (Littell et al. 2008): increases in summer precipitation and/or summer soil moisture sufficient to meet the water demand associated with increased temperature will probably increase growth, whereas decreases in moisture will lead to decreased growth. However, for the latitudes of the Northwest, the projected future changes in precipitation are less certain than projected warming (e.g., Mote and Salathé 2010).

Forest ecosystem processes, such as biogeochemical cycling, also respond to climate. Climate change will affect carbon cycling both directly and indirectly. Photosynthesis and respiration both increase with warming. However, other aspects of climate change also affect carbon cycling. Warming increases evaporative demand and decreases soil moisture, and increases in atmospheric CO_2 may stimulate plant growth provided other factors (water availability, nutrients) are not more limiting. Reduced snowpacks, earlier snowmelt, and longer growing seasons may result in reduced carbon sequestration (Hu et al. 2010). The net response of forest carbon fluxes to future climate change will thus be a balance of these factors, and vary depending on location and other factors including land use, disturbance, and variability in climate (Birdsey et al. 2007; McKinley et al. 2011). Simulated projections in the northern US Rocky Mountains suggest slight increases in net primary production and total carbon compared with recent years but a high degree of variability in the projected net ecosystem flux, with carbon sources estimated for drier climate projections (Boisvenue and Running 2010). This range of responses, including regional variability, occurs in other modeling studies as well (e.g., Smithwick et al. 2009, Rogers et al. 2011, Lenihan et al. 2008). Climate change also indirectly affects carbon sequestration: increasing disturbance may alter carbon budgets and reduce carbon stores in the Northwest (McKinley et al. 2011; French et al. 2011; Hicke et al. 2012). For example, in Washington, increasing burn area in the coming decades was projected to reduce carbon stocks by 17–37% (Raymond and McKenzie 2012). In addition to climate influences, however, the net change in NW forest carbon budgets depends significantly on forest management (McKinley et al. 2011), and in the future, bioenergy potential (Hudiburg et al. 2011). Other biogeochemical cycles also interact with climate change to affect forests. For instance, nitrogen alters the capability of forests to respond to climate change, though substantial uncertainties exist (Suddick and Davidson 2012).

In summary, climate change is likely to substantially affect the distribution, growth, and functioning of the forests of the Northwest. Statistical and mechanistic modeling studies suggest that, under projected future climate changes, the spatial distribution of suitable climate for many important NW tree species and vegetation types may change considerably by the end of the 21st century, and some vegetation types, such as subalpine forests, will become extremely limited compared to their current distribution (figs. 5.3, 5.5). Tree growth responses to future climate change will be highly variable, but

BOX 5.1

Changes in Non-forest Systems: High-Elevation Habitats, Grasslands, and Shrublands

Alpine, subalpine, and montane systems are highly diverse and are likely to respond to climate change in complex ways (Beniston 2003). Of particular concern at high elevations is the limited potential for species to move upslope or across complex, unfavorable landscapes (which present significant barriers) in response to changing climate conditions (Spies et al. 2010).

Changes to the region's alpine systems may include shifts in species composition and distribution, altered plant productivity, and increased disturbances. In some areas (e.g., the Olympic Peninsula of Washington State), suitable climates for alpine tundra and subalpine vegetation are projected to decline substantially in area or disappear by the end of the century, with temperate forest species moving into these locations (Halofsky et al. 2011). Research also suggests that climatic suitability will decline for high-elevation populations of whitebark pine, Brewer spruce (*Picea breweriana*), Engelmann spruce (*Picea engelmannii*), and subalpine fir (*Abies lasiocarpa*) in the region (e.g., Rehfeldt et al. 2006, Spies et al. 2010, Coops and Waring 2011, Rogers et al. 2011). Changes in these habitats will affect associated wildlife species and alpine biodiversity (Malanson et al. 2007; Spies et al. 2010).

Grasslands and Shrublands

Grassland and shrubland systems in the Northwest are comprised of diverse species and ecosystem types, including prairies and oak savannas west of the Cascade Range and sagebrush-steppe habitats inland (figs. 5.1, 5.6). Throughout much of the region, the extent of these systems has already declined significantly from historical levels as a result of land use changes, fire suppression, and a range of stressors associated with human development (Bachelet et al. 2011). Given the considerable diversity of species and ecosystem types represented by the region's grassland and shrubland vegetation, the effects of future climate change across the Northwest will likely be highly varied (Finch 2012).

In western Oregon and Washington, grass-dominated prairies and oak savannas are adapted to periodic drought, and warmer and drier conditions projected for the future may not significantly affect these systems (Bachelet et al. 2011). Bachelet et al. (2011) note that many prairie plant species may be able to expand under projected future increases in growing season length. Projected declines in climatic suitability for tree species (e.g., Littell et al. 2010, after Rehfeldt et al. 2006) at lower elevations as a result of climate change also may offer opportunities for grassland systems to expand upward in elevation in some areas. Similarly, several studies suggest that habitat suitability for Garry oak (*Quercus garryana*) is likely to improve in some parts of the Northwest under projected climate change, although connectivity between climatically suitable habitats is projected to decline (Bodtker et al. 2009; Pellatt et al. 2012).

The region's sagebrush-steppe systems and associated species such as greater sage-grouse (*Centrocercus urophasianus*) and pygmy rabbit (*Brachylagus idahoensis*), on the other hand, are likely to be more vulnerable to climate change (Finch 2012). These systems are sensitive to altered precipitation patterns (Bates et al. 2006; Chambers and Pellant 2008; Xian et al. 2012). Vegetation models of sagebrush-steppe systems in eastern Oregon and Washington simulate large declines in current distributions of shrublands under future climate conditions (Neilson et al. 2005; Rogers et al. 2011), with shrubs largely replaced by woodland and forest vegetation.

The response to climate change of grassland and shrubland systems throughout the Northwest will be influenced by invasive species that are currently present in these systems or may be able to expand into these systems as climate changes (Dennehy et al. 2011). Many non-native invasive

BOX 5.1 (Continued)

species have been introduced to the region and are increasingly widespread. For example, meadow knapweed (*Centaurea xmoncktonii*) is found in prairies throughout western Oregon and Washington (Dennehy et al. 2011) and yellow star-thistle (*Centaurea solstitialis*) is outcompeting native grasses and affecting rangeland forage in the interior Northwest (Roché and Roché 1988; Roché and Thill 2001). In addition to non-native invasive species, native "invasive" species also affect these systems. For example, the historical expansion of native western juniper (*Juniperus occidentalis*) into grassland and shrubland areas of eastern Oregon and Washington is well-documented (Miller and Rose 1999; Carey et al. 2012) and tree expansion in these areas is projected to continue in the future (Rogers et al. 2011).

Future Changes in Wildfire in Non-forested Systems

While many NW grassland, prairie, and woodland species are adapted to high-frequency, low-intensity fires, some species, such as sagebrush (*Artemisia* spp.) and shadscale (*Atriplex confertifolia*), are fire-intolerant (Meyer 2012). Fire suppression over the last century has altered historical fire regimes and contributed to woody vegetation (e.g., oak, juniper) expansion into grasslands and shrublands throughout the Northwest (e.g., Miller and Rose 1999). Future climate change is projected to alter fire activity across grassland and shrubland systems. For example, Rogers et al. (2011) used MC1 to simulate changes in fire regimes for the Columbia Plateau ecoregion of eastern Oregon

and Washington using future climate simulations from three GCMs (CSIRO Mk3, MIROC 3.2 medres, Hadley CM3) for the SRES-A2 emissions scenario of continued growth for 2070–2099. Biomass consumed by fire was projected to increase under all three simulations, whereas area burned increased for one simulation but decreased for the other two simulations (Rogers et al. 2011). Projected increases in future area burned by fires may decrease distributions of fire-intolerant shrubs and shrub-obligate species (e.g., greater sage-grouse).

Changes in wildfire regimes associated with climate change may exacerbate existing impacts to grassland and shrubland systems from invasive species, including exotic annual grasses such as medusahead (*Taeniatherum caput-medusae*), cheatgrass (*Bromus tectorum*), and red brome (*Bromus madritensis* subsp. *rubens*) (Chambers and Pellant 2008). Bioclimatic envelope models have been used to simulate changes in potential suitable habitat for some invasive species such as cheatgrass, which can alter fire regimes in the areas it invades (Bradley et al. 2009; Brooks et al. 2004). However, the response of invasive species to climate change varies considerably. For example, suitable habitat for cheatgrass is projected to increase in some areas of the Northwest while decreasing in others (Bradley et al. 2009). A major limitation of many vegetation modeling studies is that they often focus on the potential responses of native species to future climate change without including the potential effects of either invasive exotic species or native species that may expand into a region as climate changes.

Figure 5.6 A high-elevation meadow in the Cascade Range near Crater Lake, Oregon (left) and shrubland in eastern Oregon near Malheur Lake (right). Photos: S. Shafer

some locations are likely to experience higher growth (e.g., higher elevations) whereas other areas are likely to experienced reduced growth (e.g., the eastern Cascades). Forest ecosystem processes such as carbon cycling will very likely be affected by future climate change, though the magnitude and sign of the response will vary considerably within the region and through time. Substantial uncertainty about future responses resulting from climate models, emissions scenarios, and understanding of tree physiology and forest disturbances is associated with most findings.

5.3 Indirect Effects of Climate Change through Forest Disturbances

Forests in the Northwest also will likely be affected by climate-driven changes in disturbance regimes, such as wildfire (Littell et al. 2010), insect outbreaks (e.g., mountain pine beetle; Logan et al. 2003), disease (e.g., Swiss needle cast; Black et al. 2010), and drought (van Mantgem et al. 2009; Knutson and Pyke 2008). The response of plant species to future climate changes may also be mediated by a number of other factors, including land-use changes (e.g., grazing) and interactions with other species (e.g., invasive species) (Chambers and Wisdom 2009).

Given recent historical rates of disturbance (fig. 5.7), it seems probable that the rate of change in vegetation and biodiversity will be driven by a combination of climate-mediated disturbance and the rate and kind of vegetation response after disturbance – both disturbances and regeneration are key to understanding future vegetation trajectories (McKenzie et al. 2009; Littell et al. 2010). In this section, we describe the mechanisms by which climate affects forest disturbances (drought, fire, insects, and disease) and the past and potential future climate-related impacts.

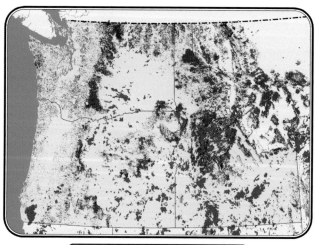

Figure 5.7 Areas of recent fire and insect disturbance in the Northwest. The total area of forest impacts due to recent disturbance is high, and climate change is expected to increase the probability and alter the spatial distribution of fire and insect outbreaks in much of the region. Therefore, future changes in forests may be controlled more by the effects of disturbance and subsequent vegetation regeneration than by direct climate effects on vegetation. Fire data: Monitoring Trends in Burn Severity (MTBS, http://www.mtbs.gov/) fire perimeter polygons (1984–2008) (Eidenshink et al. 2007). Insects and disease data: Aerial Detection Survey (ADS, http://www.fs.fed.us/foresthealth/technology/adsm.shtml, 1997–2008).

5.3.1 WILDFIRES

5.3.1.1 Climate Influence

Changes in temperature and precipitation affect fuel amount, structure, and availability over the long-term by influencing vegetation type and growth, as well as over the short-term by affecting fuel moisture during the fire season. Climate, through its influence on snowpack and summer water balance deficit, also affects the length of the fire season (e.g., Westerling et al. 2006), during which existing fuels become available to burn (e.g., Littell, McKenzie, et al. 2009). Fire history studies (evidence from trees scarred by fires or age classes of trees established after fire and independently reconstructed climate) and modern fire-climate comparisons (evidence from observed fire events and observed climate occurring in the seasons leading up to and during the fire) agree on basic mechanisms, such as growing season temperatures and precipitation, although the specifics vary with forest and region (Westerling et al. 2003; Littell, McKenzie, et al. 2009). Fire activity (occurrence, area) in most NW forests increases with higher summer temperature, lower summer precipitation (Westerling et al. 2003; McKenzie et al. 2004; Littell, McKenzie, et al. 2009), or increases in PET and water balance deficit (PET–AET) (Littell et al. 2010; Littell and Gwozdz 2011). In the western United States, a positive trend in spring and summer temperatures have been associated with longer fire seasons and an increase in area burned (Westerling et al. 2006). In some ecosystems, the maximum temperature in the warmest months (July–August) is well correlated with area burned (Littell, McKenzie, et al. 2009; Littell, Oneil, et al. 2009; Littell and Gwozdz 2011). In contrast to these ecosystems in which warming and drying promotes fire activity, in other ecosystems (typically non-forest or woodland ecosystems), the "normal" climate may limit the availability of existing fuels for fire through drought or increased temperature, which increase heat stress and limit biomass (fuel) production (Northwest: Heyerdahl et al. 2002, Hessl et al. 2004, Heyerdahl, McKenzie, et al. 2008; Northern Rockies: Heyerdahl, Morgan, et al. 2008; Western United States: Westerling et al. 2003, Littell, McKenzie, et al. 2009). Years of higher precipitation can facilitate the development of new, fine fuels through vegetation growth that subsequently becomes available fuel, increasing the likelihood of larger area burned (Swetnam and Betancourt 1998; Littell, McKenzie, et al. 2009). In some drier forests, both effects are detectable (Littell, Oneil, et al. 2009; Littell et al. 2010), but precipitation facilitation of fire is weaker in most NW forests (Littell, McKenzie, et al. 2009) than in more arid ecosystems such as drier forests, shrublands, or grasslands. Western US syntheses of pre-settlement fires (Kitzberger et al. 2007) and historical records (Collins et al. 2006) have also shown important relationships between ocean-atmosphere modes of climate variability (e.g., El Niño-Southern Oscillation, Pacific Decadal Oscillation, and Atlantic Multidecadal Oscillation), which modulate NW climate on interannual to interdecadal time scales.

5.3.1.2 Past and Projected Future Fire Activity

Despite changes in land use and the resulting effects on fuels, climate correlates with area burned and the number of large fires in both the pre-settlement period and the last few decades. The impact of climate change on NW forest fires has been assessed using statistical models that project area burned from climate variables (western US: McKenzie et al. 2004, Littell 2010; Northwest: Littell et al. 2010). Decreased summer precipitation and increased summer temperature (Mote and Salathé 2010; Chapter 2) expected in the region are the primary mechanisms for the projected increase in area burned (Littell 2010). Other seasonal effects also influence fuel moisture: earlier snowmelt leads to earlier onset of the fire season (Westerling et al. 2006) due to earlier water balance deficit (Littell and Gwozdz 2011). The area burned in NW forests is very likely to increase in response to expected future warming because warmer, drier conditions reduce moisture of existing fuels, facilitating fire (McKenzie et al. 2004; Littell, McKenzie, et al. 2009; Littell et al. 2010). The range of changes in future area burned given projected climate change in these studies is from <100% to >500% increase in median area burned depending on the time frame, methods, future emissions and climate scenario, and region.

For the Northwest, median regional area burned is projected to increase from about 0.2 million hectares (0.5 million acres, 1980–2006) to 0.3 million hectares (0.8 million acres) in the 2020s, 0.5 million hectares (1.1 million acres) in the 2040s, and 0.8 million hectares (2.0 million acres) in the 2080s (average of area burned calculated separately for climate simulated by CGCM3 and ECHAM5 GCMs run under the SRES-A1B emissions scenario of continued growth peaking at mid-century) (Littell et al. 2010). The probability of exceeding the 95th percentile area burned for the period 1916–2006 increases from 0.05 to 0.48 by the 2080s (Littell et al. 2010). Sub-regionally, the probability of exceeding the late 20th century historical (1980–2006) 95th percentile under expected future climate for the 2020s, 2040s, and 2080s ranges from 0 to 0.30 for non-forested systems, 0.01 to 0.19 for the western Cascade Range, and 0 to 0.76 for the eastern Cascade Range, Blue Mountains, and Okanogan Highlands under emissions scenarios of continued growth peaking at mid-century (SRES-A1B; 20 GCM realizations) and substantial reductions (SRES-B1; 19 GCM realizations) (after Littell et al. 2010). These increases in probabilities suggest that large fire years will become more frequent in the future. No statistical relationships could be constructed for the Oregon Coast Range and Olympic Mountains due to low fire activity in the recent record, although climate effects on fire in those ecosystems are evident in the recent paleoecological record and the consequences of rare events are extreme, with > 0.5 million hectares (>1.2 million acres) burned in single events (Henderson et al. 1989).

Although statistical models are limited in their ability to simulate the dynamic effects of climate on fire regimes, other research using DGVMs also supports the inference of increased future fire activity in much of the Northwest. Bachelet et al. (2001) showed that changes in biomass burned ranged from -80% to +500% depending on region, climate model, and emissions scenario. Rogers et al. (2011) used the MC1 DGVM to simulate fire regimes given projected climate changes and dynamic vegetation for much of Oregon and Washington for three GCM simulations (CSIRO Mk3, MIROC 3.2 medres, Hadley CM3) under the SRES-A2 emissions scenario of continued growth. The authors reported large increases in area burned (76–310%, depending on climate and fire suppression scenario) and burn severities (29–41%) by the end of the 21st century compared to 1971–2000.

Compared to area burned, there is much less quantitative information on likely responses of forest fire frequency, severity, and intensity to climate change. Fire area increases imply increases in fire frequency for any definable unit, but detecting changes in fire frequency relative to the mid- and late-21st century is difficult because natural fire return intervals vary from less than 10 to over 500 years within the Northwest. Fire severity (proportion of overstory mortality) is potentially influenced by climate (Dillon et al. 2011). However, severity may be more sensitive to landscape factors such as topography or the arrangement and availability of fuels (which affect intensity, defined as the energy release of a fire) than area burned, and so future climate effects on severity are more complex to project. To our knowledge, there are no peer-reviewed regional syntheses of climate-fire severity relationships or projections of future severity as a function of climate. The increase in extreme events associated with future climate change (Hansen et al. 2012), especially drought and heat waves, is likely to increase the fire activity in the Northwest, which, combined with fire-driven extreme weather, suggests it is plausible to expect greater fire severity at least in forest systems.

Relationships between climate variability, climate change, and wildfire interact with other factors that influence fuels (Stephens 2005). In the Northwest, regional land-use history (including timber harvest and forest clearing, fire suppression, and possibly fire exclusion through grazing) has affected the amount and structure of fuels. This effect is particularly evident for drier forests in the eastern Cascade Range, Blue Mountains, and northwestern US Rocky Mountains in Washington, Oregon, Idaho, and western Montana, where fire suppression has increased fire return intervals (Heyerdahl et al. 2002, 2008a, 2008b), in contrast to wetter forests (e.g., maritime coast of Oregon and Washington).

5.3.2 FOREST INSECTS

5.3.2.1 Climate Influence

Insects are key agents of disturbance in the forests of the Northwest. Outbreaks of bark beetles and defoliating insects have affected millions of hectares of forest regionally in the last several decades (Hicke et al. 2012; US Department of Agriculture Forest Service 2010; Meddens et al. 2012; fig. 5.7). Bentz et al. (2010) list 15 bark beetle species that have the capacity to produce mortality across western US landscapes. Climate is a major driver of insect disturbances in several ways (Sturrock et al. 2011; Bentz et al. 2009; Raffa et al. 2008; Ayres and Lombardero 2000; Bale et al. 2002). Temperature directly affects insect mortality and life stage development rates. Unseasonably low temperatures during the fall, winter, and spring can kill insects (Wygant 1940; Régnière and Bentz 2007). Year-round temperatures regulate development rates, thereby influencing the number of years required to complete an insect's life cycle and, for bark beetles, affecting population synchronization for mass-attacking host trees (Logan and Powell 2001; Hansen et al. 2001). In addition, host trees can be more vulnerable to insects due to drought (Safranyik et al. 2010; Raffa et al. 2008; Bentz et al. 2010) and increased vapor pressure deficit (Littell et al. 2010, after Oneil 2006).

5.3.2.2 Past and Projected Future Insect Outbreaks

Recent climate has been related to more intense, frequent, or severe insect outbreaks in the Northwest, and also to outbreaks in places where historical insect activity was low or unknown (Logan and Powell 2009). Higher average temperatures and drought stress

Figure 5.8 Mountain pine beetle kills several pine species native to the Northwest (left) and frequently results in substantial tree mortality across large areas of regional forests (right). Increasing temperatures have synchronized the insects' life cycles and reduced winter beetle mortality, facilitating outbreaks in places where mountain pine beetle activity was historically low or absent. Projected future climate changes suggest increasing areas of NW forests suitable for outbreaks at high elevations (Hicke et al. 2006). Photos: J. Hicke

are contributing to significant outbreaks of mountain pine beetle across pine forests of the region (fig. 5.8), increasing the frequency and levels of tree mortality (Logan and Powell 2001; Carroll et al. 2004; Oneil 2006). In British Columbia, mountain pine beetle outbreaks have been influenced by warming (both cold season and year-round temperatures; Carroll et al. 2004), and northward expansion of the beetle is occurring (Safranyik et al. 2010). In mid- and high-elevation forests in western North America, warming has facilitated prolonged outbreaks in locations considered typically too cold to support the insect, again related to both cold season and year-round temperatures (Logan and Powell 2001; Logan et al. 2010). Major outbreaks of two other important bark beetles have been linked to warming and/or drought in other regions. Anomalously warm and dry conditions were associated with outbreaks of spruce beetle in Alaska (Berg et al. 2006; Sherriff et al. 2011) and Utah and Colorado (Hebertson and Jenkins 2003; DeRose and Long 2012), further underscoring the climatic connections with insect outbreaks. Extreme drought in the Southwest in the early 2000s was tightly coupled to a population increase of pinyon ips (*Ips confusus*) that may have amplified the pinyon mortality (Raffa et al. 2008). In many outbreak locations, large numbers of stands of host trees were in structural conditions that were highly susceptible to attack by bark beetles (Hicke and Jenkins 2008; Werner et al. 2006), thereby contributing to the extensive mortality in addition to the climatic factors.

Future climate change is expected to affect the frequency and area of outbreaks of insects in the Northwest. The region of suitable year-round temperatures for outbreaks of mountain pine beetle is projected to move upslope with future warming, continuing the high level of susceptibility of high-elevation pine forests to this insect (Williams and Liebhold 2002; Hicke et al. 2006; Littell et al. 2010; Bentz et al. 2010; Evangelista et al. 2011). Similarly, at high elevations the increased probability that sufficient warmth exists for spruce beetles to complete their life cycle will lead to enhanced probability of outbreak in western North America (Bentz et al. 2010), and other bark beetle species may also expand their ranges (e.g., western pine beetle [*Dendroctonus brevicomis*]) (Evangelista et al. 2011). In contrast, simulations suggest that at the current lower elevation areas of mountain pine beetle, future conditions become too warm to support outbreaks by disrupting population synchronization and life cycles, thereby resulting in range contraction (Hicke et al. 2006). Ranges of other bark beetles will likely decrease as well (e.g., pine engraver beetle [*Ips pini*]) due to climatic conditions less favorable for outbreaks (Evangelista et al. 2011). In the Northwest, outbreaks of western spruce budworm (*Choristoneura occidentalis*), which is commercially important because of its damage to various conifer species including Douglas-fir, have been linked to warm, dry summers (Thomson et al. 1984).

Insect outbreaks also have the capacity to affect future climate via changes in atmospheric CO_2 concentrations and surface albedo, thereby creating both positive and negative feedbacks between climate change and outbreaks (Adams et al. 2010; Hicke et al. 2012). Reduced photosynthesis following attack and increased decomposition of killed trees and soil carbon flux can turn forests into carbon sources instead of sinks (Kurz et al. 2008). However, modifications to surface albedo can lead to surface cooling that may be greater than warming associated with carbon release (O'Halloran et al. 2012).

5.3.3 FOREST DISEASES

5.3.3.1 Climate Influence

Diseases are also important disturbance agents in the forests of the Northwest, and disease plays a significant role in regulating forest structure and function. Climate influences forest pathogens through temperature and foliar moisture (Sturrock et al. 2011). Foliage fungi appear to be affected by increased spring and summer precipitation (see Chmura et al. 2011 for a brief review). For example, higher average temperatures and increased spring precipitation in the Oregon Coast Range have contributed to an increase in the severity and distribution of Swiss needle cast in Douglas-fir (Stone et al. 2008; Sturrock et al. 2011). If these relationships hold, similar future trends could reasonably be expected given projected climate changes in the region (Mote and Salathé 2010; Chapter 2).

Like insect outbreaks, disease outbreaks are also indirectly affected by climate change influences on host tree stress, which in turn influences the capacity of a tree to defend itself from attack (Ayers and Lombardero 2000; Sturrock et al. 2011). Drought can decrease tree vigor (see summary in Chmura et al. 2011), and has been linked to disease epidemics (Sturrock et al. 2011). Compared with insects and fire, the relationships between pathogens and climate are unclear for many diseases, but climate influences pathogen range and survival, host vulnerability, and host-pathogen relationships.

Root rot pathogens are most likely to increase in stressed host trees (by climate or other factors, Chmura et al. 2011), and Klopfenstein et al. (2009) suggest the potential for climatically caused increases in *Armillaria* root rot in western conifers. However, as with insects and climate, the climatic influences on pathogens are likely to be species- and host-specific, such that generalizations are difficult to make (Kliejunas 2011).

5.3.3.2 Past and Projected Future Disease Outbreaks

Several pathogen epidemics have been linked to climate change. Swiss needle cast is a disease caused by *Phaeocryptopus gaeumannii*, a foliar pathogen that infests Douglas-fir in the coastal areas of the Northwest. Expansion of the area of Swiss needle cast has been associated with warming and precipitation changes, and is projected to have increased capacity to affect Douglas-fir in the future (Stone et al. 2008). Sudden oak death caused by *Phytophthora ramorum*, a virulent invasive pathogen in California and the Northwest, is affected by temperature and moisture (Venette and Cohen 2006). Projections of climate change suggest increased sudden oak death in response to climate change (Sturrock et al. 2011). Kliejunas (2011) evaluated the relative risk of increased disease damage in forests of the Northwest by combining the likelihood (probability) of increased damage and the consequences (impacts) for several diseases. The risk potential depended on disease and climate scenario (warmer wetter versus warmer drier), but by 2100, Cytospora canker of alder, dwarf mistletoes, and yellow-cedar decline were projected to have high risk and *Armillaria* root disease was projected to have very high risk if precipitation decreased. If precipitation increased, *Armillaria* and dwarf mistletoes were projected to have high risk and sudden oak death was projected to have very high risk (Kliejunas 2011).

5.3.4 DISTURBANCE INTERACTIONS AND CUMULATIVE EFFECTS

The vulnerability to climate change of the region's forest ecosystems and services is increased by the potential for synergy between multiple disturbances, including insect and disease outbreaks and wildfires. For example, synergy between white pine blister rust and mountain pine beetles has been associated with mortality in high-elevation pines (Bockino and Tinker 2012; Six and Adams 2007). Moreover, areas with severe insect or disease outbreaks and significant tree mortality may be more vulnerable to severe wildfires depending on fire characteristic and time since outbreak (Lynch et al. 2006; Jenkins et al. 2008; McKenzie et al. 2009; Hicke et al. 2012).

The cumulative effects of disturbance and the future effects of climate on species' distributions are not completely separable. For the sake of clarity, we have first outlined the literature on these mechanisms in the Northwest. But, the forests that establish after disturbance events and under different climatic conditions will be the product of disturbance and climate as well as the other conditions (such as nutrient availability, competition, etc.) that affect tree life histories and forest processes. For example, species whose individuals are resistant to fires, such as ponderosa pine, western larch (*Larix occidentalis*), or Douglas-fir, may be favored under more frequent fires, and this resistance could be a more important factor affecting these species' distributions than climatic tolerances once trees are established. Ultimately, the cumulative effects will vary with biophysical context and other stressors, suggesting difficulty in predicting future conditions.

5.4 Implications for Economics and Natural Systems

The physical changes in forest ecosystems and processes induced by climate change will have a range of impacts, both positive and negative, for the NW forest economy, forest recreation, and natural systems. For example, the risk posed by future disturbance in a changing climate is a function of the likely impacts to human and ecological systems, and there are important implications for adaptation and vulnerability. In this section, we address some potential consequences of the projected climate-mediated changes in forests (described in earlier sections) on other dimensions including timber markets, nonmarket and recreational uses of the forests, and natural systems. There are also important considerations for human health that are discussed in Chapter 7.

5.4.1 ECONOMIC CONSEQUENCES

 Forty-seven percent of the land area in the Northwest is forested, with more forested land in Washington (~52%) and Oregon (~49%) than in Idaho (~41%) (Smith et al. 2009; see also fig. 5.1). As such, forested land in the Northwest contributes substantially to the region's economy both through forest industry activities as well as recreational and tourism activities. For example, in Oregon, the forest industry contributes $12.7 billion to Oregon's economy each year and represents 6.8% of total industrial output (Oregon Forest Resources Institute 2012). In Washington, the forest industry provides approximately 15% of manufacturing jobs and about 3.2% of gross business income (Washington State Department of Natural Resources 2007). Idaho's wood and paper industries account for nearly one-fifth of all the labor income generated in the state, and more than one-tenth of the state's total employment (Idaho Forest Products Commission 2012). Publicly owned forests in the region also provide a wide range of recreational opportunities such as hiking, biking, camping, skiing, snowshoeing, and snowmobiling. Visitors to these forests generate significant economic benefits to the region's economies through visitor expenditures on food, recreation equipment, and lodging, and create a demand for tourism-related employment.

Land ownership is important when assessing the economic consequences of climate change since the uses and management of forested areas differ by land ownership. The heterogeneity of the forest ownership across Washington, Oregon, and Idaho is shown in table 5.1. Only about one-third of the forest land is privately owned, with the remainder publicly owned forests that are used and managed for a variety of timber and nontimber uses.

Tourism and recreation opportunities on publicly owned lands are important parts of the economies of the Northwest and part of the social fabric of the region. Adverse impacts of climate change on the sustainability and health of the forests will have ripple effects in recreational markets and may decrease the (nonmarket) values of the forest ecosystems. Understanding how these values and recreational experiences will change under a changing climate is critical to identifying and quantifying climate change impacts and critical to designing effective management responses and synergistic policies across the spectrum of forest land ownership.

Although the entire forest economy, including federal, state, and privately owned lands, may be affected by the sensitivities noted in this chapter, the economic

Table 5.1 Percent and totals of forest land by ownership for Washington, Oregon, and Idaho

Total (%)	Ownership (%)	Washington (%)	Oregon (%)	Idaho (%)
US National Forest	36.8	46.4	76.4	53.2
US National Grassland	0	0.2	0	0.1
US National Park Service	5.6	0.6	0.4	2.1
US Bureau of Land Management	0.3	12.4	4.1	5.6
US Fish and Wildlife Service	0.3	<0.1	0	0.1
Other Federal	0.3	<0.1	0	0.1
State	11.6	3.2	7.2	7.3
Local (county, municipal, etc.)	1.6	0.6	0	0.7
Private – noncorporate	22.0	16.8	6.2	16
Private – corporate	22.0	19.9	5.7	15
Total (hectares)	9,016,000	12,209,000	8,672,000	29,897,700
(acres)	22,279,000	30,169,000	21,430,000	73,878,000

consequences are more complex to quantify and attribute to climate change. Land ownership and management heterogeneity, and uncertainties in forest growth and biodiversity changes co-exist with a constantly changing set of federal and state policies regarding use of these forested lands. For example, the Northwest Forest Plan has decreased the level of harvest on federal lands in the Northwest. The average 1992–2010 timber volume cut on US Forest Service (USFS) and Bureau of Land Management (BLM) lands was 11% of 1965–1991 levels in Washington and 13% in Oregon (Warren 2011). The economic impacts associated with those policy changes in forest management may be as important as any subsequent economic effects attributable to climate change on forest productivity.

Changes in forest productivity and disturbance could increase or decrease the viability of forestry products, bioenergy, and carbon markets with the impacts being highly place- and economy-dependent. Access to population centers, composition and alternative uses of the forest products, and relative productivity of the forests across climatic regions are critical factors to consider in assessing the net economic impacts of climate change in the region. In the absence of future economic developments that create viable bioenergy or carbon markets, climate change factors may have less impact on the NW forest industry, because of the comparatively large historical impact of the Northwest Forest Plan. This policy, enacted in 1994, changed management priorities for a significant area of USFS and BLM lands from timber production to conservation. In the following sub-sections are a few examples from the literature of the types of regional or site-specific impacts of climate change in the Northwest.

5.4.1.1 Timber Market Effects

Changes in the growth rates of trees, increases in insects and disease, and changes in species migration patterns due to climate change have the potential to substantially impact timber markets in North America. Sohngen and Sedjo (2005) and Alig (2010) reviewed several studies that examined the potential impacts of climate change for the North American forest sector. Both reviews found that climate change would result in aggregate yield increases at the national level, leading to increased timber production and reduced prices. To determine the overall impacts on consumers and producers, these studies have used a concept of net surplus: the difference between the change in benefits to consumers from lower prices and higher yields, and the change in economic profits of forest landowners. Overall, net surpluses are likely to increase for the United States as a whole due to strong positive benefits to consumers from lower timber prices; however, on a regional basis the net surpluses will vary. For example, simulations by Sohngen and Mendelsohn (2001) suggest a potential decline in the net surpluses in timber regions of the Northwest and Southeast due to species redistribution and altered timber growth rates that may shift forestry away from these two regions that have historically provided a large share of the US timber supply. The projected decline in net surplus is due to the overall reductions in net returns to producers that are not offset by gains to consumers: the negative changes in producer surpluses are greater than the positive changes in consumer surpluses, resulting in negative net benefits overall.

5.4.1.2 Economic Effects of Disturbance

Potential economic benefits from projected timber yield increases as a result of climate change may be offset by insect and disease outbreaks, forest fires, and shifting ranges of tree species. For example, increasing severity and extent of Swiss needle cast (Black et al. 2010), which leads to loss of needles and a reduction in growth, is an important finding for the timber industry west of the Cascade Range, particularly for Douglas-fir producers. Douglas-fir is susceptible to Swiss needle cast and is one of the most common commercially important tree species west of the Cascade Range, as well as a popular Christmas tree (Oregon Department of Forestry 2009).

Within the dry forests east of the Cascade Range, increased frequency and intensity of insect outbreaks, such as the mountain pine beetle (Ryan et al. 2008), and increased incidence and extent of root diseases (such as *Armillaria*), will damage the growth and yield of ponderosa pine, which is the dominant commercial species east of the Cascade Range and ranks second in total value (Western Wood Products Association 1995). In addition to yield losses, increasing bark beetle-caused tree mortality will likely cause economic losses and costs related to management and possibly increased wildfire risk (Capalbo et al. 2010). Forest fires impose an array of economic consequences from loss of timber values and tourism dollars to loss of life and property. Since 1970, fire suppression has accounted for more than half of all USFS fire-related expenditures (Schuster et al. 1997).

5.4.1.3 Non-Timber Market Effects

The non-timber market effects include changes in recreational opportunities and changes to ecosystem services that public and private forest lands provide. On a national scale, estimates of the adverse impacts of climate change on forest-based recreational activities, such as hiking and camping, are expected to exceed $650 million by 2060, and adverse impacts on snow-based recreational industries are estimated to be $4.2 billion by 2060 (Loomis and Crespi 1999).

Some climate change impacts may result in positive as well as negative economic impacts. For example, Loomis and Richardson (2006) concluded that climate change-related impacts on recreation in the United States in the near-term (2020) would result in a net positive impact on visitation rates at alpine areas within national parks due to higher temperatures and a lengthening of the season for most recreation activities. In places like Rocky Mountain National Park, which currently has short recreation seasons because of ice, snow, and cold weather, the near-term effects of climate change may be viewed by some recreational users as beneficial due to the lengthening of the high-use summer season. Thus there are both winners and losers: the multitude of traditional park users such as hikers and campers may benefit from warmer temperatures and a longer season, while cross country skiers and other winter visitors may lose recreational opportunities. Lower snowfall has been associated with changes in skier visits. According to Burakowski and Magnusson (2012), skier visits in Washington and Oregon, as well as the Southwest (Arizona and New Mexico), are sensitive to snowfall; visitation rates decline by about 30% in lower snowfall years compared with higher snowfall years. These findings are consistent with previous climate-recreation studies (Mendelsohn and Markowski 1999; Loomis and Crespi 1999) that have found that higher benefits to reservoir, stream, golf, and beach recreation were partially offset by losses to winter-based activities.

Of particular relevance to the NW's eastern forests are findings regarding more localized impacts on resource-dependent communities. Starbuck et al. (2006) used a model of recreation behavior for New Mexico's national forests to study forest closures due to wildfires. They found that fires currently cause significant economic losses to local economies resulting from restricted visitations, inferior conditions associated with poor visibility during fires, and subsequent adverse aesthetic impacts while the forest regenerates. Recreational uses following a wildfire, however, are somewhat sensitive to the severity of fire as well as the length of time since the fire. Loomis et al. (2001) found that crown fires have a significant adverse impact on mountain biking recreation, whereas hikers were less adversely impacted by such fires.

Although site-specific recreational research that provides the basis for an economic analysis of changes in recreation uses has not been conducted in the Northwest, regional impacts are likely to be negative given the extent of forested and recreational lands in the region coupled with the projected increased risk of wildfires and decreased snowpack (Capalbo et al. 2010). The overall impact will hinge on the extent and severity of wildfires, regional influences of warmer springs and longer summers on visitation rates, and corresponding declines in winter recreational activities due to a reduced snowpack and higher temperatures.

5.4.1.4 Valuing Ecosystem Services

The value of ecosystem services is by necessity inseparable from biophysical and ecological function, and many ecosystem services are not traded in markets and thus lack observable prices. Boyd (2010) noted that, climate change will alter the amount and lo-cations of ecosystem services and goods produced by natural systems, and "any bio-physical change in delivery of ecosystem goods and services creates a corresponding economic change." Quantifying the impacts of climate change on the value of ecosystem services therefore requires understanding both the biophysical production functions and changes (biophysical agenda) and the benefits and costs of those changes (economic agenda) (Boyd 2010).

Forest ecosystem services (e.g., flood protection or water purification) and ecosys-tem goods (e.g., species habitat or forest products) create value or wealth to society and will be affected by climate change in the Northwest. To date, much of the literature has focused on categorizing the main types of ecosystem services and goods provided from private and public forest lands and, to a limited extent, assessing the total value of goods and services from forest ecosystems. The economic measure of the total value of ecosystem goods and services is reflected in benefits to consumers or households; valuing the changes in the levels of ecosystem services is measured as a change in the area under a demand curve for these goods and services, commonly referred to as a "consumer surplus" measure (National Research Council 2005). Valuing the impacts of a single factor (such as climate change) on changes in ecosystem goods and services is rarely possible—attribution requires quantifying the complex interactions that result in the affected valuation and markets. Other common factors that impact the production and value of ecosystem goods and services include land development, water demands, and air pollution, and are thus related in complex ways to climate change. This com-plexity prevents definitive conclusions regarding how much of observed changes may be attributed solely to a specific factor such as temperature or precipitation.

Several previous studies combined ecology and economics to identify and estimate the value of ecosystem services (e.g., Daily et al. 1997, Costanza et al. 1997, Wainger et al. 2001, Polasky et al. 2005, Boyd 2006, Boyd and Banzhaf 2006, 2007, Brown et al. 2007, Daily and Matson 2008). A study by Krieger (2001) reviewed estimates of forest ecosystem goods and services for the United States, from several studies (mostly from the 1990s), and summarized forest ecosystem monetary values by region and main type of good or service. For the Northwest, Krieger (2001) highlighted the following values:

- Water quality from the forest ecosystems, as reflected in the value of water puri-fication services, is $920,000 to $3.2 million per year
- Soil stabilization and erosion control, valued in terms of costs associated with sedimentation, is $5.5 million per year in Oregon's Willamette Valley
- Aesthetics, cultural, and other non-consumptive uses of the forest ecosystems are estimated at $48–$144 per household per year
- Endangered species habitat is valued at $15–$95 per household per year

These values provide a sample of the potential range of values for various goods and services provided by forest ecosystems. How each of these goods and services, and their marginal values, will change due to climate changes is an area of further research. Climate change will alter natural systems, thus leading to new values for ecosystem services. In some cases climate change will enhance productivity of certain ecosystems, and in other cases it will reduce productivity. As noted by Boyd (2010), uncertainties associated with identifying the production and delivery of ecosystem goods and services make it difficult to evaluate gains and losses in the value of ecosystem services, and these uncertainties are magnified with climate change.

5.4.2 CONSEQUENCES FOR NATURAL SYSTEMS

NW forest ecosystems support hundreds of species of fish and wildlife and play a fundamental role in ecosystem services, including clean air and water, soil stabilization, and biogeochemical regulation (e.g., carbon and nitrogen cycles). As with plant species, there is substantial evidence that future changes in climate may affect the abundance and geographical distributions of wildlife species (Root et al. 2003). Climate change may directly affect species (e.g., mortality from increased frequency of lethal temperatures) or indirectly affect species by altering habitat (e.g., shifts in the seasonality and amounts of snowpack and runoff), disturbance regimes (e.g., frequency and severity of fire), competition and predator-prey interactions with other species, and disease (Parmesan 2006; Hixon et al. 2010). Distribution changes have been documented for a number of aquatic and terrestrial species in the Northwest, such as Edith's checker-spot butterfly (*Euphydryas editha*, Parmesan 2006; see also Hixon et al. 2010, Janetos et al. 2008). Wolverines (*Gulo gulo*, Copeland et al. 2010) and pikas (*Ochotona princeps*, Beever et al. 2010) may be affected by the future loss of alpine and subalpine habitat. For example, a recent study of potential climate change impacts on wolverines suggests that reductions in spring snowpack required for denning could contribute to a decline in population connectivity in some areas (McKelvey et al. 2011). However, the study also found that contiguous areas of wolverine habitat are likely to persist through the 21st century, particularly in British Columbia, north-central Washington, northwestern Montana, and the Greater Yellowstone area, suggesting the potential for these areas to be protected as refugia.

Although it is uncertain to what degree climate change will influence high-intensity, stand-replacing fires in the Northwest, the consequences could be severe for species associated with late-seral/old growth forests such as marbled murrelets (*Brachyramphus marmoratus*) and northern spotted owls (*Strix occidentalis caurina*) (McKenzie et al. 2004). On the other hand, species that thrive in conditions after severe fires, such as the northern flicker (*Colaptes auratus*) and hairy woodpecker (*Picoides villosus*), could benefit under an altered fire regime (Smucker et al. 2005). Increases in wildfire extent and severity may have compounding effects on stream temperatures in some areas, particularly where those fires alter riparian vegetation (Dwire and Kauffman 2003; Dunham et al. 2007; Isaak et al. 2010). For example, a study in Idaho found that areas affected by major wildfires have experienced a significant decline in the extent of headwater stream lengths thermally suitable for spawning and early juvenile rearing for bull trout (*Salvelinus confluentus*) between 1993 and 2006, especially at higher elevations (Isaak et al. 2010).

The effects of climate change combined with other existing stressors on "natural" systems – those relatively unmanaged systems such as wilderness – may result in high cumulative impacts. Changes in disturbance and vegetation distribution may disproportionately affect areas dominated by natural vegetation because of the lack of land use changes in the historical record. The climatically mediated consequences (i.e., changes in habitat availability and quality) may affect native species and ecosystem processes in most places, although the degree and mechanisms of change will vary with ecological and physical context across the region. Two ecosystems with particularly high risk are subalpine forests and alpine vegetation, which may undergo almost complete conversion to temperate or maritime forests given vegetation projections (Rogers et al. 2011). Vegetation at the forest transition to grassland, woodland, or shrubland will also be likely to undergo significant change because of shifts in species distributions and the potential transition in some areas to more water-limited ecosystems. However, these impacts and consequences are likely to depend on the magnitude and persistence of increased water-use efficiency induced by increased atmospheric CO_2 concentrations (Bachelet et al. 2011). It is also plausible that increased management for adaptation (e.g., forest thinning to increase resistance to fire and insect outbreaks) will affect these lower-elevation systems in novel ways, although the impact on native species will vary substantially with the tolerances of those species for habitat changes. The indirect effects of climatically mediated forest changes in riparian areas may also present large consequences for aquatic ecosystems (e.g., sedimentation associated with fire, changes in stream temperature associated with changes in riparian forest cover).

5.5 Knowledge Gaps and Research Needs

Despite the increasing understanding of the likely impacts of climate change on the NW's forests and the consequences of those impacts for forest-derived services, there are still important knowledge gaps and research needs for both basic scientific understanding and applied uses such as adaptation to climate change.

An improved understanding of the role of climate in species distribution and ecosystem function, including the combined ecological and physical mechanisms that result in more favorable or less favorable conditions, will allow better projection of future vegetation at the species level. A substantial amount of research has focused on economically important species, such as Douglas-fir, and additional research is needed for other NW species as well, including invasive species. The forest ecosystems of the Northwest are also not limited to "east" and "west" of the Cascade Range crest, and studies of other systems (such as coastal forests) with unique but potentially climatically influenced structure and function are needed.

Although projections of disturbance probability and area are available, these are uncertain, and furthermore, there is a lack of quantitative information on the effects of climate change on disturbance frequency, severity, and intensity that limits our understanding of the future effects of disturbance on ecosystem structure and services. Improved ability to model and project climate-mediated disturbances (fire, insects, disease, windthrow, drought) and their interactions on forest landscapes at the watershed level will facilitate planning and adaptation efforts for multiple agencies, including those that manage water resources. The limits on regeneration of tree species after disturbance in "new" climates of the future are poorly understood, but are required to understand the trajectory of forests after disturbance. The development of a monitoring system that ranges from plot to regional scales and integrates physical and ecological processes will allow advances in bottom-up modeling of forest responses to climate change.

Understanding of forest impacts is often limited to mean climate conditions, but the role of climate extremes in disturbance, mortality, and plant growth may be as or more important.

We have approached impacts and their consequences primarily as outcomes of direct and indirect climate-mediated effects because impacts modeling often focuses on one or the other, with the exception of some ecosystem models such as DGVMs. Modeling realistic responses, however, should include dynamic interactions among impacts, including feedbacks.

An emphasis on studies that assess GCM capability in the region, comparability among GCMs, and include ensemble (e.g., Elsner et al. 2010) and multiple GCM simulations that "bracket" the range of plausible future conditions (e.g, Littell, McKenzie, et al. 2011, Rogers et al. 2011) will allow development of scenarios useful for developing more specific adaptation options (Littell et al. 2012).

Additional research is needed to quantify the effects of climate change on forest-related economic activity and ecosystem services, including hydrology, carbon sequestration, and habitat. Similarly, the impact of changes in forest ecosystems and health and effects on the local economies and on recreational use needs more elaboration within the Northwest, as does the role of forest land use changes in the Northwest of the future. The efficacy of public policy in mitigating and offsetting adverse impacts of declines in forest productivity and forest diversity is not well understood. Finally, an improved understanding of the economic costs and benefits of alternative management approaches that decrease forest vulnerability to climate change will greatly assist the transition from general adaptation theory to specific, local adaptation action.

However, we emphasize that scientific uncertainty in some areas is not a reason to forestall the transition from adaptation theory to adaptation action. On the contrary, an increasing amount of research is focused on incorporating climate change impacts into planning and management of forest ecosystems (e.g., Spies et al. 2010). A transition from impacts to applied adaptation research is needed. For example, where the legacy of past timber harvest has left even age, large, nearly monocultural stands, will landscape management for higher species and stand diversity mitigate the severity of subsequent disturbance? Alternatively, how do we evaluate the barriers to using fire as a management tool to hasten the transition to more fire-resilient forests in currently fire prone areas?

5.6 Adaptive Capacity and Implications for Vulnerability

Ability to prepare for the impacts of climate change on NW forest ecosystems and their services varies with ownership and management priorities. Adaptation actions that decrease forest vulnerability exist, but none are appropriate across diverse climate threats, land-use histories, and management objectives (Millar et al. 2007). The complexities of the NW's climate and forests – and the vulnerabilities of natural systems and economies – suggest there is no one-size fits-all approach to adaptation in the region (Littell et al. 2012).

Reducing the severity of disturbance through silvicultural approaches that modify stand conditions is one active approach to adaptation (Spittlehouse and Stewart 2003). Surface and canopy thinning can reduce the occurrence and effects of high-severity fire in previously low-severity fire systems (e.g., drier eastern Cascade Range forests) (Peterson, Halofsky, et al. 2011; Prichard et al. 2010), but may be ineffective in historically high-severity fire forests (e.g., western Cascade Range, Olympic Mountains, some subalpine forests). Thinning can increase water availability to remaining trees and thus may reduce tree mortality from insect outbreaks, but this approach is not feasible on the scale of the outbreaks currently extant in much of the West (Chmura et al. 2011; Littell et al. 2012). Alternatively, prescribed fire could plausibly affect the same outcomes, though with different limitations including air pollution and the risk of escaped fires (e.g., as described in Littell et al. 2012). Such adaptation strategies may reduce risk locally by decreasing the vulnerability to severe disturbances.

In the Northwest, science-management partnerships have been established to approach adaptation to climate change (Halofsky et al. 2011; Littell et al. 2012), and a guidebook on the process of institutional adaptation within the USFS, including the Pacific Northwest Region, has been developed (Peterson, Millar, et al. 2011). Peterson, Millar, et al. (2011) note: "Planning and management for the expected effects of climate change on natural resources are just now beginning in the western United States (US), where the majority of public lands are located. Federal and state agencies have been slow to address climate change as a factor in resource production objectives, planning strategies, and on-the-ground applications." These are first steps toward developing adaptation strategies and actions tailored to the region's resource management problems, including climate impacts on forests, but considerable investment of time and resources will be required to realize the goal of adaptation: increasing the resilience of NW forests and ecosystem services to the impacts of climate change.

Acknowledgements

The authors wish to thank Dominique Bachelet (Conservation Biology Institute), Don McKenzie and Dave Peterson (PNW Research Station, US Forest Service), and Stacy Vynne (Puget Sound Partnership) for their helpful comments on a previous version of the manuscript. J.S. Littell was supported by the US Geological Survey National Climate Change and Wildlife Science Center and the Department of Interior Alaska Climate Science Center. J.A. Hicke was supported by the University of Idaho, US Geological Survey Western Mountain Initiative, USDA Forest Service Western Wildland Environmental Threat Assessment Center, and Los Alamos National Laboratory. S.L. Shafer was supported by the US Geological Survey Climate and Land Use Change Research and Development Program.

References

Adams, H. D., A. K. Macalady, D. D. Breshears, C. D. Allen, N. L. Stephenson, S. R. Saleska, T. E. Huxman, and N. G. McDowell. 2010. "Climate-Induced Tree Mortality: Earth System Consequences." *EOS Transactions of the American Geophysical Union* 91 (17): 153. doi: 10.1029/2010EO170003.

Alig, R. J., tech. coord. 2010. "Economic Modeling of Effects of Climate Change on the Forest Sector and Mitigation Options: A Compendium of Briefing Papers." General Technical Report PNW-GTR-833. Portland, OR: US Department of Agriculture, Forest Service, Pacific Northwest Research Station. 169 pp.

Ayres, M. P., and M. J. Lombardero. 2000. "Assessing the Consequences of Global Change for Forest Disturbance from Herbivores and Pathogens." *Science of the Total Environment* 262 (3): 263-286. doi: 10.1016/S0048-9697(00)00528-3.

Bachelet, D., B. R. Johnson, S. D. Bridgham, P. V. Dunn, H. E. Anderson, and B. M. Rogers. 2011. "Climate Change Impacts on Western Pacific Northwest Prairies and Savannas." *Northwest Science* 85 (2): 411-429. doi: 10.3955/046.085.0224.

Bachelet, D., R. P. Neilson, J. M. Lenihan, and R. J. Drapek. 2001. "Climate Change Effects on Vegetation Distribution and Carbon Budget in the United States." *Ecosystems* 4 (3): 164 -185. doi: 10.1007/s10021-001-0002-7.

Bale, J. S., G. J. Masters, I. D. Hodkinson, C. Awmack, T. M. Bezemer, V. K. Brown, J. Butterfield, A. Buse, J. C. Coulson, J. Farrar, J. E. G. Good, R. Harrington, S. Hartley, T. H. Jones, R. L. Lindroth, M. C. Press, I. Symrnioudis, A. D. Watt, and J. B. Whittaker. 2002. "Herbivory in Global Climate Change Research: Direct Effects of Rising Temperature on Insect Herbivores." *Global Change Biology* 8 (1): 1-16. doi: 10.1046/j.1365-2486.2002.00451.x.

Bates, J. D., T. Svejcar, R. F. Miller, and R. A. Angell. 2006. "The Effects of Precipitation Timing on Sagebrush Steppe Vegetation." *Journal of Arid Environments* 64 (4): 670-697. doi: 10.1016/j.jaridenv.2005.06.026.

Beever, E. A., C. Ray, P. W. Mote, and J. L. Wilkening. 2010. "Testing Alternative Models of Climate-Mediated Extirpations." *Ecological Applications* 20 (1): 164-178. doi: 10.1890/08-1011.1.

Beniston, M. 2003. "Climatic Change in Mountain Regions: A Review of Possible Impacts." *Climatic Change* 59: 5-31. doi: 10.1023/A:1024458411589.

Bentz, B., J. Logan, J. MacMahon, C. D. Allen, M. Ayres, E. Berg, A. Carroll, M. Hansen, J. Hicke, L. Joyce, W. Macfarlane, S. Munson, J. Negrón, T. Paine, J. Powell, K. Raffa, J. Régnière, M. Reid, B. Romme, S. J. Seybold, D. Six, D. Tomback, J. Vandygriff, T. Veblen, M. White, J. Witcosky, and D. Wood. 2009. "Bark Beetle Outbreaks in Western North America: Causes and Consequences." Salt Lake City: University of Utah Press.

Bentz, B. J., J. Régnière, C. J. Fettig, M. E. Hansen, J. L. Hayes, J. A. Hicke, R. G. Kelsey, J. F. Negron, and S. J. Seybold. 2010. "Climate Change and Bark Beetles of the Western United States and Canada: Direct and Indirect Effects." *BioScience* 60 (8): 602-613. doi: 10.1525/bio.2010.60.8.6.

Berg, E. E., J. D. Henry, C. L. Fastie, A. D. De Volder, and S. M. Matsuoka. 2006. "Spruce Beetle Outbreaks on the Kenai Peninsula, Alaska, and Kluane National Park and Reserve, Yukon Territory: Relationship to Summer Temperatures and Regional Differences in Disturbance Regimes." *Forest Ecology and Management* 227 (3): 219-232. doi: 10.1016/j.foreco.2006.02.038.

Birdsey, R. A., J. C. Jenkins, M. Johnston, E. Huber-Sannwald, B. Amero, B. d. Jong, J. D. E. Barra, N. French, F. Garcia-Oliva, M. Harmon, L. S. Heath, V. J. Jaramillo, K. Johnsen, B. E. Law, E. Marín-Spiotta, O. Masera, R. Neilson, Y. Pan, and K. S. Pregitzer. 2007. "North American Forests." In *The First State of the Carbon Cycle Report (SOCCR): The North American Carbon Budget and Implications for the Global Carbon Cycle*, edited by A. W. King, L. Dilling, G. P. Zimmerman, D. M. Fairman, R. A. Houghton, G. Marland, A. Z. Rose, and T. J. Wilbanks, 117-126. National Oceanic and Atmospheric Administration, National Climatic Data Center, Asheville, NC, USA.

Black, B. A., D. C. Shaw, and J. K. Stone. 2010. "Impacts of Swiss Needle Cast on Overstory Douglas-fir Forests of the Western Oregon Coast Range." *Forest Ecology and Management* 259 (8): 1673-1680. doi: 10.1016/j.foreco.2010.01.047.

Bockino, N. K., and D. B. Tinker. 2012. "Interactions of White Pine Blister Rust and Mountain Pine Beetle in Whitebark Pine Ecosystems in the Southern Greater Yellowstone Area." *Natural Areas Journal* 32 (1): 31-40. doi: 10.3375/043.032.0105.

Bodtker, K. M., M. G. Pellatt, and A. J. Cannon. 2009. "A Bioclimatic Model to Assess the Impact of Climate Change on Ecosystems at Risk and Inform Land Management Decisions." Report for the Climate Change Impacts and Adaptation Directorate, CCAF Project A718. Western and Northern Service Centre, Vancouver.

Boisvenue, C., and S. W. Running. 2010. "Simulations Show Decreasing Carbon Stocks and Potential for Carbon Emissions in Rocky Mountain Forests over the Next Century." *Ecological Applications* 20 (5): 1302-1319. doi: 10.1890/09-0504.1.

Boyd, J. 2006. "The Nonmarket Benefits of Nature: What Should Be Counted in Green GDP?" *Ecological Economics* 61 (4): 716–723. doi: 10.1016/j.ecolecon.2006.06.016.

Boyd, J. 2010. "Ecosystem Services and Climate Adaptation." Issue Brief 10-16, Resources for the Future. http://www.rff.org/RFF/Documents/RFF-IB-10-16.pdf.

Boyd, J., and S. Banzhaf. 2006. "What are Ecosystem Services? The Need for Standardized Environmental Accounting Units." Discussion Paper RFF-DP-06-02. Washington, DC: Resources for the Future. 26 pp.

Boyd, J., and S. Banzhaf. 2007. "What are Ecosystem Services? The Need for Standardized Environmental Accounting Units." *Ecological Economics* 63: 616–626. doi: 10.1016/j.ecolecon.2007.01.002.

Bradley, B. A., M. Oppenheimer, and D. S. Wilcove. 2009. "Climate Change and Plant Invasions: Restoration Opportunities Ahead?" *Global Change Biology* 15 (6): 1511-1521. doi: 10.1111/j.1365-2486.2008.01824.x.

Brooks, M. L., C. M. D'Antonio, D. M. Richardson, J. B. Grace, J. E. Keeley, J. M. DiTomaso, R. J. Hobbs, M. Pellant, and D. Pyke. 2004. "Effects of Invasive Alien Plants on Fire Regimes." *BioScience* 54 (7): 677-688. doi: 10.1641/0006-3568(2004)054[0677:EOIAPO]2.0.CO;2.

Brown, T. C., J. C. Bergstrom, and J.B. Loomis. 2007. "Defining, Valuing, and Providing Ecosystem Goods and Services." *Natural Resources Journal.* 47 (2): 329-376. http://lawlibrary.unm.edu/nrj/47/2/04_brown_goods.pdf.

Burakowski, E. A., and M. Magnusson. 2012. "Climate Impacts on the Winter Tourism Economy in the United States." Natural Resources Defense Council and Protect Our Winters. Washington, DC. http://www.nrdc.org/globalwarming/files/climate-impacts-winter-tourism-report.pdf.

Capalbo, S., J. Julian, T. Maness, and E. Kelly. 2010. "Toward Assessing the Economic Impacts of Climate Change on Oregon." In *Oregon Climate Assessment Report*, edited by K. D. Dello and P. W. Mote, 359–391. Oregon Climate Change Research Institute, Oregon State University, Corvallis, OR. http://occri.net/ocar.

Carey, M. P., B. L. Sanderson, K. A. Barnas, and J. D. Olden. 2012. "Native Invaders – Challenges for Science, Management, Policy, and Society." *Frontiers in Ecology and Environment* 10 (7): 373-381. doi: dx.doi.org/10.1890/110060.

Carroll, A. L., S. W. Taylor, J. Régnière, and L. Safranyik. 2004. "Effects of Climate Change on Range Expansion by the Mountain Pine Beetle in British Columbia." In *Mountain Pine Beetle Symposium: Challenges and Solutions*, edited by T. L. Shore, J. E. Brooks, and J. E. Stone, 223-232. Natural Resources Canada. Canadian Forest Service. Pacific Forestry Centre. Information Report BC-X-399. Victoria, BC. 298 pp.

Case, M. J., and D. L. Peterson. 2005. "Fine-Scale Variability in Growth-Climate Relationships of Douglas-fir, North Cascade Range, Washington." *Canadian Journal of Forest Research* 35 (11): 2743-2755. doi: 10.1139/x05-191.

Cayan, D. R., S. A. Kammerdiener, M. D. Dettinger, J. M. Caprio, and D. H. Peterson. 2001. "Changes in the Onset of Spring in the Western United States." *Bulletin of the American Meteorological Society* 82: 399-415. doi: 10.1175/1520-0477(2001)082<0399:CITOOS>2.3.CO;2.

Chambers, J. C., and M. Pellant. 2008. "Climate Change Impacts on Northwestern and Inter-mountain United States Rangelands." *Society for Range Management* 30 (3): 29-33. doi: 10.2111/1551-501X(2008)30[29:CCIONA]2.0.CO.

Chambers, J. C., and M. J. Wisdom. 2009. "Priority Research and Management Issues for the Imperiled Great Basin of the Western United States." *Restoration Ecology* 17 (5): 707-714. doi: 10.1111/j.1526-100X.2009.00588.x.

Chen, P.-Y., C. Welsh, and A. Hamann. 2010. "Geographic Variation in Growth Response of Douglas-fir to Interannual Climate Variability and Projected Climate Change." *Global Change Biology* 16 (12): 3374-3385. doi: 10.1111/j.1365-2486.2010.02166.x.

Chmura, D. J., P. D. Anderson, G. T. Howe, C. A. Harrington, J. E. Halofsky, D. L. Peterson, D. C. Shaw, and B. J. St. Clair. 2011. "Forest Responses to Climate Change in the North-western United States: Ecophysiological Foundations for Adaptive Management." *Forest Ecology and Management* 261 (7): 1121-1142. doi: 10.1016/j.foreco.2010.12.040.

Churkina, G., and S. W. Running. 1998. "Contrasting Climatic Controls on the Estimated Produc-tivity of Global Terrestrial Biomes." *Ecosystems* 1 (2): 206-215. doi: 10.1007/s100219900016.

Churkina, G., S. W. Running, A. L. Schloss and The Participants of the Potsdam NPP Model Intercomparison. 1999. "Comparing Global Models of Terrestrial Net Primary Productivity (NPP): The Importance of Water Availability." *Global Change Biology* 5 (Suppl. 1): 46-55. doi: 10.1046/j.1365-2486.1999.00009.x.

Collins, B. M., P. N. Omi, and P. L. Chapman. 2006. "Regional Relationships Between Climate and Wildfire-Burned Area in the Interior West, USA." *Canadian Journal of Forest Research* 36 (3): 699-709. doi: 10.1139/x05-264.

Coops, N. C., and R. H. Waring. 2011. "Estimating the Vulnerability of Fifteen Tree Species under Changing Climate in Northwest North America." *Ecological Modelling* 222 (13): 2119-2129. doi: 10.1016/j.ecolmodel.2011.03.033.

Copeland, J. P., K. S. McKelvey, K. B. Aubry, A. Landa, J. Persson, R. M. Inman, J. Krebs, E. Lofroth, H. Golden, J. R. Squires, A. Magoun, M. K. Schwartz, J. Wilmot, C. L. Copeland, R. E. Yates, I. Kojola, and R. May. 2010. "The Bioclimatic Envelope of the Wolverine (*Gulo gulo*): Do Climatic Constraints Limit its Geographic Distribution?" *Canadian Journal of Zoology* 88 (3): 233-246. doi: 10.1139/Z09-136.

Costanza, R., R. d'Arge, R. de Groot, S. Farber, M. Grasso, B. Hannon, K. Limburg, S. Naeem, R. O'Meill, J. Paruelo, R. G. Raskin, P. Sutton, and M. van den Belt. 1997. "The Value of the World's Ecosystem Services and Natural Capital." *Nature* 387: 253-260. http://www.nature.com/nature/journal/v387/n6630/pdf/387253a0.pdf.

Daily, G. C., S. Alexander, P. R. Ehrlich, L. Goulder, J. Lubchenco, P. A. Matson, H. A. Mooney, S. Postel, S. H. Schneider, D. Tilman, G. M. Woodwell. 1997. "Ecosystem Services: Benefits Supplied to Human Societies by Natural Ecosystems." *Issues in Ecology* 2: 1-16. http://www.esa.org/science_resources/issues/FileEnglish/issue2.pdf.

Daily, G. C., and P. A. Matson. 2008. "Ecosystem Services: From Theory to Implementa-tion." *Proceedings of the National Academy of Sciences* 105 (28): 9455–9456. doi: 10.1073/pnas.0804960105.

Davis, M. B., and R. G. Shaw. 2001. "Range Shifts and Adaptive Responses to Quaternary Climate Change." *Science* 292 (5517): 673-679. doi: 10.1126/science.292.5517.673.

Dennehy, C., E. R. Alverson, H. E. Anderson, D. R. Clements, R. Gilbert, and T. N. Kaye. 2011. "Management Strategies for Invasive Plants in Pacific Northwest Prairies, Savannas, and Oak Woodlands." *Northwest Science* 85 (2): 329-351. doi: 10.3955/046.085.0219.

DeRose, R. J., and J. N. Long. 2012. "Drought-Driven Disturbance History Characterizes a Southern Rocky Mountain Subalpine Forest." *Canadian Journal of Forest Research* 42 (9): 1649–1660. doi: 10.1139/x2012-102.

Dillon, G. K., Z. A. Holden, P. Morgan, M. A. Crimmins, E. K. Heyerdahl, and C. H. Luce. 2011. "Both Topography and Climate Affected Forest and Woodland Burn Severity in Two Regions of the Western US, 1984 to 2006." *Ecosphere* 2 (12): 130. doi: 10.1890/ES11-00271.1

Dunham, J. B., A. E. Rosenberger, C. H. Luce, and B. E. Rieman. 2007. "Influences of Wildfire and Channel Reorganization on Spatial and Temporal Variation in Stream Temperature and the Distribution of Fish and Amphibians." *Ecosystems* 10 (2): 335-346. doi: 10.1007/s10021-007-9029-8.

Dwire, K. A., and J. B. Kaufman. 2003. "Fire and Riparian Ecosystems in Landscapes of the Western USA." *Forest Ecology and Management* 178: 61-74. doi: 10.1016/S0378-1127(03)00053-7.

Eidenshink, J., B. Schwind, K. Brewer, Z. Zhu, B. Quayle, and S. Howard. 2007. "A Project for Monitoring Trends in Burn Severity." *Fire Ecology* 3 (1): 3-21. doi: 10.4996/fireecology.0301003.

Elsner, M. M., L. Cuo, N. Voisin, J. S. Deems, A. F. Hamlet, J. A. Vano, K. E. B. Mickelson, S. Y. Lee, and D. P. Lettenmaier. 2010. "Implications of 21st Century Climate Change for the Hydrology of Washington State." *Climatic Change* 102: 225-260. doi: 10.1007/s10584-010-9855-0.

Evangelista, P. H., S. Kumar, T. J. Stohlgren, and N. E. Young. 2011. "Assessing Forest Vulnerability and the Potential Distribution of Pine Beetles under Current and Future Climate Scenarios in the Interior West of the US." *Forest Ecology and Management* 262 (3): 307-316. doi: 10.1016/j.foreco.2011.03.036.

Finch, D.M., ed. 2012. "Climate Change in Grasslands, Shrublands, and Deserts of the Interior American West: A Review and Needs Assessment." General Technical Report RMRS-GTR-285. Fort Collins, CO: US Department of Agriculture, Forest Service, Rocky Mountain Research Station. 139 pp.

French, N. H. F., W. J. de Groot, L. K. Jenkins, B. M. Rogers, E. Alvarado, B. Amiro, B. de Jong, S. Goetz, E. Hoy, E. Hyer, R. Keane, B. E. Law, D. McKenzie, S. G. McNulty, R. Ottmar, D. R. Pérez-Salicrup, J. Randerson, K. M. Robertson, and M. Turetsky. 2011. "Model Comparisons for Estimating Carbon Emissions from North American Wildland Fire." *Journal of Geophysical Research: Biogeosciences* 116: G00K05. doi: 10.1029/2010JG001469.

Halofsky, J. E., D. L. Peterson, K. A. O'Halloran, and C. H. Hoffman (eds.) 2011. "Adapting to Climate Change at Olympic National Forest and Olympic National Park." General Technical Report PNW-GTR-844. Portland, OR: US Department of Agriculture, Forest Service, Pacific Northwest Research Station, 130pp. http://www.fs.fed.us/pnw/pubs/pnw_gtr844.pdf.

Hansen, M. E., B. J. Bentz, and D. L. Turner. 2001. "Temperature-Based Model for Predicting Univoltine Brood Proportions in Spruce Beetle (Coleoptera: Scolytidae)." *Canadian Entomologist* 133 (6): 827-841. doi: 10.4039/Ent133827-6.

Hansen, J., M. Sato, and R. Ruedy. 2012. "Perception of Climate Change." *Proceedings of the National Academy of Sciences* 109 (37): E2415-E2423. doi: 10.1073/pnas.1205276109.

Hebertson, E. G., and M. J. Jenkins. 2003. "Historic Climate Factors Associated with Major Avalanche Years on the Wasatch Plateau, Utah." *Cold Regions Science and Technology* 37 (3): 315-332. doi: 10.1016/S0165-232X(03)00073-9.

Henderson, J. A., D. H. Peter, R. D. Lesher, and D. C. Shaw. 1989. "Forested Plant Associations of the Olympic National Forest." US Department of Agriculture, Forest Service, Pacific Northwest Region. R6-ECOL-TP 001-88. 502 pp.

Hessl, A. E., D. McKenzie, and R. Schellhaas. 2004. "Drought and Pacific Decadal Oscillation Linked to Fire Occurrence in the Inland Pacific Northwest." *Ecological Applications* 14 (2): 425-442. doi: 10.1890/03-5019.

Heyerdahl, E. K., L. B. Brubaker, and J. K. Agee. 2002. "Annual and Decadal Climate Forcing of Historical Regimes in the Interior Pacific Northwest, USA." *The Holocene* 12 (5): 597-604. doi: 10.1191/0959683602hl570rp.

Heyerdahl, E. K., D. McKenzie, and L. D. Daniels, A. E. Hessl, J. S. Littell, and N. J. Mantua. 2008. "Climate Drivers of Regionally Synchronous Fires in the Inland Northwest (1651–1900)." *International Journal of Wildland Fire* 17 (1): 40-49. doi: 10.1071/WF07024.

Heyerdahl, E. K., P. Morgan, and J. P. Riser. 2008. "Multi-Season Climate Synchronized Historical Fires in Dry Forests (1650-1900), Northern Rockies, U.S.A." *Ecology* 89 (3): 705-16. doi: 10.1890/06-2047.1.

Hicke, J. A., and J. C. Jenkins. 2008. "Mapping Lodgepole Pine Stand Structure Susceptibility to Mountain Pine Beetle Attack across the Western United States." *Forest Ecology and Management* 255: 1536-1547. doi: 10.1016/j.foreco.2007.11.027.

Hicke, J. A., M. C. Johnson, J. L. Hayes, and H. K. Preisler. 2012. "Effects of Bark Beetle-Caused Tree Mortality on Wildfire." *Forest Ecology and Management* 271: 81-90. doi: 10.1016/j.foreco.2012.02.005.

Hicke J. A., J. A. Logan, J. Powell, and D. S. Ojima. 2006. "Changing Temperatures Influence Suitability for Modeled Mountain Pine Beetle (Dendroctonus ponderosae) Outbreaks in the Western United States." *Journal of Geophysical Research* 111: G02019. doi: 10.1029/2005JG000101.

Hixon, M. A., S. V. Gregory, W. D. Robinson. 2010. "Oregon's Fish and Wildlife in a Changing Climate." In *Oregon Climate Assessment Report*, edited by K. D. Dello and P. W. Mote, 268-360. Oregon Climate Change Research Institute, Oregon State University, Corvallis, Oregon. http://occri.net/ocar.

Holman, M. L., and D. L. Peterson. 2006. "Spatial and Temporal Variability in Forest Growth in the Olympic Mountains, Washington: Sensitivity to Climatic Variability." *Canadian Journal of Forest Research* 36 (1): 92-104. doi: 10.1139/x05-225.

Hu, J., D. J. P. Moore, S. P. Burns, and R. K. Monson. 2010. "Longer Growing Seasons Lead to Less Carbon Sequestration by a Subalpine Forest." *Global Change Biology* 16 (2): 771-783. doi: 10.1111/j.1365-2486.2009.01967.x.

Hudiburg, T. W., B. E. Law, C. Wirth, and S. Luyssaert. 2011. "Regional Carbon Dioxide Implications of Forest Bioenergy Production." *Nature Climate Change* 1: 419-423. doi: 10.1038/nclimate1264.

Idaho Forest Products Commission. 2012. "Economics." Accessed May 8. http://www.idaho forests.org/money1.htm.

Isaak, D. J., C. H. Luce, B. E. Rieman, D. E. Nagel, E. E. Peterson, D. L. Horan, S. Parkes, G. L. Chandler. 2010. "Effects of Climate Change and Wildfire on Stream Temperature and Salmonids Thermal Habitat in a Mountain River Network." *Ecological Applications* 20 (5): 1350-1371. doi: 10.1890/09-0822.1.

Janetos, A., L. Hansen, D. Inouye, B. P. Kelly, L. Meyerson, B. Peterson, and R. Shaw. 2008. "Biodiversity." In *The Effects of Climate Change on Agriculture, Land Resources, Water Resources, and Biodiversity in the United States*, edited by M. Walsh, 151-181. US Climate Change Science Program Synthesis and Assessment Product 4.3, US Department of Agriculture, Washington, DC.

Jenkins, M. J., E. Hebertson, W. Page, and C. A. Jorgensen. 2008. "Bark Beetles, Fuels, Fires, and Implications for Forest Management in the Intermountain West." *Forest Ecology and Management* 254 (1): 16-34. doi: 10.1016/j.foreco.2007.09.045.

Kitzberger, T., P. M. Brown, E. K. Heyerdahl, T. W. Swetnam, and T. T. Veblen. 2007. "Contingent Pacific-Atlantic Ocean Influence on Multi-Century Wildfire Synchrony over Western North America." Proceedings of the National Academy of Sciences 104 (2): 543-548. doi: 10.1073/pnas.0606078104.

Kliejunas, J. T. 2011. "A Risk Assessment of Climate Change and the Impact of Forest Diseases on Forest Ecosystems in the Western United States and Canada." General Technical Report PSW-GTR-236. Albany, CA: US Department of Agriculture, Forest Service, Pacific Southwest Research Station. 70 pp.

Kliejunas, J. T., B. W. Geils, J. M. Glaeser, E. M. Goheen, P. Hennon, M.-S. Kim, H. Kope, J. Stone, R. Sturrock, and S. J. Frankel. 2009. "Review of Literature on Climate Change and Forest Diseases of Western North America." General Technical Report PSW-GTR-234. Albany, CA: US Department of Agriculture, Forest Service, Pacific Southwest Research Station.

Klopfenstein, N. B., M. S. Kim, J. W. Hanna, B. A. Richardson, A. Bryce, and J. E. Lundquist. 2009. "Approaches to Predicting Potential Impacts of Climate Change on Forest Disease: An Example with Armillaria Root Disease." Res. Pap. RMRS-RP-76. Fort Collins, CO: US Department of Agriculture, Forest Service, Rocky Mountain Research Station. 10 pp.

Knutson, K. C., and D. A. Pyke. 2008. "Western Juniper and Ponderosa Pine Ecotonal Climate-Growth Relationships across Landscape Gradients in Southern Oregon." *Canadian Journal of Forest Research* 38 (12): 3021-3032. doi: 10.1139/X08-142.

Krieger, D. J. 2001. "Economic Value of Forest Ecosystems: A Review." Analysis prepared for the Wilderness Society. http://www.cfr.washington.edu/classes.esrm.465/2007/readings/WS_valuation.pdf

Kurz, W. A., C. C. Dymond, G. Stinson, G. J. Rampley, E. T. Neilson, A. L. Carroll, T. Ebata, and L. Safranyik. 2008. "Mountain Pine Beetle and Forest Carbon Feedback to Climate Change." *Nature* 452 (7190): 987-990. doi: 10.1038/nature06777.

Lenihan, J., D. Bachelet, R. Neilson, and R. Drapek. 2008. "Simulated Response of Conterminous United States Ecosystems to Climate Change at Different Levels of Fire Suppression, CO_2 Emission Rate, and Growth Response to CO_2." *Global and Planetary Change* 64: 16-25. doi: 10.1016/j.gloplacha.2008.01.006.

Littell, J. S. 2010. "Impacts in the Next Few Decades and Coming Centuries." In *Climate Stabilization Targets: Emissions, Concentrations, and Impacts over Decades to Millennia*, Board of Atmospheric Sciences and Climate, National Research Council, 159-216. http://www.nap.edu/catalog.php?record_id=12877.

Littell, J. S., M. M. Elsner, G. S. Mauger, E. Lutz, A. F. Hamlet, and E. Salathé. 2011. "Regional Climate and Hydrologic Change in the Northern US Rockies and Pacific Northwest: Internally Consistent Projections of Future Climate for Resource Management." Project Report for USFS JVA 09-JV-11015600-039. Prepared by the Climate Impacts Group, University of Washington, Seattle. April, 2011.

Littell, J. S., and R. Gwozdz. 2011. "Climatic Water Balance and Regional Fire Years in the Pacific Northwest, USA: Linking Regional Climate and Fire at Landscape Scales." In *The Landscape Ecology of Fire, Ecological Studies*, edited by D. McKenzie, C. M. Miller, and D. A. Falk, 213. doi 10.1007/978-94-007-0301-8_5, ©Springer Science+Business Media B.V. 2011.

Littell, J. S., D. McKenzie, B. K. Kerns, S. Cushman, and C. G. Shaw. 2011. "Managing Uncertainty in Climate-Driven Ecological Models to Inform Adaptation to Climate Change." *Ecosphere* 2 (9): 102. doi: 10.1890/ES11-00114.1.

Littell, J. S., D. McKenzie, D. L. Peterson, and A. L. Westerling. 2009. "Climate and Wildfire Area Burned in Western U.S. Ecoprovinces, 1916-2003." *Ecological Applications* 19 (4): 1003-21. doi: 10.1890/07-1183.1.

Littell, J. S., E. E. Oneil, D. McKenzie, J. A. Hicke, J. Lutz, R. A. Norheim, and M. M. Elsner. 2009. "Forest Ecosystems, Disturbance, and Climatic Change in Washington State, USA." In *The Washington Climate Change Impacts Assessment*, edited by M. McGuire Elsner, J. Littell, and L Whitely Binder, 255-284. Center for Science in the Earth System, Joint Institute for the Study of the Atmosphere and Oceans, University of Washington, Seattle, Washington. http://www.cses.washington.edu/db/pdf/wacciareport681.pdf.

Littell, J. S., E. E. Oneil, D. McKenzie, J. A. Hicke, J. Lutz, R. A. Norheim, and M. M. Elsner. 2010. "Forest Ecosystems, Disturbance, and Climatic Change in Washington State, USA." *Climatic Change* 102: 129-158. doi: 10.1007/s10584-010-9858-x.

Littell, J. S., D. L. Peterson, C. I. Millar, and K. A. O'Halloran. 2012. "U.S. National Forests adapt to climate change through science-management partnerships." *Climatic Change* 110: 269-296. doi: 10.1007/s10584-011-0066-0.

Littell, J. S., D. L. Peterson, and M. Tjoelker. 2008. "Water Limits Tree Growth from Stand to Region: Douglas-fir Growth-Climate Relationships in Northwestern Ecosystems." *Ecological Monographs* 78 (3): 349-368. doi: 10.1890/07-0712.1.

Logan, J. A., W. W. Macfarlane, and L. Willcox. 2010. "Whitebark Pine Vulnerability to Climate-Driven Mountain Pine Beetle Disturbance in the Greater Yellowstone Ecosystem." *Ecological Applications* 20 (4): 895-902. doi: 10.1890/09-0655.1.

Logan J. A., and J. A. Powell. 2001. "Ghost Forests, Global Warming, and the Mountain Pine Beetle (Coleoptera: Scolytidae)." *American Entomologist* 47 (3): 160-173. http://www.entsoc.org/PDF/Pubs/Periodicals/AE/AE-2001/fall/feature-logan.pdf.

Logan, J. A. and J. A. Powell. 2009. "Ecological Cconsequences of Cclimate Cchange Aaltered Fforest Insect Ddisturbance Rregimes." In *Climate Change in Western North America: Evidence and Environmental Effects*, edited by F. H. Wagner, 98-109. University of Utah Press.

Logan, J. A., J. Régnière, and J. A. Powell. 2003. "Assessing the Impacts of Global Warming on Forest Pest Dynamics." *Frontiers in Ecology and the Environment* 1 (3): 130-137. doi: 10.1890/1540-9295(2003)001[0130:ATIOGW]2.0.CO;2.

Loomis, J., and J. Crespi. 1999. "Estimated Effects of Climate Change on Selected Outdoor Recreation Activities in the United States." In *The Impact of Climate Change on the United States Economy*, edited by R. Mendelsohn and J. E. Neumann, 289-314. Cambridge, UK: Cambridge University Press.

Loomis, J., A. Gonzalez-Caban, and J. Englin. 2001. "Testing for Differential Effects of Forest Fires on Hiking and Mountain Biking Demand and Benefits." *Journal of Agricultural and Resource Economics* 26 (2): 508-522. http://purl.umn.edu/31049.

Loomis, J. B. and Richardson, R. B. 2006. "An External Validity Test of Intended Behavior: Comparing Revealed Preference and Intended Visitation in Response to Climate Change." *Journal of Environmental Planning and Management* 49 (4): 621-630. doi: 10.1080/09640560600747562.

Lynch, H. J., R. A. Renkin, R. L. Crabtree, and P. R. Moorcroft. 2006. "The Influence of Previous Mountain Pine Beetle (Dendroctonus ponderosae) Activity on the 1988 Yellowstone Fires." *Ecosystems* 9 (8): 1318-1327. doi: 10.1007/s10021-006-0173-3.

Malanson, G. P., D. R. Butler, D. B. Fagre, S. J. Walsh, D. F. Tomback, L. D. Daniels, L. M. Resler, W. K. Smith, D. J. Weiss, D. L. Peterson, A. G. Bunn, C. A. Hiemstra, D. Lipzin, P. S. Bourgeron, Z. Shen, and C. I. Millar. 2007. "Alpine Treeline of Western North America:Linking Organism-To-Landscape Dynamics." *Physical Geography* 28 (5): 378-396. doi: 10.2747/0272-3646.28.5.378.

McKelvey, K. S., J. P. Copeland, M. K. Schwartz, J. S. Littell, K. B. Aubry, J. R. Squires, S. A. Parks, M. M. Elsner, and G. S. Mauger. 2011. "Climate Change Predicted to Shift Wolverine Distributions, Connectivity, and Dispersal Corridors." *Ecological Applications* 21 (8): 2882-2897. doi: 10.1890/10-2206.1.

McKenney, D. W., J. H. Pedlar, and K. Lawrence. 2007. "Potential Impacts of Climate Change on the Distribution of North American Trees." *BioScience* 57: 939. doi: 10.1641/B571106.

McKenney, D. W., J. H. Pedlar, R. B. Rood, and D. Price. 2011. "Revisiting Projected Shifts in the Climate Envelopes of North American Trees Using Updated General Circulation Models." *Global Change Biology* 17 (8): 2720-2730. doi: 10.1111/j.1365-2486.2011.02413.x.

McKenzie, D., Z. Gedalof, D. L. Peterson, and P. Mote. 2004. "Climatic Change, Wildfire, and Conservation." *Conservation Biology* 18 (4): 890-902. doi: 10.1111/j.1523-1739.2004.00492.x.

McKenzie, D., D. L. Peterson, and J. S. Littell. 2009. "Global Warming and Stress Complexes in Forests of Western North America." In *Developments in Environmental Science 8:, Wildland Fires and Air Pollution*, edited by A. Bytnerowicz, M. Arbaugh, A. Riebau, and C. Anderson, 319-337. Elsevier Science, Ltd., Amsterdam, The Netherlands.

McKenzie, D., D. W. Peterson, D. L. Peterson, and P. E. T. Thornton. 2003. "Climatic and Biophysical Controls on Conifer Species Distributions in Mountain Forests of Washington State, USA." *Journal of Biogeography* 30 (7): 1093-1108. doi: 10.1046/j.1365-2699.2003.00921.x.

McMahon, S. M., S. P. Harrison, W. S. Armbruster, P. J. Bartlein, C. M. Beale, M. E. Edwards, J. Kattge, G. Midgley, X. Morin, and I. C. Prentice. 2011. "Improving Assessment and Modeling of Climate Change Impacts on Global Terrestrial Biodiversity." *Trends in Ecology and Evolution* 26 (5): 249-259. doi: 10.1016/j.tree.2011.02.012.

McKinley, D. C., M. G. Ryan, R. A. Birdsey, C. P. Giardina, M. E. Harmon, L. S. Heath, R. A. Houghton, R. B. Jackson, J. E. Morrison, B. C. Murray, D. E. Pataki, and K. E. Skog. 2011. "A Synthesis of Current Knowledge on Forests and Carbon Storage in the United States." *Ecological Applications*. 21(6): 1902-1924. doi: 10.1890/10-0697.1.

Meddens, A. J. H., J. A. Hicke, and C. A. Ferguson. 2012. "Spatiotemporal Patterns of Observed Bark Beetle-Caused Tree Mortality in British Columbia and the Western US." *Ecological Applications* 22 (7): 1876-1891. doi: 10.1890/11-1785.1.

Mendelsohn, R., and M. Markowski. 1999. "The Impact of Climate Change on Outdoor Recreation." *The Impact of Climate Change on the United States Economy*, edited by R. Mendelsohn and J.E. Neumann, 267–288. Cambridge, UK: Cambridge University Press.

Meyer, S. E. 2012. "Restoring and Managing Cold Desert Shrublands for Climate Change Mitigation." In *Climate Change in Grasslands, Shrublands, and Deserts of the Interior American West: A Review and Needs Assessment*, edited by D. M. Finch. General Technical Report RMRS-GTR-285. Fort Collins, CO: US Department of Agriculture, Forest Service, Rocky Mountain Research Station. 139 pp.

Millar, C. I., R. Neilson, D. Bachelet, R. Drapek, and J. Lenihan. 2006. "Climate Change at Multiple Scales." In *Forests, Carbon, and Climate Change: A Synthesis of Science Findings*, edited by H. Salwasser and M. Cloughesy. Oregon Forest Resources Institute Publication, 182 pp.

Millar, C. I., N. L. Stephenson, and S. L. Stephens. 2007. "Climate Change and Forests of the Future: Managing in the Face of Uncertainty." *Ecological Applications* 17 (8): 2145-2151. doi: 10.1890/06-1715.1.

Millar, C. I., R. D. Westfall, D. L. Delany, J. C. King, and L. J. Graumlich. 2004. "Response of Subalpine Conifers in the Sierra Nevada, California, U.S.A., to 20th-Century Warming and

Decadal Climate Variability." *Arctic, Antarctic, and Alpine Research* 36 (2): 181-200. doi: 10.1657/1523-0430(2004)036[0181:ROSCIT]2.0.CO;2.

Miller, R. F., and J. A. Rose. 1999. "Fire History and Western Juniper Encroachment in Sagebrush Steppe." *Journal of Range Management* 52 (6): 550-559. http://oregonstate.edu/dept/eoarc/sites/default/files/publication/430.pdf.

Mote, P. W., and E. P. Salathé. 2010. "Future Climate in the Pacific Northwest." *Climatic Change* 102: 29-50. doi: 10.1007/s10584-010-9848-z.

Nakićenović, N., O. Davidson, G. Davis, A. Grübler, T. Kram, E. Lebre La Rovere, B. Metz, T. Morita, W. Pepper, H. Pitcher, A. Sankovski, P. Shukla, R. Swart, R. Watson, and Z. Dadi. 2000. *Special Report on Emissions Scenarios: A Special Report of Working Group III of the Intergovernmental Panel on Climate Change,* Cambridge University Press, Cambridge, UK, 599 pp. http://www.grida.no/climate/ipcc/emission/index.htm.

National Center for Earth Resources Observations and Science, US Geological Survey. 2002. North American Land Cover Characteristics – 1-Kilometer Resolution: National Atlas of the United States, http://nationalatlas.gov/mld/landcvi.html.

National Research Council. 2005. *Valuing Ecosystem Services: Toward Better Environmental Decision-making* National Academies Press, Washington DC.

Neilson, R. P., J. M. Lenihan, D. Bachelet, and R. J. Drapek. 2005. "Climate Change Implications for Sagebrush Ecosystems." *Transactions of the North American Wildlife and Natural Resources Conference* 70: 145-159.

Nemani, R. R., C. D. Keeling, H. Hashimoto, M. Jolly, S. W. Running, S. C. Piper, C. J. Tucker, and R. Myneni. 2003. "Climate Driven Increases in Terrestrial Net Primary Production from 1982 to 1999." *Science* 300 (5625): 1560-1563. doi: 10.1126/science.1082750.

O'Halloran, T. L., B. E. Law, M. L. Goulden, Z. Wang, J. G. Barr, C. Schaaf, M. Brown, J. D. Fuentes, M. Göckede, A. Black, and V. Engel. 2012. "Radiative Forcing of Natural Forest Disturbances." *Global Change Biology* 18 (2): 555-565. doi: 10.1111/j.1365-2486.2011.02577.x.

Oneil, E. E. 2006. "Developing Stand Density Thresholds to Address Mountain Pine Beetle Susceptibility in Eastern Washington Forests." PhD diss., University of Washington.

Oregon Department of Forestry. 2009. "Forest Facts, Oregon's Forests: Some Facts and Figures." Last modified September. http://www.oregon.gov/odf/pubs/docs/forest_facts/ffforestryfacts figures.pdf.

Oregon Forest Resources Institute. 2012. "The 2012 Forest Report: An Economic Assessment of Oregon's Forest and Wood Products Manufacturing Sector." http://oregonforests.org/sites/default/files/publications/pdf/OFRI_Forest_Report_2012_0.pdf.

Parmesan, C. 2006. "Ecological and Evolutionary Responses to Recent Climate Change." *Annual Review of Ecological and Evolutionary Systems* 37: 637-639. doi: 10.1146/annurev.ecolsys.37.091305.110100.

Pellatt, M. G., S. J. Goring, K. M. Bodtker, and A. J. Cannon. 2012. "Using a Down-Scaled Bioclimate Envelope Model to Determine Long-Term Temporal Connectivity of Garry Oak (Quercus garryana) Habitat in Western North America: Implications for Protected Area Planning." *Environmental Management* 49 (4): 802-815. doi: 10.1007/s00267-012-9815-8.

Peterson, D. L., J. E. Halofsky, and M. C. Johnson. 2011. "Managing and Adapting to Changing Fire Regimes in a Warmer Climate." In *The Landscape Ecology of Fire*, edited by D. McKenzie, C. Miller, and D. A. Falk, 249-268. New York, NY: Springer Verlag.

Peterson, D. L., C. I. Millar, L. A. Joyce, M. J. Furniss, J. E. Halofsky, R. P, Neilson, and T. L. Morelli. 2011. "Responding to Climate Change in National Forests: A Guidebook for Developing Adaptation Options." General Technical Report PNW-GTR-855. Portland, OR: US Department of Agriculture, Forest Service, Pacific Northwest Research Station. 109 pp.

Peterson, D. W., and D. L. Peterson. 2001. "Mountain Hemlock Growth Responds to Climatic Variability at Annual and Decadal Time Scales." *Ecology* 82 (12): 3330. doi: 10.1890/0012-9658(2001)082[3330:MHGRTC]2.0.CO;2.

Polasky S., E. Nelson, E. Lonsdorf , P. Fackler, and A. Starfield. 2005. "Conserving Species in a Working Landscape: Land Use with Biological and Economic Objectives." *Ecological Applications* 15 (4): 1387–1401. doi: 10.1890/03-5423.

Prichard, S. J., D. L. Peterson, and K. Jacobson. 2010. "Fuel Treatments Reduce the Severity of Wildfire Effects in Dry Mixed Conifer Forest, Washington, USA." *Canadian Journal of Forest Research* 40: 1615-1626. doi: 10.1139/X10-109.

Raffa, K. F., B. H. Aukema, B.J. Bentz, A. L. Carroll, J. A. Hicke, M. G. Turner, and W. H. Romme. 2008. "Cross-Scale Drivers of Natural Disturbances Prone to Anthropogenic Amplification: The Dynamics of Bark Beetle Eruptions." *BioScience* 58 (6): 501-517. doi: 10.1641/B580607.

Raymond, C. L., and D. McKenzie. 2012. "Carbon Dynamics of Forests in Washington, USA: 21st Century Projections Based on Climate-Driven Changes in Fire Regimes." *Ecological Applications* 22 (5): 1589-1611. doi: 10.1890/11-1851.1.

Régnière, J., and B. Bentz. 2007. "Modeling Cold Tolerance in the Mountain Pine Beetle, *Dendroctonus ponderosae*." *Journal of Insect Physiology* 53 (6): 559-572, doi: 10.1016/j.jinsphys.2007.02.007.

Rehfeldt, G. E. 2006. "A Spline Climate Model for the Western United States." General Technical Report 165. Fort Collins, CO: US Department of Agriculture, Forest Service, Rocky Mountain Research Station.

Rehfeldt, G. E., N. L. Crookston, M. V. Warwell, and J. S. Evans. 2006. "Empirical Analyses of Plant-Climate Relationships for the Western United States." *International Journal of Plant Sciences* 167 (6): 1123-1150. doi: 10.1086/507711.

Rehfeldt, G. E., D. E. Ferguson, and N. L. Crookston. 2008. "Quantifying the Abundance of Co-Occurring Conifers along Inland Northwest (USA) Climate Gradients." *Ecology* 89 (8): 2127-39. doi: 10.1890/06-2013.1.

Roché, C. T., and B. F. Roché, Jr. 1988. "Distribution and Amount of Four Knapweed (*Centaurea* L.) Species in Eastern Washington." *Northwest Science* 62 (5): 242-253. https://research.wsulibs.wsu.edu:8443/xmlui/bitstream/handle/2376/1708/v62%20p242%20Roche%20and%20Roche%20Jr..PDF?sequence=1.

Roché, C. T., and D. C. Thill. 2001. "Biology of Common Crupina and Yellow Starthistle, Two Mediterranean Winter Annual Invaders in Western North America." *Weed Science* 49 (4): 439-447. doi: 10.1614/0043-1745(2001)049[0439:BOCCAY]2.0.CO;2.

Rogers, B. M., R. P. Neilson, R. Drapek, J. M. Lenihan, J. R. Wells, D. Bachelet, and B. E. Law. 2011. "Impacts of Climate Change on Fire Regimes and Carbon Stocks of the U.S. Pacific Northwest." *Journal of Geophysical Research* 116 (G03037): 1-13. doi: 10.1029/2011JG001695.

Root, T. L., J. T. Price, K. R. Hall, S. H. Schneider, C. Rosenzweig, and J. A. Pounds. 2003. "Fingerprints of Global Warming on Wild Animals and Plants." *Nature* 421 (6918): 57-60. doi: 10.1038/nature01333.

Ryan, M. G., S. R. Archer, R. Birdsey, C. Dahm, L. Heath, J. Hicke, D. Hollinger, T. Huxman, G. Okin, R. Oren, J. Randerson, and W. Schlesinger. 2008. "Land Resources." In *The Effects of Climate Change on Agriculture, Land Resources, Water Resources, and Biodiversity in the United States*, edited by M. Walsh, 75-120. US Climate Change Science Program Synthesis and Assessment Product 4.3, US Department of Agriculture, Washington, DC.

Safranyik, L., A. L. Carroll, J. Régnière , D. W. Langor, W. G. Riel, T. L. Shore, B. Peter, B. J. Cooke, V. G. Nealis, and S. W. Taylor. 2010. "Potential for Range Expansion of Mountain

Pine Beetle into the Boreal Forest of North America." *Canadian Entomologist* 142 (5): 415-442. doi: 10.4039/n08-CPA01.

Schuster, E. G., D. A. Cleaves, and E. F. Bell. 1997. "Analysis of USDA Forest Service Fire-Related Expenditures 1970-1995." Res. Paper PSW-RP-230. Albany, CA: US Department of Agriculture, Forest Service, Pacific Southwest Research Station, 29 pp.

Shafer, S. L. M. E. Harmon, R. P. Neilson, R. Seidl, B. St. Clair, and A. Yost. 2010. "The Potential Effects of Climate Change on Oregon's Vegetation." In *Oregon Climate Assessment Report*, edited by K. D. Dello and P. W. Mote, 175–210. Oregon Climate Change Research Institute, Oregon State University, Corvallis, OR. http://occri.net/ocar.

Sherriff, R. L., E. E. Berg, and A. E. Miller. 2011. "Climate Variability and Spruce Beetle (Dendroctonus rufipennis) Outbreaks in South-Central and Southwest Alaska." *Ecology* 92 (7): 1459-1470. doi: 10.1890/10-1118.1.

Six, D. L., and J. Adams. 2007. "White Pine Blister Rust Severity and Selection of Individual Whitebark Pine by the Mountain Pine Beetle (Coleoptera: Curculionidae, Scolytinae)." *Journal of Entomological Science* 42 (3): 345-353.

Smith, W. B., tech. coord.; Miles, P. D., data coord.; Perry, C. H., map coord.; Pugh, S. A., Data CD coord. 2009. "Forest Resources of the United States, 2007." General Technical Report WO-78. Washington, DC: US Department of Agriculture, Forest Service, Washington Office. 336 pp. (Table 1 starting on page 151). http://www.fs.fed.us/nrs/pubs/gtr/gtr_wo78.pdf.

Smithwick, E. A., M. G. Ryan, D. M. Kashian, W. H. Romme, D. B. Tinker, and M. G. Turner. 2009. "Modeling the Effects of Fire and Climate Change on Carbon and Nitrogen Storage in Lodgepole Pine (*Pinus contorta*) Stands." *Global Change Biology* 15 (3): 535-548. doi: 10.1111/j.1365-2486.2008.01659.x.

Smucker, K. M., R. L. Hutto, and B. M. Steele. 2005. "Changes in Bird Abundance after Wildfire: Importance of Fire Severity and Time since Fire." *Ecological Applications* 15 (5): 1535-1549. doi: 10.1890/04-1353.

Sohngen, B., and R. Mendelsohn. 2001. "Timber: Ecological-Economic Analysis." In *Global Warming and the American Economy: A Regional Assessment of Climate Change Impacts*, edited by R. Mendelsohn, 52pp. Northhampton, MA: Edward Elgar. doi: 10.4337/1840645938 .00012.

Sohngen, B., and R. Sedjo. 2005. "Impacts of Climate Change on Forest Product Markets: Implications for North American Producers." *The Forestry Chronicle* 81 (5): 669-674. http://pubs. cif-ifc.org/doi/pdf/10.5558/tfc81669-5.

Spies, T. A., T. W. Giesen, F. J. Swanson, J. F. Franklin, D. Lach, and K. N. Johnson. 2010. "Climate Change Adaptation Strategies for Federal Forests of the Pacific Northwest, USA: Ecological, Policy, and Socio-Economic Perspectives." *Landscape Ecology* 25 (8): 1185-1199. doi: 10.1007/ s10980-010-9483-0.

Spittlehouse, D. L., and R. B. Stewart. 2003. "Adaptation to Climate Change in Forest Management." *BC Journal of Ecosystems and Management (Perspectives)* 4 (1): 1-11. http://www.forrex .org/sites/default/files/publications/jem_archive/ISS21/vol4_no1_art1.pdf.

Starbuck, C. M., R. P. Berrens, and M. McKee. 2006. "Simulating Changes in Forest Recreation Demand and Associated Economic Impacts Due to Fire and Fuels Management Activities." *Forest Policy and Economics* 8 (1): 52-66. doi: 10.1016/j.forpol.2004.05.004.

Stephens, S. L. 2005. "Forest Fire Causes and Extent on United States Forest Service Lands." *International Journal of Wildland Fire* 14 (3): 213-222. doi: 10.1071/WF04006.

Stephenson, N. L. 1990. "Climatic Control of Vegetation Distribution: The Role of the Water Balance." *American Naturalist* 135 (5): 649-670. http://www.jstor.org/stable/2462028.

Stone, J. K., L. B. Coop, and D. K. Manter. 2008. "Predicting the Effects of Climate Change on Swiss Needle Cast Disease Severity in Pacific Northwest Forests." *Canadian Journal of Plant Pathology* 30 (2): 169-176. doi: 10.1080/07060661.2008.10540533.

Sturrock, R. N., S. J. Frankel, A. V. Brown, P. E. Hennon, and J. T. Kliejunas. 2011. "Climate Change and Forest Diseases." *Plant Pathology* 60 (1): 133-149. doi: 10.1111/j.1365-3059.2010.02406.x.

Suddick, E. C. and E. A. Davidson. 2012. "The Role of Nitrogen in Climate Change and the Impacts of Nitrogen-Climate Interactions on Terrestrial and Aquatic Ecosystems, Agriculture, and Human Health in the United States." A Technical Report Submitted to the US National Climate Assessment. Falmouth, MA: North American Nitrogen Center of the International Nitrogen Initiative.

Swetnam, T. W., and J. L. Betancourt. 1998. "Mesoscale Disturbance and Ecological Response to Decadal Climatic Variability in the American Southwest." *Journal of Climate* 11 (12): 3128-3147. doi: 10.1175/1520-0442(1998)011<3128:MDAERT>2.0.CO;2.

Thomson, A. J., R. F. Shepherd, J. W. E. Harris, and R. H. Silversides. 1984. "Relating Weather to Outbreaks of Western Spruce Budworm, *Choristoneura occidentalis* (Lepidoptera: Tortricidae), in British Columbia." *Canadian Entomologist* 116 (3): 375-381. doi: 10.4039/Ent116375-3.

US Department of Agriculture Forest Service. 2010. "Major Forest Insect and Disease Conditions in the United States: 2009 Update." FS-952, Washington, DC.

van Mantgem, P. J., N. L. Stephenson, J. C. Byrne, L. D. Daniels, J. F. Franklin, P. Z. Fulé, M. E. Harmon, A. J. Larson, J. M. Smith, A. H. Taylor, and T. T. Veblen. 2009. "Widespread Increase of Tree Mortality Rates in the Western United States." *Science* 323 (5913): 521-524. doi: 10.1126/science.1165000.

Venette, R. C., and S. D. Cohen. 2006. "Potential Climatic Suitability for Establishment of *Phytophthora ramorum* within the Contiguous United States." *Forest Ecology and Management* 231: 18-26. doi: 10.1016/j.foreco.2006.04.036.

Wainger, L. A., D. M. King, J. Salzman, and J. Boyd. 2001. "Wetland Value Indicators for Scoring Mitigation Trades." *Stanford Environmental Law Journal* 20 (2): 413-478.

Waring, R. H., and J. F. Franklin. 1979. "Evergreen Coniferous Forests of the Pacific Northwest." Science 204 (4400): 1380-1386. doi: 10.1126/science.204.4400.1380.

Warren, D. D. 2011. "Harvest, Employment, Exports, and Prices in Pacific Northwest Forests, 1965-2010." General Technical Report PNW-GTR-857. Portland, OR: US Department of Agriculture, Forest Service, Pacific Northwest Research Station. 17 pp.

Washington State Department of Natural Resources. 2007. "The Future of Washington Forests." Washington State Department of Natural Resources. http://www.dnr.wa.gov/Research-Science/Topics/ForestResearch/Pages/futureofwashingtonsforest.aspx.

Werner, R. A., E. H. Holsten, S. M. Matsuoka, and R. E. Burnside. 2006. "Spruce Beetles and Forest Ecosystems in South-Central Alaska: A Review of 30 Years of Research." *Forest Ecology and Management* 227: 195-206. doi: 10.1016/j.foreco.2006.02.050.

Westerling, A. L., T. J. Brown, A. Gershunov, D. R. Cayan, and M. D. Dettinger. 2003. "Climate and Wildfire in the Western United States." *Bulletin of the American Meteorological Society* 84 (5): 595-604. doi: 10.1175/BAMS-84-5-595.

Westerling, A. L., H. G. Hidalgo, D. R. Cayan, and T. W. Swetnam. 2006. "Warming and Earlier Spring Increase Western US Forest Wildfire Activity." *Science* 313 (5789): 940-943. doi: 10.1126/science.1128834.

Western Wood Products Association. 1995. "Ponderosa Pine." http://www2.wwpa.org/SPECIES PRODUCTS/PonderosaPine/tabid/298/Default.aspx.

Williams, D. W., and A. M. Liebhold. 2002. "Climate Change and the Outbreak Ranges of Two North American Bark Beetles." *Agricultural and Forest Entomology* 4 (2): 87-99. doi: 10.1046/j.1461-9563.2002.00124.x.

Wygant, N. D. 1940. "Effects of Low Temperature on the Black Hills Beetle (*Dendroctonus ponderosae* Hopkins)." PhD diss., State University of New York, Syracuse, NY.

Xian, G., C. G. Homer, and C. L. Aldridge. 2012. "Effects of Land Cover and Regional Climate Variations on Long-Term Spatiotemporal Changes in Sagebrush Ecosystems." *GIScience & Remote Sensing* 49 (3): 378-396. doi: 10.2747/1548-1603.49.3.378.

Chapter 6

Agriculture
Impacts, Adaptation, and Mitigation

AUTHORS
Sanford D. Eigenbrode, Susan M. Capalbo, Laurie L. Houston,
Jodi Johnson-Maynard, Chad Kruger, Beau Olen

6.1 Introduction

Agriculture is critical to the environment, economy, and cultural identity of the Northwest (NW) region. Approximately 24% of the land area of Washington, Oregon, and Idaho is devoted to agricultural crops or rangeland and pastureland (US Department of Agriculture [USDA] Census of Agriculture 2010). Agricultural commodities not only contribute directly to the GDP of the Northwest, but also support food system economies of the region and provide the economic and cultural foundation for rural populations. The principal crops are wheat, potatoes, tree fruit, sugarbeets, legumes, and forage crops, but approximately 300 minor crops are also grown (USDA National Agricultural Statistics Service [NASS] 2012b). The region has significant rangeland and confined animal operations for beef and dairy (USDA Census of Agriculture 2010). These agricultural industries will be affected by projected warming and changes in the amount and seasonal distribution of precipitation in the Northwest (see Chapter 2).

Projected effects of climate changes on NW agriculture depend upon the specific agricultural sector, geographic location, global climate models (GCMs), and emission scenarios used. In parts of the region, warmer, drier summers will potentially cause yield reductions due to heat and drought stress or increase demands for irrigation water (Stöckle et al. 2010). Warmer conditions in cool seasons could affect production of fruits and wine grapes that require chilling for fruit set and quality (Jones 2005). Heat and drought stress can negatively affect forage production or directly affect the health of livestock. On the other hand, warmer, wetter winters could be advantageous for some cropping systems, reducing cold stress and providing opportunities for diversification. Furthermore, expected increasing atmospheric CO_2 concentrations are beneficial to some plants and could offset climate-related yield losses at least to mid-21st century in several major crops grown in the region (Stöckle et al. 2010).

Available studies of the effects of climate change on NW agriculture are limited to a subset of agricultural commodities, and are also limited in geographic extent, the specific impacts assessed (e.g., average yield), and the climate scenarios considered. Most results depend upon assumptions regarding availability of irrigation water, and the studies rely upon simplified scenarios of production technology, market, and policy, and do not consider climate effects on pressure from pests, weeds, and diseases. Nonetheless, a review of this literature will help delineate the extent of current knowledge and identify needs for additional study to understand the challenges and opportunities that climate change presents for NW agriculture.

6.2 Environmental, Economic, and Social Importance

Agriculture in the Northwest comprises several major subsectors (fig. 6.1), which together contribute substantively to the region's economy, cultural identity, and social well-being. The influence of agriculture on the environments of the Northwest is substantial because of the large areas devoted to it and the intensity of management of these lands. The value of agricultural commodities for Washington, Oregon, and Idaho totals $17 billion (USDA Census of Agriculture 2010), or 3% of the region's GDP. Farming

Figure 6.1 Northwest agricultural systems are diverse, ranging from extensive rain-fed livestock grazing to intensive horticultural production systems dependent on irrigation. Major production systems include (clockwise, from top left): extensive livestock grazing (courtesy of Jodi Johnson-Maynard, Oct. 11 2012); center top, irrigated field and row crop systems (processed vegetable, forages, grains, etc.) (courtesy of Soil Science, May 18, 2009); tree fruit (courtesy of Peggy Greb, USDA-ARS); confined animal feeding operations (dairy and feedlot) (courtesy of Dana Pride, August 1, 2011); irrigated row crops (potato) (courtesy of Brad King, USDA-ARS, Kimberly ID); rain-fed cereal grain systems (courtesy of Dag Endresen, July 28, 2009); vineyards (courtesy of Dana Pride, August 1, 2011); more livestock grazing (courtesy of Dana Pride, August 1, 2011). All Creative Commons Attribution.

income alone comprises 1.5–6.3% of the total private earnings, and combined food and agriculture industries are major employers and contributors to the states' economies. For example, Washington State's $46 billion food and agriculture industry employs approximately 160,000 people and contributes 13% to the state's economy (Washington State Department of Agriculture 2011). In Oregon, agriculture provides over 234,000 jobs (Oregon State Board of Agriculture 2011) and is connected to more than 15% of all economic activity in the state (Sorte et al. 2011). In Idaho, agriculture generated 11% of total sales and employed 6% of the state's workforce, generating 6% of Idaho's GDP (Watson et al. 2008). Agriculture is very diverse in the region and includes approximately 300 commodities. Among these, the most important economically are cattle and dairy, wheat, potato, hay (e.g., alfalfa), greenhouse and nursery, tree fruit, and vineyard. These major industries are significant on a national scale. Idaho is ranked #1 for milk and dairy products, and Washington and Oregon are ranked #3 and 4, respectively for both vegetable and fruit production (USDA Census of Agriculture 2010). Washington, Idaho, and Oregon are ranked #4, 5, and 8 in wheat production and together produce 17% of the nation's wheat (USDA NASS 2012b). Idaho and Washington are ranked #1 and 2 in potato production and Idaho ranks #3 in sugarbeet production (USDA NASS 2012b). The region's more than 900,000 dairy cows account for 11% of US milk production and Oregon is ranked #1 nationally for cut Christmas trees (USDA Census of Agriculture 2010).

Agriculture provides the cultural fabric and sense of identity for many rural communities of the Northwest. Ranching and farming have been ways of life for generations and many agricultural families continue to farm land homesteaded by their ancestors. Many communities throughout the region are built around agricultural service industries. Agriculture remains predominant in shaping new cultures of these regions, even though the face of agriculture has changed in parts of the Northwest through farm consolidation, the advent and expansion of irrigation-dependent sectors such as wine, tree fruit, and diversified crops, and the immigration of farmworkers (Mackun 2009).

The average proportion of the total land area devoted to crops and pastureland for the three states of the Northwest is 24% (USDA Census of Agriculture 2010) (fig. 6.2). While the land area devoted to agriculture nationally has declined by 12% since 1982, that decline has been only 3.5% in the Northwest in the same period (USDA 2009). Although the reasons for this are complex, the trend suggests a sustained importance of agriculture for the economy and environment of the region.

The extensive agricultural lands and rangelands in the Northwest are key components of the complex ecology of the region. As is true throughout the world, agriculturally managed lands and native or less disturbed habitat are intermingled on landscapes and must be considered as complex interacting systems (Daily 1997; Farina 2000; Diekötter et al. 2008). In the Northwest, grasslands now managed for grazing still contain significant native flora and fauna (Tisdale 1986; Mancuso and Moseley 1994; Lichthardt and Mosely 1997). Elsewhere, such as in the Palouse of eastern Washington and northern Idaho (Looney and Eigenbrode 2012), networks of remnant patches of native ecosystems are embedded in landscapes mostly converted to production agriculture. In these settings worldwide, native habitats are responsible for sustaining ecosystem services like biological control and pollination (Losey and Vaughan 2006; Turner and Daily 2008). Although studies documenting the impacts are lacking, ecosystem services from native habitats are likely important in NW agricultural systems.

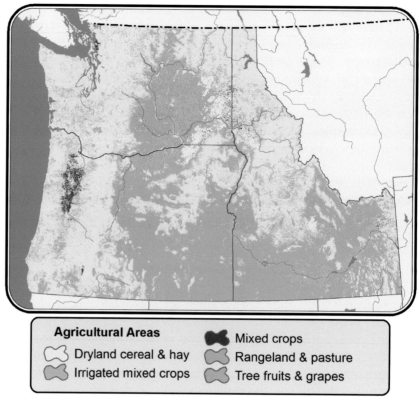

Agricultural Areas
- Dryland cereal & hay
- Irrigated mixed crops
- Mixed crops
- Rangeland & pasture
- Tree fruits & grapes

Figure 6.2 Agricultural areas of the Northwest region. Much of this heterogeneity reflects climatic constraints, primarily temperature and precipitation regimes, and availability of water for irrigation. Production systems are diverse and heterogeneous within each zone. For example, irrigation allows production of potatoes, vegetables, fruits, corn, seed crops, and other commodities. Within any area local conditions or producer preferences produce heterogeneity of practice not shown here. For example, "mixed crops" in the Willamette Valley and elsewhere can include irrigation and rainfed systems in close proximity. (Figure prepared by Rick Rupp, Washington State University. Data sources: USDA NASS CropScape (http://nassgeodata.gmu.edu/CropScape/), USGS, US Geological Survey, ESRI - Environmental Systems Research Institute, TANA – TeleAtlas, North America, AND – Automotive Navigation Data (Rotterdam, Netherlands)

Ultimately, considerations of the impacts of climate change on NW agriculture must include potential effects on the large-scale processes that link agriculture, human communities, and natural ecosystems of the region.

6.3 Vulnerabilities to Projected Climate Change

The potential impacts of climate change on agriculture worldwide are a serious concern because they threaten the capacity of humanity to meet the food and fiber needs of a continuously growing population (Smith et al. 2007). The severity of projected impacts tends to be greater in the subtropics and tropics than in temperate zones within which the NW region is located (Rosenzweig and Parry 1994; Parry et al. 2004; Parry et al. 2005; Schlenker and Roberts 2009). Tubiello et al. (2002), in an evaluation of climate change impacts on US crop production, concluded that climate change generally favors northern areas and can worsen conditions in southern areas. Nonetheless, NW agriculture is potentially vulnerable to projected climate change because of its dependence on reliable annual and seasonal precipitation or irrigation supplies from annual surface water sources, adequate temperatures and growing seasons, and the sensitivity of crops to temperature extremes, all of which are projected to change, albeit with different levels of uncertainty (see Chapter 2), during the coming century. According to the IPCC (Easterling et al. 2007) warming that exceeds 4.5–5 °C (8–9 °F) in higher latitude regions worldwide will tend to overwhelm autonomous adaptation causing declining yields. Northwest temperatures are projected to approach this level of warming by late 21st century under the Special Report on Emissions Scenarios (SRES)-A2 emissions scenario of continued growth (Kunkel et al. 2013; Nakićenović et al. 2000). Models are also in relatively good agreement that heat extremes (days with maximum temperature greater than 32 °C [90 °F]) and precipitation extreme events (days with more than ~2.5–10.2 cm [1–4 in] of precipitation) will increase in the NW region (see Chapter 2), which can adversely affect agriculture. Changes in the timing of spring planting and late spring freezes potentially expose crops to greater risks of frost injury. For example, freezing temperatures in early June reduced yield in spring wheat in the Northwest in 2012 and 2002. Despite these recent examples, observations show a pronounced shift toward earlier dates of last freeze, and models project a continued decrease in the frequency of late freeze events in the Northwest (J. Abatzoglou, pers. comm.).

The principal climate change drivers leading to impacts on NW agriculture differ among subsectors, but some general patterns are noteworthy (table 6.1). Projected warming trends will bring increases in the probability of heat-related stress and water shortages to field crops and tree fruit, but will also be associated with longer growing seasons and, perhaps, shifts in precipitation that can benefit some crops (Littell et al. 2009; Stöckle et al. 2010). Thus, net effects will be complex. Furthermore, increasing atmospheric CO_2 concentrations are expected to be beneficial for most NW commodities due to CO_2 fertilization at least until mid-21st century, offsetting climate-related reductions in productivity (Tubiello et al. 2007; Stöckle et al. 2010; Hatfield et al. 2011). In addition, increases in CO_2 increase water use efficiency, which could mitigate the effects of drought (Hatfield et al. 2011). In this chapter, whenever CO_2 fertilization effects have been included in published projections, we present those results since GCMs projecting warming are premised on increasing atmospheric CO_2 concentrations. Although CO_2 fertilization effects are typically included in climate change projections (e.g., Thomson et al. 2005; Hatfield et al. 2008), there is some controversy about whether CO_2 fertilization effects might be transient. We include them because experimental results on transience are equivocal (Long et al. 2004) and, at least for annual crops, transience seems less likely to be important for agricultural systems.

Climate-related changes in pressure from plant diseases, pests, and weeds are also difficult to project. Generally, warmer temperatures are coupled with greater pressure from insect pests, stemming from changes in geographic ranges, dates of spring arrival, and shorter generation times (Parmesan 2006; Trumble and Butler 2009). Some pests of

Table 6.1 Climate change drivers and their implications for Northwest agriculture

Climate Driver	Possible effects on NW agriculture
Increase in mean summer temperature	Heat stress-related reductions in yields and yield stability of major NW crops and livestock; changes in pressure from pests, diseases, and invasive species
Increase in mean cool-season temperature	Greater productivity or survival of winter crops and cold-sensitive perennials; changes in pressure from pests, diseases, and invasive species
Increase in length of growing season	More flexibility in crops that can be grown and cropping system design; changes in pressure from pests, diseases, and invasive species
Increase in growing degree days	Faster maturation of some crops; changes in pressure from pests, diseases, and invasive species
Increase in mean evapotranspiration	Greater risk of drought stress
Decrease in summer soil moisture	Greater risk of drought stress of rain-fed crops and those dependent on surface water irrigation
Decrease in mean summer precipitation	Greater risk of drought stress of rain-fed crops and those dependent on surface water irrigation
Increase in mean winter precipitation	Greater available soil moisture for establishing spring crops; wetter soils in spring potentially impede spring planting operations in some systems.
Increased atmospheric CO_2 (not a climate variable per se, but models projecting warming depend upon increased greenhouse gases, including CO_2)	Potentially increases productivity of annual and perennial crops

NW crops are potentially limited by climatic conditions and could become more serious or more difficult to manage with warming and changes in precipitation. Increases in the generations or intrinsic growth rates of aphids potentially increase pressure from these pests, whose development is dependent upon accumulated degree days (Clement et al. 2010). Many pests are kept in check by natural enemies and warming could alter these natural controls, either offsetting or exacerbating potential temperature-related increases in pest pressure (Thomson et al. 2010). Whereas insects are generally responsive to temperature, pathogens respond more to humidity and rainfall (Coakley et al. 1999). The severity of diseases caused by fungal pathogens of cereals can change with climate depending upon the requirements of the fungi for soil moisture levels and temperature (T. Paulitz, pers. comm.). Soil-borne fungi can only actively grow and infect plants when soil moisture is adequate and temperatures are optimum. Under extremely dry, cold, or hot conditions, fungi cease growth and form resistant structures to survive until conditions are suitable. Thus, changes in climate may have a profound effect on the distribution of fungal diseases. Increased temperatures could increase pressure from weed species that are drought tolerant and respond well to increased temperatures, notably Downy Brome (*Bromus tectorum*) (Ball 2004; I. Burke, pers. comm.). All these responses are species-specific and cannot be projected reliably in general terms. Rather, vulnerabilities to biotic stresses must be assessed for specific cropping systems and their respective pest, disease, and weed complexes.

6.4 Potential Impacts of Climate Change on Selected Subsectors

The most comprehensive assessment published to date of the effects of climate change on NW agriculture currently available was conducted as part of the Washington Climate Change Impacts Assessment (Littell et al. 2009; Stöckle et al. 2010). This chapter relies on those publications, but supplements with additional information, including prior and subsequent publications. Each of the agricultural sectors across the region has distinct vulnerabilities and capacities for adaptation. The market values of major agricultural sectors in the Northwest are shown in figure 6.3, along with potential effects of climate change on these sectors. Where published studies are available we treat these sectors separately, but many sectors have not been studied specifically. In all cases, the impacts will depend upon the degree of climate change, local conditions, policies, markets, and other factors.

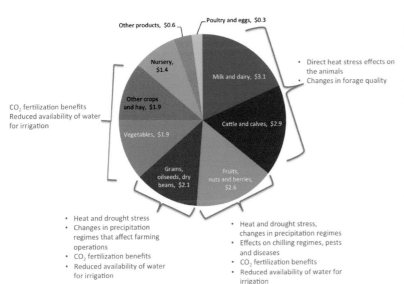

Figure 6.3 Northwest agricultural commodities with market values shown in $ (billion) in 2007. Potential effects of climate change on these sectors, if any have been projected, are shown. Detailed discussion of these potential effects is provided in the text of this chapter. Total value of commodities is $16.8 billion (source: http://www.agcensus.usda.gov/Publications/2007/Full_Report/Census_by_State/)

6.4.1 ANNUAL CROPS

6.4.1.1 Dryland Cereal Cropping Systems

The semiarid portions of central Washington and the Columbia Plateau in Washington, Oregon, and Idaho support cereal-based cropping systems without irrigation. The region can be subdivided into agroclimatic zones (Douglas et al. 1992) ranging from a warm, dry zone (located in the dryland cereal and hay production areas, fig. 6.2) where winter wheat-fallow production predominates, to cooler, wet zones (located in the non-irrigated mixed crops areas, fig. 6.2) where continuous cropping incorporates cool season legumes in rotation with spring and winter cereals. Depending upon emission scenarios and projected dates, these dryland regions are vulnerable to projected reductions in summer precipitation and warming, which potentially reduce yields or exacerbate production challenges on marginal lands, as is currently the case in the western portions of the dryland cereal areas of central Washington. Projected increases in mean temperature and warm weather episodes and decreases in summer soil moisture levels would likely reduce yields of wheat and other cereals in all zones. Wheat is vulnerable to heat stress, which can accelerate wheat senescence (the period between maturity and death of a plant or plant part) and reduce leaf and ear photosynthesis, which impedes grain-filling (Ferris et al. 1998) or causes grain shriveling, negatively impacting grain quality (Ortiz et al. 2008). Warmer, drier conditions could exacerbate soil erosion by wind and reduce early stand establishment of winter wheat on summer fallow. Based on crop models, potential yield losses from projected climate change alone would be severe by end of 21st century (Stöckle et al. 2010), but these effects are offset by CO_2 fertilization. Tubiello et al. (2002) projected US West Coast (California, Oregon, and Washington) non-irrigated winter wheat production to increase 10–30% by 2030, relative to baseline climate (1951–1994). In an analysis of the Yakima Basin in Washington, Thomson et al. (2005), using an agricultural production model (EPIC), projected non-irrigated winter wheat yield increases of 19–23% with warming of 1 °C (1.8 °F) over the baseline (1961–1990) and CO_2 concentration of 560 ppm, depending upon which of three separate GCMs (BMRC, UIUC, UIUC + Sulfates) was employed. Using a different crop model (CropSyst), four GCMs (PCM1, CCSM3, ECHAM5, CGCM3), and the SRES-A1B scenario, Stöckle et al. (2010) projected dryland winter wheat yield increases of 13–15% by the 2020s, 13–25% by the 2040s, and 23–35% by the 2080s for a range of locations across Washington State, relative to baseline climate (1975–2005) when warming and CO_2 fertilization were included. In the same study, dryland spring wheat yields for a range of locations across Washington State were projected to change by +7% to +8% by the 2020s, -7% to +2% by the 2040s, and -11% to +0% by the 2080s. The range of values obtained depend upon the production zone and planting date, with lower increases or deficits in lower rainfall zones and better performance occurring if planting is adjusted earlier in the season, avoiding higher temperatures during vulnerable stages. The models were based on changes in mean temperatures and did not consider the frequency of extreme heat events, which could negatively affect yields but for which projections are less certain.

Cool-season (October–March) precipitation is generally projected to increase in the region, but models are not in close agreement on this projection (see Chapter 2).

Depending on the timing of that precipitation, planting of spring wheat could be hampered, reducing yields or causing a shift toward more winter cropping. On the other hand, greater soil moisture entering the growing season could mitigate the effects of projected reduced summer precipitation. Winter wheat is produced throughout the region and could benefit from warmer winters, but may be challenged if drier summers impede late summer and fall planting of these crops, reducing germination and stand establishment.

Several species of cereal aphids are periodically pests of wheat in the region. Flights of these aphids are influenced by weather and have been occurring earlier (Halbert et al. 1985; Davis et al., in preparation). Some of these species are vectors of *Barley yellow dwarf virus* and the match between phenology of their flights and vulnerable stages of spring or winter wheat could shift with climate, affecting prevalence of the disease caused by this virus (Halbert et al. 1985). Projected temperature and precipitation patterns for the Northwest are more favorable for the Cereal leaf beetle (CLB) (*Oulema melanopus*), a recent invasive pest of the Northwest, than are current climates; on the other hand, the phenological overlap between this pest and a successful biological control agent, the parasitoid wasp (*Terastichus julis*) (Roberts and Rao 2012), is largely unaffected or increases under projected climates (Eigenbrode and Abatzoglou, in preparation). Evans et al. (2012) found that warmer springs were associated with lower rates of parasitism by *T. julis*, using a 10-year record of surveys in Utah, suggesting that longer term warming trends could hamper CLB control. Thus, the severity of CLB may be relatively unchanged or increase, depending upon the seasonal patterns of warming in the NW. Reproduction by the parasitic wasp, *Cotesia marginiventris*, which attacks many species of pest caterpillars including those affecting NW crops, was found to be reduced drastically by a 3 °C (5.4 °F) increase in summer temperatures, potentially disrupting biological control and allowing pest numbers to increase (Trumble and Butler 2009).

6.4.1.2 Irrigated Annual Cropping Systems

Much of the Columbia Basin proper, the greater Columbia-Snake River Valley, many river valleys along the eastern slopes of the Cascade Mountain Range, and all of the Snake River Plain in southern Idaho and the Klamath Basin in Oregon have low annual and summer precipitation. The agriculture in these areas is dependent upon irrigation, much of which comes from the rivers of the greater Columbia Basin (Snake, Columbia, and Yakima Rivers) and the Klamath Basin (fig. 6.2). River flows of the region are projected to be altered, with reduced summer flows when irrigation demands are highest (Mote 2006; Elsner et al. 2010; Chapter 3). These changes, coupled with projected warming trends that could increase demands, potentially exacerbate water shortages for irrigation in some locations. For a rise in temperature of 1 °C (1.8 °F), irrigation demands are projected to increase by at least 10% in arid and semi-arid regions (Fischer et al. 2002), such as those that prevail in much of the NW region. A study focusing on the Yakima Basin (Vano et al. 2010) projected water shortage years to increase from a historic baseline of 14% of years (1979–1999) to 27% by the 2020s and to 68% by the 2080s based on an emissions scenario with substantial reductions (SRES-B1) and historical records of water shortages. In the Snake River Basin, regional warming has been associated with significant reductions in annual water diversion trends (Hoekema and Sridhar 2011). Precipitation in that region has trended to increasing spring flows and diversions to agriculture, but soil moisture in spring remains relatively low. These trends suggest increasing water demands in the future (Hoekema and Sridhar 2011). The Washington State Department of Ecology (2011) indicates that future economic conditions may moderate the impact of future climate-driven effects on irrigation water supply, projecting only a 2.2% increase in irrigation demand in the Columbia River Basin under average flow conditions for business-as-usual economic and emissions scenarios by the 2030s. Crops affected throughout the Northwest include irrigated wheat, potatoes, sugarbeets, forages, corn, tree fruit, high value vegetable crops, and others. Because of their prominence in the Northwest, potatoes and tree fruit have been studied specifically for the potential effects of climate change on their production.

Potatoes are a principal crop grown under irrigation in central Washington and the Snake River valleys of Idaho. Potato yields or quality are susceptible to warming in three important ways. First, rising temperatures accelerate plant development and leaf senescence, effectively reducing growing season length (Timlin et al. 2006). Second, warming impedes translocation of carbohydrates from plant tissue to tubers, resulting in reduced tuber-bulking. Third, higher temperatures (in excess of 30 °C [86 °F]) for extended periods during tuberization (formation and expansion of tubers) can contribute to lower tuber quality (Alva et al. 2002). Attempts to project effects of climate change on potatoes have used different scenarios, baselines, and models, but there is general agreement that higher temperatures will reduce potato yields. Without the effects of CO_2 fertilization, crop models project substantial yield losses due to these negative effects of warming (Rosenzweig et al. 1996; Stöckle et al. 2010). If CO_2 fertilization is included in these projections, projections vary. Rosenzweig et al. (1996) projected yield increases for Yakima, Washington, and Boise, Idaho, ranging from 3–7% over 1951–1980 baselines, depending on CO_2 concentrations and temperature increases. Stöckle et al. (2010) concluded that yield losses by end of the 21st century would be only 2–3% for all scenarios when the effect of CO_2 fertilization is included. In contrast, Tubiello et al. (2002) found potato yields to decline by 10–15% in Pendleton and Medford, Oregon, and 30–40% in Boise, Idaho, by the year 2090 relative to baseline climate (1951–1994) and based on CO_2 fertilization from a 550 ppm projected atmospheric CO_2 concentration. Tubiello et al. (2002) used two separate GCMs (HCGS and CCGS) and a crop growth model (DSSAT). The large differences in projections are attributable to choice of GCM and scenario and location modeled.

Potato pests in Washington State, including aphids and the Colorado potato beetle (*Leptinotarsa decemlineata*) are partly kept in check by a suite of natural enemies (Crowder et al. 2010). Ongoing work at Washington State University (D. W. Crowder, pers. comm.) indicates that predator diversity and effectiveness declines with increasing seasonal degree days, with implications for managing these pests under future climates. The potato psyllid, *Bactericera cockerelli*, has become established in the region and is a threat as a vector of *Candidatus* Liberibacter *solanacearum* responsible for Zebra chip disease in potatoes. Some authors have suggested that invasion by the psyllid was facilitated by warming trends (Trumble and Butler 2009; Liu and Trumble 2007). It is unknown, however, if future warming trends would exacerbate problems caused by this pest.

6.4.2 PERENNIAL CROPS

6.4.2.1 Tree Fruit and Small Fruit

Tree fruit production occurs largely in Payette County, Idaho, the Willamette Valley of Oregon, and central Washington, and other small, localized areas like the Hood River Valley of Oregon. Central Washington is one of the most important tree fruit growing areas in the world (Washington State University Tree Fruit Research and Extension Center). In all of these regions, fruit production requires irrigation. These systems may be affected by heat stress and by changes in seasonal temperature regimes important for their phenology. Fruit and nut trees require chilling periods in order to ensure uniform flowering and fruit set. Every fruit and nut tree species and cultivar has unique winter chill requirements that are necessary for them to break seasonal dormancy in spring and to achieve uniform flowering (Saure 1985). Insufficient chilling can result in late or staggered bloom, decreased fruit set, and poor fruit quality, which will decrease the marketable yield of these commodities (Weinberger 1950). Projected warmer temperatures could disrupt chilling, potentially reducing fruit set for tree fruits that are currently productive in parts of the Northwest. On the other hand, these trends could also allow some species and varieties of tree fruit and nuts that are cold sensitive to be grown successfully in the region, leading to net increases in fruit production and profitability of the operations. Luedeling et al. (2011) concluded that as climate warming decreases winter chill accumulation, "the ecological niche of many fruits and nuts in the Western United States is likely to move north, from California's Central Valley towards Northern California, Oregon, and Washington." Winter chill accumulation is projected to decline across the West Coast as climate change progresses, but the Northwest may experience a less severe decline than tree fruit growing areas in California over the 21st century (Luedeling and Brown 2011; Luedeling et al. 2011).

In addition to the changes in chilling conditions, warming trends can affect productivity after fruit set and may require adaptations in crop load management (e.g., fruit thinning). For one important tree fruit currently produced in the region, namely irrigated apples, future climate change is projected to decrease production by 1%, 3%, and 4% for the 2020s, 2040s, and 2080s, respectively, under an emissions scenario with continued growth peaking at mid-century (SRES-A1B) and four GCMs (PCM1, CCSM3, ECHAM5, CGCM3) relative to 1975–2005. When the fertilizing effect of CO_2 is considered, yields are projected to increase by 6% (2020s), 9% (2040s), and 16% (2080s) (Stöckle et al. 2010). These results assume current crop load management practices remain the same. Similar effects could be anticipated for these and other fruits in Washington, Oregon, and Idaho. Some of the effects result from warmer spring temperatures eliciting early budding, making the trees more vulnerable to frost as has occurred more frequently in the eastern United States (Gu et al. 2008). In addition, fruit and nut trees are relatively water intensive crops (Stöckle et al. 2010) and across much of the Northwest they are dependent upon irrigation, which may pose production risks as water supplies become scarce in locations where there are competing demands for water (Washington State Department of Ecology 2011).

Tree fruits are susceptible to numerous pests and diseases, but there is little research examining the implications of climate change for managing them. The principal pest affecting apples in Washington State is the codling moth. Projected warming under the SRES-A1B scenario (continued growth peaking at mid-century) causes adult moths to appear 6, 9, and 14 days earlier and increases the fraction of third generation hatch by 36%, 55%, and 81%, for the 2020s, 2040s, and 2080s, with implications for management of this pest (Stöckle et al. 2010). A significant pathogen affecting tree fruit is powdery mildew. Cherry powdery mildew is projected to increase under the CCSM3 (2020 only) and the CGCM3 projected climates (Stöckle et al. 2010).

6.4.2.2 Wine Grapes and Wines

A majority of US premium grape production currently exists on the West Coast, primarily in California but also notably in western Oregon and in the Columbia River Basin of Oregon and Washington. Vineyards in most of the region are dependent upon irrigation, except for the Willamette and Umpqua Valleys of Oregon. Changes in important measures of climate are already observed in wine-growing areas of the Northwest, though most published studies have focused on Oregon. Oregon's wine regions have seen the length of the frost-free period increase by 17 to 35 days in the past century (Coakley et al. 2010). Climate warming could cause shifts in which varieties are produced in specific localities (Jones 2005, 2006a, 2006b, 2007), since each wine grape varietal has an optimal growing season temperature range for production (Jones 2006a; White et al. 2006). As temperature deviates from this range wine quality and style is affected, resulting in lower sales value or higher production costs to maintain wine attributes. Certain important varieties in the Northwest could become nonviable where currently grown (e.g., Pinot Noir and Pinot Gris in Oregon). The Climate Leadership Initiative (2009a) used the results of Jones (2006b) and the very high growth emission scenario (SRES-A1FI) to reach the conclusion that the 2 °C (3.6 °F) temperature thresholds for Pinot Noir and Pinot Gris would be exceeded by 2040, resulting in annual lost production value of $24 million. On the other hand, as the climate warms, new varieties of grape will become viable in the Northwest providing opportunities for the industry (Jones 2007; Diffenbaugh et al. 2011). Many wine varieties that are currently grown in California are anticipated to become viable in the cooler climates of Oregon and Washington as climate change takes hold (Diffenbaugh et al. 2011; Jones 2007). Nonetheless, costs associated with replacing the long-lived vines must be taken into consideration. The long productive lives (25–30 years) and maturation times (4–6 years) of grapevines makes planning for future production difficult in face of the uncertainty of climate change. There is evidence that producers are experimenting with warmer climate grape varieties in the northern Willamette Valley, such as Syrah. However, there is no record of producers doing so on a large scale, in part because these varieties are not able to ripen consistently. The 2011 Oregon Vineyard Report (NASS 2012) states that the north Willamette Valley produced just 37 tons of Syrah, which is less than 1% of total production for the north Willamette Valley. The more likely near term scenario will be for producers to move toward warmer end cool climate grapes such as Riesling rather than invest in varieties that are in a new heat bracket.

Economically, NW wine producers have experienced significant growth in terms of acres planted and wine revenue in recent years. In 2010, the value of production for wine grapes in Washington exceeded $166 million, up from $149 million in 2008 (USDA

NASS 2011a). There were 43,849 planted acres in 2011 versus 31,000 planted acres in 2006 (USDA NASS 2011b). In Washington, red varieties comprised more than half the acreage, including 10,294 acres of Cabernet Sauvignon and 7,654 acres of Chardonnay. In Oregon, the value of production for wine grapes exceeded $81 million in 2011, up from $42 million in 2005. There were 20,400 planted acres in 2011 versus 14,100 planted acres in 2005 (USDA NASS 2006, 2012a). Vineyard establishment is a long-term investment with the potential life of the vineyard in excess of 50 years and establishment costs approaching $20,000 per acre (Julian et al. 2008). Sixteen varietals predominate in the Northwest including Riesling, Chardonnay, Pinot Grigio, Cabernet Savignon, Merlot and Pinot Noir. Given the expected long life of vineyards and high costs to establish and maintain, the projected temperature increases associated with climate change could have significant impacts on NW wine industries. If temperatures increase as projected, vineyards at lower elevations may no longer have the appropriate microclimate for premium wine production. This would force growers to choose between producing lower quality grapes, on average, or starting over with a wine grape better suited for the vineyard's location. In addition to these direct effects on the vines, there may be effects on pests and pathogens.

6.4.3 ANIMAL PRODUCTION SYSTEMS

6.4.3.1 Rangeland

Grazing lands provide important ecosystem services for production of beef, dairy, and sheep, regulation of water supplies, genetic resources (plant materials used for restoration), wildlife habitat, and climate regulation through carbon sequestration (Follett and Reed 2010; Brown and MacLeod 2011). Based on the land area occupied by grazing lands, as well as the ecosystem services they provide, knowledge of the impacts of climate change on these systems is critical. Rangeland systems are known to be heterogeneous and sensitive to extreme events and variation in weather, suggesting that these systems may be extremely vulnerable to climate change (Polley et al. 2010; Abatzoglou and Kolden 2011; Brown and MacLeod 2011).

Major issues related to climate change impacts in rangeland systems include invasive species and changes in plant productivity and nutritional value. These impacts result from a combination of the effects of warming, increased CO_2 levels, and changes in precipitation amount and timing (table 6.1 and fig. 6.3). Increased temperatures, as projected for the NW region, generally lengthen the growing season and change plant phenology. Temperature effects, however, will be more beneficial in rangeland located in cooler climates within the Northwest as long as nutrient resources are not limiting. Rangeland systems located in warmer, drier climatic zones may experience a decrease in productivity if temperatures exceed the optimum range of the plant species in the area (Izaurralde et al. 2011). Wan et al. (2005) reported that experimental warming of 2 °C (3.6 °F) in a tallgrass prairie extended the growing season by an average of 19 days, increased green above-ground biomass in spring and autumn, but only significantly reduced soil moisture in one out of three years. Grasses grown under enriched CO_2 conditions generally show an increase in growth with the rate depending on photosynthetic pathway (C3 versus C4, defined in the following section) and species (Poorter 1993; Hatfield et al. 2011). Competitive advantages under enriched CO_2 conditions may lead to species replacement (Dukes et al. 2011) and potential undesirable changes in both productivity and forage quality. Morgan et al. (2004) reported enhanced productivity (41% greater above-ground biomass) in a CO_2 enrichment study with three shortgrass steppe species. The enhanced production, however, was mostly due to effects on one species (*Stipa comata*). The authors of this same study reported decreased digestibility of all three species when grown under enriched CO_2 conditions. Enrichment of CO_2 may also lead to greater water use efficiency (Izaurralde et al. 2011). The ecological impacts of increased soil water storage will likely depend on the type of rangeland system (Parton et al. 2001), with moisture-limited desert and shrub-grass systems benefitting more than rangelands occurring in wetter environments.

A warming climate could potentially exacerbate pressure from invasive species that can degrade rangeland and abundance and feeding by insect pests. For example, cheatgrass (*Bromus tectorum*) is adapted for rapid early growth to avoid drought, and yellow starthistle (*Centauria solsticialis*) possesses an effective taproot system, attributes that can help these species outcompete native grasses and other higher quality forage species. Above-ground biomass of *C. solstitialis* increased more than six-fold across warming, precipitation, nitrate, and burning treatments when CO_2 was elevated (+300 ppm) (Dukes et al. 2011). Cold-intolerant annual invasive grasses may also be favored by warmer temperatures, longer growing seasons, and shifts in annual precipitation patterns that favor the invasion-fire cycle in desert systems (Abatzoglou and Kolden 2011). Similar climate drivers could exacerbate plant invasions of western rangelands.

Grasshoppers, the best-studied insect pests of rangeland, can cause significant economic loss in outbreak years when their populations are large. Although beneficial through their contributions to nutrient cycling, they also can become pests by competing with livestock for biomass (Branson et al. 2006; Branson 2008). These insects are sensitive to temperature (Joern and Gaines 1990). Some northern species of grasshoppers have a multi-year lifecycle, spending two winter periods in the egg stage (Fielding 2008; Fielding and Defoliart 2010). Fielding and Defoliart (2010) reported that increasing soil temperature by 2, 3, or 4 °C (3.6–7.2 °F) moved up egg hatch of *Melanopus borealis* and *M. sanguinipes* by 3, 5, or 7 days. The authors suggested that this range of temperature increases, along with other factors, could cause an increase in univoltinism (reproducing every year instead of alternate years), with implications for population sizes. Grasshoppers are kept in check by predators and fungal pathogens, but studies specifically examining effects of climate on these beneficial organisms are lacking or inconclusive.

The ability of rangeland systems to adapt to expected changes in climate within the Northwest has not been adequately quantified. The degree of change required will most likely depend on the type of system. Grazing systems in arid to semiarid zones may experience declines in productivity and profitability due to the impacts of climate change. Current estimates from Wyoming, for example, suggest that increased variability in summer precipitation may reduce profitability by a maximum of 23% (Ritten et al. 2010). A comparable study is not available for NW rangeland, but rangelands of the region could have similar vulnerability. Factors that impart resiliency in rangeland systems include availability of supplemental feed during dry periods and the availability of a variety of grazing grounds with plant communities that differ in growth, soil moisture stores, and nutrition (i.e., adaptive foraging) (Fynn 2012). Projected changes may improve grazing in wetter zones while being detrimental to drier zones. Options for integrating crop and animal production where feasible should be evaluated as an adaptation strategy for grazing management.

6.4.3.2 Pasture and Forage

The impacts of climate change on pasture and forage crop production are difficult to quantify due to a relative lack of research on pasture grasses as compared to crop species and the complex interactions between environmental conditions and factors such as plant competition and seasonal shifts in productivity (Izaurralde et al. 2011). The results of published studies, however, generally suggest a relatively greater influence in biomass production by C3 plants (plants in which the CO_2 is first fixed into a compound containing three carbon atoms before entering the Calvin cycle of photosynthesis; most temperate broadleaf plants and many temperate grasses are C3 plants) than C4 plants (Greer et al. 1995).

Similar to rangeland systems, forage and pasture will be impacted by climate change in multiple ways. Elevated CO_2 concentrations will diminish forage nutritional quality by reducing plant protein and nitrogen concentrations and can reduce the digestibility of forages that are already of poor quality. In partial support of these projections, a CO_2 enrichment experiment in a shortgrass prairie caused the protein concentration of autumn forage to fall below critical maintenance levels for livestock in three out of four years and reduced the digestibility of forage by 14% and 10% in mid-summer and autumn, respectively (Milchunas et al. 2005).

Adams et al. (2001) econometrically estimated crop yield changes, including alfalfa and hay, in relation to baseline climate (1972–2000) in four regions of California (Sacramento Valley and delta, San Joaquin Valley and desert, northeast and mountain, and coastal) at three points in time (2010, 2060, and 2100) using two separate GCMs (PCM and HCGS). This study accounted for CO_2 fertilization and assumed an annual rate of technological progress of 0.25%. Across the 24 scenarios, alfalfa hay yields were estimated to increase by 3–22%.

Lee et al. (2009) incorporated the fertilizing effect of CO_2 and projected California alfalfa hay yields to increase by 4% by 2025, 4–5% by 2050, and 0–7% by 2075, expressed as percentage deviations from mean 2000 yields. These results correspond to average yields generated for a scenario of substantially reduced emissions (SRES-B1) and a continued growth (SRES-A2) emissions scenario, as determined by six separate GCMs coupled with a crop-ecosystem model (DAYCENT).

Caution must be taken when trying to extrapolate the results of these studies to the Northwest. Even within California, Lee et al. (2009) reported that modeled alfalfa yields do not show a consistent response to climate change across counties ranging from a 3% decrease to a 14% increase under SRES-A2 (continued growth) (Lee et al. 2009). Nonetheless results of the California studies can help assess potential impacts on on NW forage systems. Thomson et al. (2005) reported greater increases in alfalfa yields in the upper Columbia Basin in Washington as compared to the California studies. Yield increases in this study range from 33–43% with warming of 1 °C (1.8 °F) above the baseline (1961–1990) and CO_2 concentration of 560 ppm, according to results from three separate

GCMs coupled with an agricultural production model (EPIC). Warming of 2.5 °C (4.5 °F) and CO_2 concentration of 560 ppm are projected to result in alfalfa yield gains of 27–45%. Overall results of climate and crop-growth models suggest an increase in alfalfa production in the Northwest, as long as water is not limiting. Research on other important forages grown in the Northwest should be conducted to determine potential changes in yield and nutritional content.

6.4.3.3 Dairy and Other Confined Animal Operations

Confined animal operations will be impacted by changes in the availability of forage and hay described in the previous sections. Similar to the other livestock production systems, direct impacts of climate on animal health will also need to be considered. Non-optimum temperature, for example, impacts animal immunological, physiological, and digestive functions (Nienaber and Hahn 2007; Mader 2009).

Frank (2001) used two GCMs (CGCM1 and HCGS) and livestock production-response models for confined swine and cattle and for milk-producing dairy operations in six regions of the United States. In the region that includes the seven western states, a doubling of atmospheric CO_2 reduces dairy cow yields, as measured by kilograms of fat-corrected milk, by 0.1–0.2%. Doubling CO_2 also increases the number of days for beef cattle to achieve finish weights by 2.2–2.5%, while a tripling of atmospheric CO_2 increases the time to achieve finish weights by 15%. The Climate Leadership Initiative (2009a, 2009b) used these results to estimate potential economic losses in the beef cattle industries of Oregon and Washington as a result of climate change. They estimate the value of reduced beef production in Oregon to be $7, $11, and $67 million (1.5%, 2.4%, and 14.8%) for the 2020s, 2040s, and 2080s, respectively, based on projected levels of CO_2 under the very high growth emissions scenario (SRES-A1FI). A recent study of the effects of heat stress projected increased losses in dairy production due to heat stress from 0–6% in decades around 2075 in the Northwest (Bauman et al., in review).

Adaptation in confined animal systems may take a variety of forms depending on the degree of change and the specific type of system. Facility modifications including shades, sprinklers, and evaporative cooling systems, as well as changes in genetic stocks and breeding may be necessary to maintain economic sustainability (Mader 2009).

6.4.4 OTHER NORTHWEST AGRICULTURE SUBSECTORS

Climates west of the Cascades differ from much of the rest of the Northwest. Temperatures are moderated by the Pacific Ocean and annual precipitation is generally greater than that received east of the Cascades (see fig. 2.1 in Chapter 2). The Willamette Valley in Oregon, and parts of western Washington, support a diversified agriculture with over 170 crops, including tree and small fruits, grains, grass and hay, ornamentals, turfgrass, vegetables, and viticulture. Increases in annual or cool season precipitation in this already wet region (Chapter 2), or more frequent and higher intensity rainfall events, could exacerbate erosion and flooding, hamper planting and harvesting, or injure crops. As is true for the rest of the Northwest, projected warming may increase vulnerability to invasion risk by pests, diseases, and invasive weeds (Coakley et al. 2010). The diversity of production systems already present within the Willamette Valley suggests flexibility that would impart resilience to climate change, but constraints include costs of changing production systems and availability of markets.

In several parts of the Northwest, including the Willamette Valley and southwestern and northern Idaho, vegetable seed, alfalfa seed, bluegrass, and other grass seed production are important industries. For example, Willamette Valley growers currently enjoy a specialty seed niche that has grown to a farm-gate value estimated at nearly $50 million for the 2012 crop year (Willamette Valley Specialty Seed Association 2012). No studies have examined the vulnerability of these industries to projected climate change.

Throughout the Northwest, canola or other brassica oilseed crops are produced mostly for edible oils. There is a growing interest and potential for substantial increases in acreage of oilseed brassicas as energy crops. In rotation with cereal crops, oilseed brassica provide several advantages for breaking disease cycles and improving soil tilth and weed and pest management (Haramoto et al. 2004; Collins et al. 2006; Matthiesson and Kirkegaard 2006). If this industry expands, varieties must be developed that are suitable for projected climates (Salisbury and Barbetti 2011) or new species, such as *Camelina* spp, with potential tolerance to high temperatures and low precipitation, need to be considered. Furthermore, compatibility of these oilseed crops with brassica vegetable seed in diversified landscapes in which cross-pollination can occur must be considered (WVSSA 2012).

Small-acreage, direct-market farming throughout the Northwest is an important area of growth in the agricultural sector (Diamond and Soto 2009). Many of these small-acreage farms have diversified operations and specialize in niche markets such as hay, direct-market fresh produce, and livestock products (Ostrom and Jussaume 2007). There is no available scientific literature that assesses the vulnerability of these more complex cropping systems to climate change in the Northwest, though anecdotal evidence indicates that the diversity of these cropping systems may be an important advantage for adapting to climate-induced changes (du Toit and Alcala 2009).

6.5 Potential to Adapt to Changing Climates

Farming and ranching are inherently and necessarily flexible and responsive to variable weather conditions, and therefore global agriculture is likely as well positioned as any economic sector to adapt to climate change. That flexibility, coupled with relatively moderate projected impacts for the NW region, indicates that NW agriculture is also well positioned for adaptation. The diversity of agriculture in the NW region in part illustrates the capacity of these industries to be productive under climatic conditions ranging from the higher rainfall, moderate seasonality, and diverse agriculture of the Pacific coastal regions and the Willamette Valley to the cold winters, low rainfall, and wheat fallow production systems of the Columbia Plateau and the arid rangelands of southern Idaho. As NW climates change, a significant amount of autonomous adaptation is expected to occur through adjustments in the timing of farming operations and selection of crop varieties, to shifts in the crops grown and the transformation of entire cropping systems and land use patterns (IPCC 2001). But the resilience, economic impacts, cropping system-specific infrastructure, and inherent adaptability of NW systems vary considerably. Diversified systems of Pacific coastal regions with greater access to urban markets may be more adaptable than agriculture in semi-arid interiors, where wheat production or rangeland grazing predominate. Moreover, throughout the region, the potential for agricultural adaptation is constrained by the availability of resources,

Box 6.1

Mitigating Greenhouse Gas Emissions from Agricultural Systems

Although this report focuses on the impacts and requirements for adaptation to climate change, agriculture is also implicated as a producer of greenhouse gases (GHGs) that are drivers of climate change. Changes in agricultural management practices have the potential to mitigate these emissions. In the Northwest, the principal opportunities are (1) reducing tillage of annual crops which can increase storage of carbon in soils, (2) improving the efficiency of nitrogen fertilization practices in order to limit nitrous oxide (N_2O) production, and (3) adopting manure management technologies that reduce methane emissions from confined animal feeding (CAF) operations. Brown and Huggins (2012) conducted an exhaustive review of the published studies of the effects of reduced tillage on soil carbon stocks in dryland agriculture within the NW region. Although data are variable, the authors concluded that adoption of no-tillage methods on acreage previously farmed using conventional tillage results in carbon increases in the surface 5 cm (2 in) of the soil profile and declining with depth to near zero at 20 cm (7.9 in). Depending upon the production zone, increases range from 0.11–1.04 megatons CO_2 equivalents of soil carbon per acre per year. Reduced tillage practices other than no-tillage have no measurable effect. Stöckle et al. (2012) simulated representative cropping systems of the Northwest and provide a more modest estimate ranging from 0.13–0.24 megatons CO_2 equivalents of soil carbon per acre per year. Since no-tillage has not been widely adopted in the region, there is potential for annual production systems to store more carbon than they do presently. No-tillage can be profitable, but presents several challenges that have slowed adoption.

Nitrogen applied as nitrate and not taken up by crops is available for metabolism by soil microbes and one of the products, N_2O, is a significant greenhouse gas (Robertson et al. 2000). Worldwide, there are opportunities to improve the efficiency of nitrogen use through a combination of practices and technologies (Snyder et al. 2009). In NW agriculture there are opportunities to improve nitrogen use efficiency through precision application technology, modifications in irrigation schedules, diversification in which legumes contribute more to nitrogen budgets, and eventually, more nitrogen efficient crop varieties (Huggins and Pan 2003; Cogger et al. 2006; Coakley et al. 2010; Huggins 2010).

Methane, released from animal manure, is a powerful greenhouse gas (IPCC 2007). Anaerobic digestion technology is available and is being adopted on CAF operations in the Northwest where research at Washington State University has shown these systems to be profitable (Yorgey et al. 2011). The extent of these industries and the cultivated land provide opportunities for GHG mitigation through improved dairy manure and food processing waste management, improved efficiency of nitrogen fertilizer use, and increased soil carbon sequestration (Kruger et al. 2010).

It appears that many of the strategies that will help with mitigation of greenhouse gas emissions will also help with adaptation and overall sustainability and profitability of farming. Further research is needed to determine best management practices in cereal production systems, animal systems, and others in the Northwest to minimize production of CO_2, N_2O, and methane.

including water, fertilizer, machinery, processing infrastructure, capital, knowledge, and management expertise.

Some specific examples illustrate these constraints on agriculture's capacity to adapt to changing climates.

- Transitioning to new varieties of perennial crops such as wine grapes and tree fruit, if indicated, is necessarily slow and expensive. And, as mentioned earlier, uncertain projections hamper decisions about making transitions in these long-lived perennial crops based on climate projections relative to consumer demands.

- Risk aversion and reliance on traditional practices among many farmers, although prudent under typical circumstances, could hamper responsiveness if climates change rapidly.

- Irrigation-dependent agricultural regions, notably central Washington, the Magic and Treasure Valleys of Idaho, and north central Oregon (fig. 6.2), possess flexibility in what crops can be produced, but may be disproportionately affected by climate-related changes in the availability of irrigation water from rivers and reservoirs (see Chapter 3).

- Parts of the Northwest have marginally viable agriculture because of low rainfall and unavailability of affordable irrigation water. These areas have few options should warmer and drier summers become the norm.

- In addition to the challenges it poses at the production level, climate change may have implications for the food processing and transportation infrastructure beyond the farm gate, including design and location of storage facilities, changes in geographic range and type of food pathogens, and impacts of mitigation and energy policies on the economics of our domestic food systems (Antle and Capalbo 2010). In the 20th century and early 21st century, public sector investment has played a substantial role in the success of US agriculture and seems likely to continue to do so as part of agriculture's response to climate change.

Some examples of existing policies and their possible effects on climate change adaptation include:

- *Agricultural subsidy programs for commodity crops (wheat, corn, rice) and trade policies.* Such policies as the import quota on sugar reduce flexibility and have unintended consequences for global markets. A common feature is that these policies encourage farmers to grow subsidized crops rather than adapting to changing conditions, including climate (Antle and Capalbo 2010).

- *Disaster assistance and production and income insurance policies.* While providing some protection against climate variability and extreme events, these policies may also reduce the incentive for farmers and ranchers to take adaptive actions.

- *Soil, water, and ecosystem conservation policies.* These policies protect water quality and enhance ecosystem services such as wildlife habitat. However, they may also limit a producer's options when responding to climate change or extreme events by reducing the ability to adapt land use to changing conditions.

- *Environmental policies affect agricultural land use and management.* Regulations for locating confined animal production and the disposal of waste from these facilities are an important example that has implications for adaptation. Changes in climate and climate extremes may significantly impact the viability of these operations in locations where waste ponds become vulnerable to extreme rainfall events and floods. Environmental regulations raise the cost of relocating facilities (Nene et al. 2009).

- *Tax policies affect agriculture in many ways, including the taxation of income and depreciable assets.* Tax rules could be utilized to facilitate adaptation for example by accelerating depreciation of assets. However, using tax policies to target incentives for adaptation may prove difficult since other types of technological and economic change will also lead to capital obsolescence.

- *Energy policies are likely to have many impacts on agricultural sectors as both consumers and producers of energy.* The increased cost of fossil fuels associated with GHG mitigation policies will likely adversely affect farmer income, however, they also provide an incentive for technology changes that are likely to improve adaptability of renewable and alternative energy systems.

- *Direct public sector investment and policies that support continued investment in new technologies.* Such policies have always played a key role in the success of US agriculture to provide benefits to society; these types of policies can continue to play an important role in encouraging adaptation under a changing climate.

Thus, despite agriculture's generally high potential for adaptation to climate change, some regions and subsectors will be particularly challenged, and outcomes will depend upon not only production systems per se, but the economics, infrastructure, and policy scenarios. Even under conducive scenarios, success will require flexibility, ingenuity, and strong partnerships between the public sector (research institutions and government agencies), private businesses, such as breeding companies, commodity organizations, and private producers. These powerful partnerships and investments have helped ensure agriculture remained strong in the preceding century and will be required in the future.

Finally, adaptation to changing climate must be sustainable in the long term. Measures taken to maintain farm incomes under climate change in the short term may exacerbate long-standing issues of sustainability in the region. Some NW agricultural sectors face threats to sustainability including uncertainties in water supplies for irrigation (Washington State Department of Ecology 2011), and loss of productivity longer-term due to soil erosion (Pruski and Nearing 2002; Mullan et al. 2012). Key to long-term sustainability is a better understanding of how increased climate variability impacts production and investment decisions and the interactions among climate, energy, and other regulatory policies that impact adaptation rates and outcomes (Antle and Capalbo 2010).

Changes in agricultural profitability would affect the region's rural economies, with associated effects on human health and well-being. Adaptation will require not only changes in technology, land use, and productivity, but community-level responses, such as public and private investment in adaptive infrastructure for transportation, processing, and storage. Most rural communities in the region are not yet coming to grips with the complex decisions and new investments that may be required (Cone et al. 2011). Continued effort by university extension personnel and others is needed to assist communities and producers to assess available information and make value assessments that are needed to determine the best responses to climate change.

6.6 Knowledge Gaps and Research Needs

Considering the extent and diversity of agriculture in the Northwest, projections for effects of climate change on agriculture remain limited. Most studies have used a limited range of climate projections (GCMs and emissions scenarios) or focused on specific commodities or geographic areas. They depend upon assumptions regarding availability of irrigation water, policy scenarios, and production practices. Most do not consider changing pressure from pests, weeds, and diseases. Although climate projections include interannual variations, projected impacts on agriculture rarely consider corresponding interannual variations in yield, which can be critical for agricultural viability.

This relative paucity of studies in part reflects the availability of necessary funding. Adaptation will depend on continued public and private investments in research to improve crops, reduce uncertainty in management outcomes and costs, and address crop protection needs, but in recent decades support for this research has been diminishing (Kruger et al. 2011). Recently, the USDA, through its National Institute of Food and Agriculture (NIFA) and Agricultural Research Service (ARS), and the National Science Foundation have initiated coordinated research efforts to address these needs. Partnerships among the NW land-grant universities and ARS, augmented by USDA competitive funding, have helped ensure agriculture remained strong in the preceding century and will be required in the future.

Climate-related agricultural projects underway or being initiated in the region include: Regional Approaches to Climate Change for Pacific Northwest Agriculture (reacchpna.org), Site-Specific Climate Friendly Farming, Climate Friendly Farming, the Columbia River Supply and Demand Forecast (http://www.ecy.wa.gov/programs/wr/cwp/forecast/forecast.html), the Cook Agronomy Farm Long-Term Agricultural Research site of the USDA ARS, Regional Earth System Modeling Project (BioEarth) (http://www.cereo.wsu.edu/bioearth/), the Watershed Integrated Systems Dynamics Modeling project (WISDM) (http://www.cereo.wsu.edu/wisdm/), Willamette Water 2100, Idaho EPSCoR Project: Water Resources in a Changing Climate (http://idahoepscor.org). Although these and similar projects are achieving great gains in fundamental understanding of the challenges faced by agriculture and many are well connected to stakeholders, they still leave significant gaps. Key needs for a research and extension agenda required to address potential effects of climate change on NW agriculture include:

- *Improved understanding of socioeconomic factors and policies that mediate adaptation and mitigation practices relevant to climate change across all agricultural subsectors.* Adaptation and mitigation will require changes in producer practices. Effective policies and successful communication with producers will depend upon a thorough understanding of the social and economic conditions that influence their business decisions and farming practices.

- *Extended effort to include neglected commodities or production systems.* Studies are limited or absent for many minor or specialty crops important for the region, including hops, sugarbeets, and small fruit. There also have been no studies that consider climate change impacts on small, diversified farms.

- *Extended effort to consider organic production systems.* There is no research examining effects of climate change on small and large producers of organic commodities in the Northwest. These systems may be uniquely challenged, or their practices may have applicability for other sectors under changing climates.

- *Responsive development of integrated weed, pest, and disease management (IWM, IPM, IDM).* Existing programs may need adjustment as climates change. Effectiveness of biological control of weeds and insects, timing of insecticide or herbicide treatments, and the overall severity or complexity of managing of certain pests, weeds, and diseases may be altered as climates change.

- *Rapid development and adoption of crop varieties adapted to changing climatic conditions (e.g., heat, drought, pest resistance).* A large NIFA-funded project (Triticeae Cooperative Agricultural Project; lead institution, University of California, Davis) is focused on accelerating development of drought and heat stress (among other traits) in wheat, with applicability for NW systems. Similar efforts for other key commodities will be needed as part of adaptation to changing climates.

- *Improved cropping system and hydrologic modeling approaches that incorporate climate change on a regionally and temporally distributed scale for multiple crops.* Cropping system responses to climatic factors are exceedingly complex and require sophisticated modeling to generate projection scenarios. Ongoing projects are addressing this gap, but models can be improved and expanded.

- *Alternative cropping systems with resilience to projected climate changes.* As climates change, opportunities may arise to diversify or otherwise modify cropping systems (crops, rotations, integration) so they are better suited to projected climates in the Northwest.

- *Improved consideration of climatic and weather variability.* Extreme events (e.g., heat, flooding) and interannual and seasonal variation in weather are difficult to project but are potentially critical for understanding effects of future climates on agriculture.

- *Improved understanding of the social and economic dimensions of local and regional decision making in rural communities in response to climate change.* Although much of the response to climate change in NW agriculture involves decisions by individual producers, communities may be involved in confronting climate change impacts.

- *More thorough interdisciplinary research that considers the effects of climate change on NW agricultural production systems as a whole, rather than focusing on agronomic, economic, or social factors in isolation.* NW agriculture is a complex system involving interacting biological and human dimensions. Interdisciplinary research considers the system wide complexities and emergent properties, ensuring that individual efforts are not based on erroneous assumptions.

- *Educating cohorts of scientists better prepared to work on complex problems relating to agricultural sustainability under drivers of change, including climate.* Climate change will present ongoing challenges to the agricultural sector, which will need to be addressed by future scientists.

- *Fostering an enhanced collaborative environment among scientists throughout the Northwest to improve and sustain efforts to address effects of climate change on the region's agriculture.* Climate change and agriculture's responses to climate change will take place across the entire NW region and require decades. Collaborations among scientists, educators and institutions will be required to address these processes at an appropriate scale.

- *Research that considers mitigation of greenhouse gas production in conjunction with adaptation to projected climates.* This could include improved technology and accelerated adoption of reduced tillage and improved nitrogen use efficiency practices for all crops.

- *Research that considers the implications of climate change for landscapes that include agricultural systems.* Agricultural systems rely upon ecosystem services including pollination, water, biological control, and regional resistance to invasive species. Similarly, agriculture can contribute to regional biodiversity conservation by providing habitats for native species. These landscape scale processes are potentially susceptible to a changing climate, but no studies are available to our knowledge that consider impacts at this scale in the Northwest.

Acknowledgments

The Agriculture chapter authors wish to acknowledge the following grants for support while working on their chapter: National Institute for Food and Agriculture competitive grant, award number: 2011-68002-30191; and US Department of Energy Earth System Modeling (ESM) Program, award number: 20116700330346. The authors also thank Dave Huggins (USDA Agricultural Research Service), Stacy Vynne (Puget Sound Partnership), and two anonymous reviewers for their thoughtful review and comments on an earlier version of this chapter.

References

Abatzoglou, J. T., and C. A. Kolden. 2011. "Climate Change in Western US Deserts: Potential for Increased Wildfire and Invasive Annual Grasses." *Rangeland Ecology & Management* 64 (5): 471-478. doi: 10.2111/REM-D-09-00151.1.

Adams, R., J. Wu, and L. L. Houston. 2001. "Changes in Crop Yields and Irrigation Demand." In *The Impact of Climate Change on Regional Systems: A Comprehensive Analysis of California*, edited by J. B. Smith and R. Mendelsohn. Cheltenham: Edward Elgar.

Alva, A. K., T. Hodges, R. A. Boydston, and H. P. Collins. 2002. "Effects of Irrigation and Tillage Practices on Yield of Potato under High Production Conditions in the Pacific Northwest." *Communications in Soil Science and Plant Analysis* 33: 1451–1460. doi: 10.1081/CSS-120004293.

Antle, J. M., and S. M. Capalbo. 2010. "Adaptation of Agricultural and Food Systems to Climate Change: An Economic and Perspective." *Applied Economics Perspectives and Policy* 32 (3): 386-416. doi: 10.1093/aepp/ppq015.

Ball, D. A., S. M. Frost, and A. I. Gitelman. 2004. "Predicting Timing of Downy Brome (Bromus tectorum) Seed Production Using Growing Degree Days." *Weed Science* 52 (4): 518-524. doi: 10.1614/WS-03-067.

Bauman, Y., G. Mauger, and E. P. Salathé. "Climate Change Impacts on Dairy Production in the Pacific Northwest." Submitted to *Weather Climate and Society*. In review.

Branson, D. H. 2008. "Influence of a Large Late Summer Precipitation Event on Food Limitation and Grasshopper Population Dynamics in a Northern Great Plains Grassland." *Environmental Entomology* 37 (3): 686-695. http://esa.publisher.ingentaconnect.com/content/esa/envent/2008/00000037/00000003/art00010.

Branson, D. H., A. Joern, and G. A. Sword. 2006. "Sustainable Management of Insect Herbivores in Grassland Ecosystems: New Perspectives in Grasshopper Control." *Bioscience* 56 (9): 743-755. doi: 10.1641/0006-3568(2006)56[743:SMOIHI]2.0.CO;2.

Brown, J., and N. MacLeod. 2011. "A Site-Based Approach to Delivering Rangeland Ecosystem Services." *Rangeland Journal* 33 (2): 99-108. doi: 10.1071/RJ11006.

Brown, T. T., and D. R. Huggins. 2012. "Soil Carbon Sequestration in the Dryland Cropping Region of the Pacific Northwest." *Journal of Soil and Water Conservation* 67 (5): 406-415. doi: 1.2489/jswc.67.5.406.

Clement, S. L., D. S. Husebye, and S. D. Eigenbrode. 2010. "Ecological Factors Influencing Pea Aphid Outbreaks in the US Pacific Northwest." In *Aphid Biodiversity under Environmental Change: Patterns and Processes*, edited by P. Kindlemann, A. F. G. Dixon, and J. P. Michaud, 108-128. Dordrecht: Springer.

Climate Leadership Initiative. 2009a. "An Overview of Potential Economic Costs to Oregon of a Business-As-Usual Approach to Climate Change." http://www.theresourceinnovationgroup.org/storage/economicreport_oregon.pdf.

Climate Leadership Initiative. 2009b. "An Overview of Potential Economic Costs to Washington of a Business-As-Usual Approach to Climate Change." http://www.theresourceinnovationgroup.org/storage/economicreport_washington.pdf.

Coakley, S. M., G. V. Jones, S. Page, and K. D. Dello. 2010. "Climate Change and Agriculture in Oregon." In *Oregon Climate Assessment Report*, edited by K. D. Dello, and P. W. Mote, 151-172. Oregon Climate Change Research Institute, Oregon State University, Corvallis, Oregon. http://occri.net/ocar.

Coakley, S. M., H. Scherm, and S. Chakraborty. 1999. "Climate Change and Plant Disease Management." *Annual Review of Phytopathology* 37: 399–426. doi: 10.1146/annurev.phyto.37.1.399.

Cogger, C. G., T. A. Forge, and G. H. Neilsen. 2006. "Biosolids Recycling: Nitrogen Management and Soil Ecology." *Canadian Journal of Soil Science* 86 (4): 613-620. doi: 10.4141/S05-117.

Collins, H. P., A. Alva, R. A. Boydston, R. L. Cochran, P. B. Hamm, A. McGuire, and E. Riga. 2006. "Soil Microbial, Fungal and Nematode Responses to Soil Fumigation and Cover Crops under Potato Production." *Biological Fertility and Soils* 42 (3): 247-257. doi: 10.1007/s00374-005-0022-0.

Cone, J., J. Borberg, and M. Russo. 2011. "Classical and Jazz: Two Approaches to Supporting Rural Community Preparation for Climate Change." *Rural Connections,* June, http://wrdc.usu.edu/files/uploads/Rural%20Connections/RCJUN11w.pdf.

Crowder, D. W., T. D. Northfield, M. R. Strand, and W. E. Snyder. 2010. "Organic Agriculture Promotes Evenness and Natural Pest Control." *Nature* 466 (7302): 109-112. doi: 10.1038/nature09183.

Daily, G. C., ed. 1997. *Nature's Services.* Washington, DC: Island Press.

Davis, T. S., J. Abatzoglou, W. Price, N. Bosque-Pérez, S. E. Halbert, S. D. Eigenbrode. "Biogeography and Climate Drivers Of Cereal Aphid Phenology In The Northwestern USA." MS in preparation

Diamond, A., and R. Soto. 2009. "Facts on Direct-to-Consumer Food Marketing: Incorporating Data from the 2007 Census of Agriculture." USDA Agricultural Marketing Service. http://naldc.nal.usda.gov/catalog/46726.

Diekötter, T., R. Billeter, and T. O. Crist. 2008. "Effects of Landscape Connectivity on the Spatial Distribution of Insect Diversity in Agricultural Mosaic Landscapes." *Basic and Applied Ecology* 9: 298-307. doi: 10.1016/j.baae.2007.03.003.

Diffenbaugh, N. S., M. A. White, G. V. Jones, and M. Ashfaq. 2011. "Climate Adaptation Wedges: A Case Study of Premium Wine in the Western United States." *Environmental Research Letters* 6 (2): 024024. doi: 10.1088/1748-9326/6/2/024024

Douglas, C. L., R. W. Rickman, B. L. Klepper, J. F. Zuzel, and D. J. Wysocki. 1992. "Agroclimatic Zones for Dryland Winter Wheat Producing Areas of Idaho, Washington, and Oregon." *Northwest Science* 66 (1): 26-34. http://hdl.handle.net/2376/1603.

Dukes, J. S., N. R. Chiariello, S. R. Loarie, and C. B. Field. 2011. "Strong Response of an Invasive Plant Species (*Centaurea solstitialis* L.) to Global Environmental Changes." *Ecological Applications* 21 (6): 1887-1894. doi: 10.1890/11-0111.1.

du Toit, L. and A. V. Alcala. 2009. "Management of Seedling Blights in Organic Vegetable Production in the Pacific Northwest." *Progress Report: Organic Cropping Research For The Northwest.* CSANR, Washington State University: http://www.tfrec.wsu.edu/pdfs/P2271.pdf.

Easterling, W.E., P. K. Aggarwal, P. Batima, K. M. Brander, L. Erda, S. M. Howden, A. Kirilenko, J. Morton, J.-F. Soussana, J. Schmidhuber and F. N. Tubiello. 2007. "Food, Fibre and Forest Products". In *Climate Change 2007: Impacts, Adaptation and Vulnerability. Contribution of Working Group II to the Fourth Assessment Report of the Intergovernmental Panel on Climate Change,* edited by M. L. Parry, O. F. Canziani, J. P. Palutikof, P. J. van der Linden, and C. E. Hanson, 273-313. Cambridge University Press, Cambridge, UK.

Eigenbrode, S. D., and J. T. Abatzoglou. "Cereal Leaf Beetle in Projected Climates of the PNW." In preparation.

Elsner, M. M., L. Cuo, N. Voisin, J. S. Deems, A. F. Hamlet, J. A. Vano, K. E. B. Mickelson, S. Y. Lee, and D. P. Lettenmaier. 2010. "Implications of 21st Century Climate Change for the Hydrology of Washington State." *Climatic Change* 102: 225-260. doi: 10.1007/s10584-010-9855-0.

Evans, E. W., N. R. Carlile, M. B. Innes, and N. Pitigala. 2012. "Warm Springs Reduce Parasitism of the Cereal Leaf Beetle Through Phenological Mismatch." *Journal of Applied Entomology.* doi: 10.1111/jen.12028. In press.

Farina, A. 2000. *Landscape Ecology in Action.* Dordrecht: Kluwer Academic Publishers.

Ferris, R., R. H. Ellis, T. R. Wheeler, and P. Hadley. 1998. "Effect of High Temperature Stress at Anthesis on Grain Yield and Biomass of Field-Grown Crops of Wheat." *Annals of Botany* 82 (5): 631-639. doi: 10.1006/anbo.1998.0740.

Fielding, D. J. 2008. "Diapause Traits of Melanoplus sanguinipes and Melanoplus borealis (Orthoptera: Acrididae)." *Annals of the Entomological Society of America* 101: 439-448. doi: 10.1603/0013-8746(2008)101[439:DTOMSA]2.0.CO;2.

Fielding, D. J., and L. S. Defoliart. 2010. "Embryonic Developmental Rates of Northern Grasshoppers (Orthoptera: Acrididae): Implications for Climate Change and Habitat Management." *Environmental Entomology* 39: 1643-1651. doi: 10.1603/EN09356.

Fischer, G., M. Shah, and H. van Velthuizen. 2002. "Impacts of Climate on Agro-Ecology." In *Climate Change and Agricultural Vulnerability*, 38–91. Vienna, Austria: IIASA.

Follett, R. F., and D. A. Reed. 2010. "Soil Carbon Sequestration in Grazing Lands: Societal Benefits and Policy Implications." *Rangeland Ecology and Management* 63 (1): 4-15. doi: 10.2111/08-225.1.

Frank, K. L. 2001. "Potential Effects of Climate Change on Warm Season Voluntary Feed Intake and Associated Production of Confined Livestock in the United States." MS thesis, Kansas State University, Manhattan.

Fynn, R. W. S. 2012. "Functional Resource Heterogeneity Increases Livestock and Rangeland Productivity." *Rangeland Ecology & Management* 65 (4): 319-329. doi: 10.2111/REM-D-11-00141.1.

Greer, D. H., W. A. Laing, and B. D. Campbell. 1995. "Photosynthetic Responses of 13 Pasture Species to Elevated CO_2 and Temperature." *Australian Journal of Plant Physiology* 22 (5): 713-722. doi: 10.1071/PP9950713.

Gu, L., P. Hanson, W. MacPost, D. P. Kaiser, B. Yang, R. Nemani, S. Pallardy, and T. Meyers. 2008. "The 2007 Eastern US Spring Freeze: Increased Cold Damage in a Warming World?" *BioScience* 58 (3): 253–262. doi: 10.1641/B580311.

Halbert, S. E., and K. S. Pike. 1985. "Spread of Barley Yellow Dwarf Virus and Relative Importance of Local Aphid Vectors in Central Washington." *Annals of Applied Biology* 107 (3): 387-395. doi: 10.1111/j.1744-7348.1985.tb03155.x.

Haramoto, E. R., and E. R. Gallandt. 2004. "Brassica Cover Cropping for Weed Management: a Review." *Renewable Agriculture and Food Systems* 19 (4):187-198. doi: 10.1079/RAFS200490.

Hatfield, J., K. J. Boote, P. Fay, L. Hahn, C. Izaurralde, B. A. Kimball, T. Mader, J. Morgan, D. Ort, W. Polley, A. Thomson, and D. Wolfe. 2008. "Agriculture." In *The Effects of Climate Change on Agriculture, Land Resources, Water Resources, and Biodiversity in the United States*, edited by M. Walsh, 21-74. US Climate Change Science Program Synthesis and Assessment Product 4.3, US Department of Agriculture, Washington, DC.

Hatfield, J. L., K. J. Boote, B. A. Kimball, L. H. Ziska, R. C. Izaurralde, D. Ort, A. M. Thomson, and D. Wolfe. 2011. "Climate Impacts on Agriculture: Implications for Crop Production." *Agronomy Journal* 103 (2): 351-370. doi: 10.2134/agronj2010.0303.

Hoekema, D. J., and V. Sridhar. 2011. "Relating Climatic Attributes and Water Resources Allocation: A Study Using Surface Water Supply and Soil Moisture Indices in the Snake River Basin, Idaho." *Water Resources Research* 47: W07536. doi: 10.1029/2010WR009697.

Huggins, D. R. 2010. "Site-Specific N Management for Direct-Seed Cropping Systems." In *Climate Friendly Farming: Improving the Carbon Footprint of Agriculture in the Pacific Northwest*, edited by C. Kruger, G. Yorgey, S. Chen, H. Collins, C. Feise, C. Frear, D. Granatstein, S. Higgins, D. Huggins, C. MacConnell, K. Painter, C. Stöckle. CSANR Research Report 2010-001. Washington State University: http://csanr.wsu.edu/pages/Climate_Friendly_Farming_Final_Report/.

Huggins, D. R., and W. L. Pan. 2003. "Key Indicators for Assessing Nitrogen Use Efficiency in Cereal-Based Agroecosystems." *Journal of Crop Production* 8: 57-86. doi: 10.1300/J144v08n01_07.

IPCC 2001. *Climate Change 2001: The Scientific Basis. Contribution of Working Group I to the Third Assessment Report of the Intergovernmental Panel on Climate Change*, edited by J. T. Houghton, Y. Ding, D. J. Griggs, M. Noguer, P. J. van der Linden, X. Dai, K. Maskell, and C. A. Johnson, and D. Xiaosu. New York: Cambridge University Press, 881 pp. http://www.ipcc.ch /ipccreports/tar/wg1/.

IPCC 2007. *Climate Change 2007: Synthesis Report. Contribution of Working Groups I, II and III to the Fourth Assessment Report of the Intergovernmental Panel on Climate Change*, edited by Core Writing Team, R. K. Pachauri, and A. Reisinger. IPCC, Geneva, Switzerland, 104 pp.

Izaurralde, R. C., A. M. Thomson, J. A. Morgan, P. A. Fay, H. W. Polley, and J. L. Hatfield. 2011. "Climate Impacts on Agriculture: Implications for Forage and Rangeland Production." *Agronomy Journal* 103 (2): 371-381. doi: 10.2134/agronj2010.0304.

Joern, A., and S. B. Gaines. 1990. "Population Dynamics and Regulation in Grasshoppers." In *Biology of Grasshoppers*, edited by R. F. Chapman and A. Joern, 415-482. Wiley Press.

Jones, G. V. 2005. "Climate Change in the Western U.S. Grape Growing Regions." *Acta Horticulturae (ISHS)*, 689: 41-60. http://www.actahort.org/books/689/689_2.htm.

Jones, G. V. 2006a. "Climate and Terroir: Impacts of Climate Variability and Change on Wine." *In Fine Wine and Terrior – The Geoscience Perspective*, edited by R. W. Macqueen, and L. D. Meinert. Geoscience Canada Reprint Series Number 9, Geological Association of Canada, St. John's, Newfoundland, 247 pages. http://www.sou.edu/envirostudies/gjones_docs /GJones%20Climate%20Change%20Geoscience%20Canada.pdf.

Jones, G. V. 2006b. "Climate Change and Wine: Observations, Impacts, and Future Implications." *Australia and New Zealand Wine Industry Journal* 21 (4): 21-26.

Jones, G. V. 2007. "Climate Change: Observations, Projections, and General Implications for Viticulture and Wine Production." *Practical Winery and Vineyard* July/August: 44-64.

Julian, J. W., C. F. Seavert, P. A. Skinkis, P. VanBuskirk, and S. Castagnoli. 2008. "Vineyard Economics: Establishing and Producing Pinot Noir Wine Grapes in Western Oregon." *Oregon State University Extension Service Bulletin EM 8969-E*, August: 1-17. http://arec.oregon state.edu/oaeb/files/pdf/EM8969-E.pdf.

Kruger, C., G. Yorgey, S. Chen, H. Collins, C. Feise, C. Frear, D. Granatstein, S. Higgins, D. Huggins, C. MacConnell, K. Painter, C. Stöckle. 2010. "Climate Friendly Farming: Improving the Carbon Footprint of Agriculture in the Pacific Northwest." CSANR Research Report 2010-001. Washington State University: http://csanr.wsu.edu/pages/Climate_Friendly _Farming_Final_Report/.

Kruger, C., G. Yorgey, and C. Stöckle. 2011. "Climate Change and Agriculture in the Pacific Northwest." *Rural Connections* June: 51-54. http://wrdc.usu.edu/files/uploads/Rural %20Connections/RCJUN11w.pdf.

Kunkel, K. E., L. E. Stevens, S. E. Stevens, L. Sun, E. Janssen, D. Wuebbles, K. T. Redmond, and J. G. Dobson. 2013. "Regional Climate Trends and Scenarios for the U.S. National Climate Assessment. Part 6. Climate of the Northwest U.S." NOAA Technical Report NESDIS 142-6, 75 pp. http://scenarios.globalchange.gov/report/regional-climate-trends-and-scenarios-us-national-climate-assessment-part-6-climate-northwest.

Lee, J., S. De Gryze, and J. Six. 2009. "Effect of Climate Change on Field Crop Production in the Central Valley of California." California Climate Change Center, CEC-500-2009-041-D.

Lichthardt, J., and R. K. Mosely. 1997. "Status and Conservation of the Palouse Grassland in Idaho." Idaho Department of Fish and Game, US Fish and Wildlife Service, Purchase Order No. 14420-5-0395.

Littell, J. S., M. McGuire, M. M. Elsner, L. C. Whitely Binder, and A. K. Snover, eds. 2009. "Executive Summary." In *Washington Climate Change Impacts Assessment: Evaluating Washington's Future in a Changing Climate*, edited by M. M. Elsner, J. Littell, and L. Whitely Binder. Climate Impacts Group, University of Washington, Seattle, Washington. www.cses.washington.edu /db/pdf/wacciaexecsummary638.pdf.

Liu, D., and J. T. Trumble. 2007. "Comparative Fitness of Invasive and Native Populations of the Potato Psyllid (*Bactericera cockerelli*)." *Entomologia Experimentalis et Applicata* 123 (1): 35-42. doi: 10.1111/j.1570-7458.2007.00521.x.

Long, S. P., E. A. Ainsworth, A. Rogers, and D. R. Ort. 2004. "Rising Atmospheric Carbon Dioxide: Plants FACE the Future." *Annual Review of Plant Biology* 55: 591-628. doi: 10.1146/ annurev.arplant.55.031903.141610.

Looney, C., and S. D. Eigenbrode. 2012. "Characteristics and Distribution of Palouse Prairie Remnants: Implications for Conservation Planning." *Natural Areas Journal* 32 (1): 75-85. doi: 10.3375/043.032.0109.

Losey, J. E., and M. Vaughan. 2006. "The Economic Value of Ecological Services Provided by Insects." *BioScience* 56 (4): 311-323. doi: 10.1641/0006-3568(2006)56[311:TEVOES]2.0.CO;2.

Luedeling, E., and P. H. Brown. 2011. "A Global Analysis of the Comparability of Winter Chill Models for Fruit and Nut Trees." *International Journal of Biometeorology* 55 (3): 411–421. doi: 10.1007/s00484-010-0352-y.

Luedeling, E., E. H. Girvetz, M.A. Semenov, and P.H. Brown. 2011. "Climate Change Affects Winter Chill for Temperate Fruit and Nut Trees." *PLoS ONE* 6 (5): e20155. doi: 10.1371 /journal.pone.0020155.

Mackun, P. J. 2009. "Population Change in Central and Outlying Areas of Metropolitan Statistical Areas: 2000 to 2007." *Current Population Reports* P25-1136. US Department of Commerce, Economics and Statistics Administration, US Census Bureau, Washington, DC.

Mader, T. L., K. L. Frank, J. A. Harrington, G. L. Hahn, and J. A. Nienaber. 2009. "Potential Climate Change Effects on Warm-Season Livestock Production in the Great Plains." *Climatic Change* 97: 529-541. doi: 10.1007/s10584-009-9615-1.

Mancuso, M., and R. Moseley. 1994. "Vegetation Description, Rare Plant Inventory, and Vegetation Monitoring for Craig Mountain, Idaho." *Bonneville Power Administration Project* 92-069. Idaho Department of Fish and Game, Boise, Idaho.

Matthiessen, J. N., and J. A. Kirkegaard. 2006. "Biofumigation and Enhanced Biodegradation: Opportunity and Challenge in Soilborne Pest and Disease Management." *Critical Reviews in Plant Science* 25 (3): 235–265. doi: 10.1080/07352680600611543.

Milchunas, D. G., A. R. Mosier, J. A. Morgan, D. R. LeCain, J. Y. King, and J. A. Nelson. 2005. "Elevated CO_2 and Defoliation Effects on a Shortgrass Steppe: Forage Quality Versus Quantity for Ruminants." *Agriculture Ecosystems & Environment* 111: 166-184. doi: 10.1016/j. agee.2005.06.014.

Morgan, J. A., A. R. Mosier, D. G. Milchunas, D. R. LeCain, J. A. Nelson, and W. J. Parton. 2004. "CO_2 Enhances Productivity, Alters Species Composition, and Reduces Digestibility of Shortgrass Steppe Vegetation." *Ecological Applications* 14 (1): 208-219. doi: 10.1890/ 02-5213.

Mote, P. W. 2006. "Climate-Driven Variability and Trends in Mountain Snowpack in Western North America." *Journal of Climate* 19 (23): 6209-6220. doi: 10.1175/JCLI3971.1.

Mullan, D., D. Favis-Mortlock, and R. Fealy. 2012. "Addressing Key Limitations Associated with Modeling Soil Erosion under the Impacts of Future Climate Change." *Agricultural and Forest Meteorology* 156 (15): 18-30. doi: 10.1016/j.agrformet.2011.12.004.

Nakićenović, N., O. Davidson, G. Davis, A. Grübler, T. Kram, E. Lebre La Rovere, B. Metz, T. Morita, W. Pepper, H. Pitcher, A. Sankovshi, P. Shukla, R. Swart, R. Watson, and Z. Dadi. 2000. *Special Report on Emissions Scenarios: A Special Report of Working Group III of the Intergovernmental Panel on Climate Change*, Cambridge University Press, Cambridge, UK, 599 pp. http://www.grida.no/climate/ipcc/emission/index.htm.

Nene, G., A. M. Azzam, and K. Schoengold. 2009. "Environmental Regulations and the Structure of U.S. Hog Farms." Paper presented at the Agricultural and Applied Economics Association Joint Annual Meeting, Milwaukee, Wisconsin, July 26-29. http://ageconsearch.umn.edu /bitstream/49395/2/AAEA%20final%20paper.pdf.

Nienaber, J. A., and G. L. Hahn. 2007. "Livestock Production System Management Responses To Thermal Challenges." *International Journal of Biometeorology* 52 (2): 149-157. doi: 10.1007/ s00484-007-0103-x.

Oregon State Board of Agriculture. 2011. *State of Oregon Agriculture: Industry Report from the State Board of Agriculture*, 2009-11, State of Oregon Agriculture, 36 pp. http://www.oregon.gov/ ODA/docs/pdf/bd_rpt.pdf.

Ortiz, R., K. D. Sayre, B. Govaerts, R. Gupta, G. V. Subbarao, T. Ban, D. Hodson, J. M. Dixon, J. I. Ortiz-Monasterio, and M. Reynolds. 2008. "Climate Change: Can Wheat Beat the Heat?" *Agriculture, Ecosystems, and Environment* 126: 46-58. doi: 10.1016/j.agee.2008.01.019.

Ostrom, M. R., and R. A. Jussaume, Jr. 2007. "Assessing the Significance of Direct Farmer-Consumer Linkages as a Change Strategy in Washington State: Civic or Opportunistic?" In *Remaking the North American Food System: Strategies for Sustainability*, edited by C.C. Hinrichs and T. Lyson. Lincoln: University of Nebraska Press.

Parmesan, C. 2006. "Ecological and Evolutionary Responses to Recent Climate Change." *Annual Review of Ecology Evolution and Systematics* 37: 637-669. doi: 10.1146/annurev.ecolsys.37 .091305.110100.

Parry, M., C. Rosenzweig, A. Iglesias, M. Livermore, G. Fischer. 2004. "Effects of Climate Change on Global Food Production under SRES Emissions and Socio-Economic Scenarios." *Global Environmental Change* 14 (1): 53-67. doi: 10.1016/j.gloenvcha.2003.10.008.

Parry, M., C. Rosenzweig, and M. Livermore. 2005. "Climate Change, Global Food Supply and Risk Of Hunger." *Philosophical Transactions of the Royal Society, Series B* 360 (1463): 2125–2138. doi: 10.1098/rstb.2005.1751.

Parton, W. J., J. A. Morgan, R. H. Kelly, and D. Ojima. 2001. "Modeling Soil C Responses to Environmental Change in Grassland Systems." In *The Potential of U.S. Grazing Lands to Sequester Carbon and Mitigate the Greenhouse Effect*, edited by R. F. Follet, J. M. Kimble and R. Lal, 371-398. Boca Raton, FL: CRC Press.

Polley, H. W., W. Emmerich, J. A. Bradford, P. L. Sims, D. A. Johnson, N. Z. Sallendra, T. J. Svejcar, R. F. Angell, A. B. Frank, R. L. Phillips, K. A. Snyder, J. A. Morgan, J. Sanabria, P. C. Mielnick, and W. A. Dugas. 2010. "Precipitation Regulates the Response of Net Ecosystem CO_2 Exchange to Environmental Variation on United States Rangelands." *Rangeland Ecology & Management* 63 (2): 176-186. doi: 10.2111/REM-D-09-00015.1.

Poorter, H. 1993. "Interspecific Variation in the Growth Response of Plants to an Elevated Ambient CO_2 Concentration." *Vegetatio* 104-105 (1): 77-97. doi: 10.1007/BF00048146.

Pruski, F. F., and M. A. Nearing. 2002. "Climate-Induced Changes in Erosion During the 21st Century for Eight U.S. Locations." *Water Resource Research* 38: 1298. doi: 10.1029/2001WR000493.

Ritten, J. P., W. M. Frasier, C. T. Bastian, and S. T. Gray. 2010. "Optimal Rangeland Stocking Decisions Under Stochastic and Climate-Impacted Weather." *American Journal of Agricultural Economics* 92 (4): 1242-1255. doi: 10.1093/ajae/aaq052.

Ritten, J. P., B. S. Rashford, and C. T. Bastian. 2011. "Can Rangeland Carbon Sequestration Help Livestock Producers and Rural Economies Adapt to Climate Change?" *Rural Connections* June: 27-32. http://wrdc.usu.edu/files/uploads/Rural%20Connections/RCJUN11w.pdf

Roberts, D., and S. Rao. 2012. "Extension Leads Multi-Agency Team in Suppressing a Pest in the West." *Journal of Extension* 50 (2): 2FEA10. http://www.joe.org/joe/2012april/a10.php.

Robertson, G. P., E. A. Paul, and R. R. Harwood. 2000. "Greenhouse Gases in Intensive Agriculture: Contributions of Individual Gases to the Radiative Forcing of the Atmosphere." *Science* 289 (5846): 1922-1925. doi: 10.1126/science.289.5486.1922.

Rosenzweig, C., and M. L. Parry. 1994. "Potential Impact of Climate Change on World Food Supply." *Nature* 367 (6459): 133-138. doi: 10.1038/367133a0.

Rosenzweig, C., J. Phillips, R. Goldberg, J. Carroll, and T. Hodges. 1996. "Potential Impacts of Climate Change on Citrus and Potato Production in the US." *Agricultural Systems* 52 (4): 455-479. doi: 10.1016/0308-521X(95)00059-E.

Salisbury, P. A., and M. J. Barbetti. 2011. "Breeding Oilseed Brassica for Climate Change." In *Crop Adaptation to Climate Change*, edited by S. D. Yadav, R. J. Redden, J. L. Hatfield, H. Lotze-Campen, and A. E. Hall. 448-463. John Wiley & Sons, Inc.

Saure, M. C. 1985. "Dormancy Release in Deciduous Fruit Trees." *Horticultural Reviews* 7: 239–300. doi: 10.1002/9781118060735.ch6.

Schlenker, W., and M. J. Roberts. 2009. "Nonlinear Temperature Effects Indicate Severe Damages to U.S. Crop Yields under Climate Change." *Proceedings of the National Academy of Sciences* 106 (37): 15594-15598. doi: 10.1073/pnas.0906865106.

Smith, P., D. Martino, Z. Cai, D. Gwary, H. Janzen, P. Kumar, B. McCarl, S. Ogle, F. O'Mara, C. Rice, B. Scholes, and O. Sirotenko. 2007. "Agriculture." In *Climate Change 2007: Mitigation. Contribution of Working Group III to the Fourth Assessment Report of the Intergovernmental Panel on Climate Change*, edited by B. Metz, O. Davidson, P. Bosch, R. Dave, and L. Meyer. Cambridge University Press, Cambridge, United Kingdom and New York, NY, USA.

Snyder, C. S., T. W. Bruulsema, T. L. Jensen, and P. E. Fixen. 2009. "Review of Greenhouse Gas Emissions From Crop Production Systems And Fertilizer Management Effects." *Agriculture, Ecosystems and Environment* 133: 247–266. doi: 10.1016/j.agee.2009.04.021.

Sorte, B., P. Lewin, and P. Opfer. 2011. "Oregon Agriculture and the Economy: An Update." Oregon State University Extension Service, Rural Studies Program, http://ruralstudies.oregonstate.edu/sites/default/files/pub/pdf/OregonAgEconomyAnUpdate.pdf.

Stöckle, C., S. Higgins, A. Kemanian, R. Nelson, D. Huggins, J. Marcos, and H. Collins. 2012. "Carbon Storage and Nitrous Oxide Emissions of Cropping Systems in Eastern Washington: A Simulation Study." *Journal of Soil and Water Conservation* 67 (5): 365-377. doi: 10.2489/jswc.67.5.365.

Stöckle, C. O., R. L. Nelson, S. Higgins, J. Brunner, G. Grove, R. Boydston, M. Whiting, and C. Kruger. 2010. "Assessment of Climate Change Impact on Eastern Washington Agriculture." *Climatic Change* 102: 77-102. doi: 10.1007/s10584-010-9851-4.

Sutherst R. W., F. Constable, K. J. Finlay, R. Harrington, J. Luck, and M. P. Zalucki. 2011. "Adapting to Crop Pest and Pathogen Risks under a Changing Climate." *Wiley Interdisciplinary Reviews: Climate Change* 2 (2): 220-237.

Thomson, A. M., R. A. Brown, N. J. Rosenberg, R. C. Izaurralde, and V. Benson. 2005. "Climate Change Impacts for the Conterminous USA: An Integrated Assessment. Part 3: Dryland Production of Grain and Forage Crops." *Climatic Change* 69 (1): 43–65. doi: 10.1007/s10584-005-3612-9.

Thomson, L. J., S. Macfadyen, and A. A. Hoffmann. 2010. "Predicting the Effects of Climate Change on Natural Enemies of Agricultural Pests." *Biological Control* 52 (3): 296-306. doi: 10.1016/j.biocontrol.2009.01.022.

Timlin, D., S. M. L. Rahman, J. Baker, V. R. Reddy, D. Fleisher, and B. Quebedeaux. 2006. "Whole Plant Photosynthesis, Development, and Carbon Partitioning in Potato as a Function of Temperature." *Agronomy Journal* 98 (5): 1195-1203. doi: 10.2134/agronj2005.0260.

Tisdale, E. W. 1986. "Native Vegetation of Idaho." *Rangelands* 8 (5): 202-207. http://www.jstor.org/stable/3901016.

Trumble, J. T., and C. D. Butler. 2009. "Climate Change Will Exacerbate California's Insect Pest Problems." *California Agriculture* 63 (2): 73-78. doi: 10.3733/ca.v063n02p73.

Tubiello, F. N., C. Rosenzweig, R. A. Goldberg, S. Jagtap, and J. W. Jones. 2002. "Effects of Climate Change on US Crop Production: Simulation Results Using Two Different GCM Scenarios. Part I: Wheat, Potato, Maize, and Citrus." *Climate Research* 20 (3): 259–270. doi: 10.3354/cr020259.

Tubiello, F. N., J.-F. Soussana, and S. M. Howden. 2007. "Crop and Pasture Response to Climate Change." *Proceedings of the National Academy of Sciences* 104 (50): 19686-19690. doi: 10.1073/pnas.0701728104.

Turner, R. K., and G. C. Daily. 2008. "The Ecosystem Services Framework and Natural Capital Conservation." *Environmental and Resources Economics*: 39 (1): 25-35. doi: 10.1007/s10640-007-9176-6.

US Department of Agriculture. 2009. *Summary Report: 2007 National Resources Inventory*, Natural Resources Conservation Service, Washington, DC, and Center for Survey Statistics and Methodology, Iowa State University, Ames, Iowa. 123 pp. http://www.nrcs.usda.gov/Internet/FSE_DOCUMENTS/stelprdb1041379.pdf.

US Department of Agriculture Census of Agriculture. 2010. "Production Fact Sheet." http://www.agcensus.usda.gov/Publications/2007/Online_Highlights/Fact_Sheets/Production/.

US Department of Agriculture National Agricultural Statistics Service. 2006. "2005 Oregon Vineyard and Winery Report." February. http://www.nass.usda.gov/Statistics_by_State/Oregon/Publications/Vineyard_and_Winery/vw-2006.pdf.

US Department of Agriculture National Agricultural Statistics Service. 2011a. "2010 Washington Wine Grape Production up 3 Percent; Cabernet Sauvignon up 16 Percent; Chardonnay down 14 Percent." January 21. http://www.nass.usda.gov/Statistics_by_State/Washington/Publications/Fruit/grape11.pdf.

US Department of Agriculture National Agricultural Statistics Service. 2011b. "Washington Vineyard Acreage Report 2011." August 16. http://www.nass.usda.gov/Statistics_by_State/Washington/Publications/Fruit/VineyardAcreage2011.pdf.

US Department of Agriculture National Agricultural Statistics Service. 2012a. "2011 Oregon Vineyard and Winery Report." April. http://www.nass.usda.gov/Statistics_by_State/Oregon/Publications/Vineyard_and_Winery/v_2011_final.pdf.

US Department of Agriculture National Agricultural Statistics Service. 2012b. *Crop Production, 2011 Summary*, ISSN: 1057-7823.

Vano, J. A., M. J. Scott, N. Voisin, C. O. Stöckle, A. F. Hamlet, K. E. B. Mickelson, M. M. Elsner, and D. P. Lettenmaier. 2010. "Climate Change Impacts on Water Management and Irrigated Agriculture in the Yakima River Basin, Washington, USA." *Climatic Change* 102: 287-317. doi: 10.1007/s10584-010-9856-z.

Wan, S. Q., D. F. Hui, L. Wallace, and Y. Q. Luo. 2005. "Direct and Indirect Effects of Experimental Warming on Ecosystem Carbon Processes in a Tallgrass Prairie." *Global Biogeochemical Cycles* 19: GB2014. doi: 10.1029/2004GB002315.

Washington State Department of Agriculture. 2011. "Agriculture - A Cornerstone of Washington's Economy". http://agr.wa.gov/AgInWa/docs/126-CropProductionMap11-11.pdf.

Washington State Department of Ecology. 2011. "Columbia River Basin Long-Term Water Supply and Demand Forecast." Publication No. 11-12-011. http://www.ecy.wa.gov/programs/wr/cwp/forecast/reports.html.

Washington State University Tree Fruit Research and Extension Center. 2012. "The Campaign for Tree Fruit." http://treefruit.wsu.edu/campaign/.

Watson, P., G. Taylor, and S. Cooke. 2008. "The Contribution of Agriculture to Idaho's Economy: 2006." *CIS Bulletin* 1144, University of Idaho Extension, http://www.cals.uidaho.edu/edComm/pdf/CIS/CIS1144.pdf.

Weinberger, J. H. 1950. "Chilling Requirements of Peach Varieties." *Proceedings of the American Society of Horticultural Science* 56: 122-128.

White, M. A., N. S. Diffenbaugh, G. V. Jones, J. S. Pal, and F. Giorgi. 2006. "Extreme Heat Reduces and Shifts United States Premium Wine Production in the 21st Century." *Proceedings of the National Academy of Sciences* 103 (30): 11217-11222. doi: 10.1073/pnas.0603230103.

Willamette Valley Specialty Seed Association). 2012. http://www.thewvssa.org/documents/Position_Rapeseed.pdf.

Yorgey, G., C. Kruger, R. Frear, R. Shumway, C. Bishop, S. Chen, and C. MacConnell. 2011. "Anaerobic Digestion in the Pacific Northwest." *Rural Connections* June: 33-38. http://whatcom.wsu.edu/ag/documents/anaerobic/AnaerobidDigestion.pdf.

Chapter 7

Human Health
Impacts and Adaptation

AUTHORS
Jeffrey Bethel, Steven Ranzoni, Susan M. Capalbo

7.1 Introduction

Climate scientists strongly agree that climate is changing (Confalonieri et al. 2007). Key elements of projected future climate change in the Northwest include increasing year-round temperatures and rising sea level (high confidence), changes in precipitation that include decreases in summer (medium confidence) and increases during the other seasons (low confidence), and increases in some kinds of extreme weather events (Chapter 2; Chapter 4). These changes will significantly affect natural and managed ecosystems and built environments in the Northwest, which in turn will have significant impacts on all aspects of society, including human health.

While some health outcomes associated with climate change are relatively direct (e.g., exposure to extreme heat), others are more complicated and involve multiple pathways (fig. 7.1). Incidents of extreme weather (e.g., floods, droughts, severe storms, heat waves, and wildfires) can directly affect human health and cause serious environmental and economic impacts. Indirect impacts can occur when climate change alters or disrupts natural and social systems. This can give rise to the spread or emergence of vector-, water-, and food-borne diseases in areas where they either have not existed, or where their presence may have been limited (Colwell et al. 1998; Gubler et al. 2001; Haines and Patz 2004; Reiter 2001). Respiratory conditions would be exacerbated by exposures to smoke from increased wildfires (Delfino et al. 2009; Spracklen et al. 2009; Littell et al. 2010; McKenzie et al. 2004). Air pollution and increases in pollen counts (and a prolonged pollen-producing season) would increase cases of allergies, asthma, and other respiratory conditions among susceptible people. Climate change can also impact mental health directly (e.g., acute or traumatic effects of extreme weather events), indirectly (e.g., threats to emotional well-being based on observation of impacts of climate change), and psychosocially (e.g., large-scale social and community effects of climate change) (Berry et al. 2010; Doherty et al. 2011). It is important to note that geographic regions in the Northwest will not experience climate change uniformly. The Northwest's climate is broadly divided by the Cascade mountain range into east and west regions that will have different levels of risk for wildfires, drought, extreme heat, and other climate-related changes.

This chapter synthesizes what is currently understood regarding the key impacts of climate change on human health in the Northwest. While the chapter focuses on the

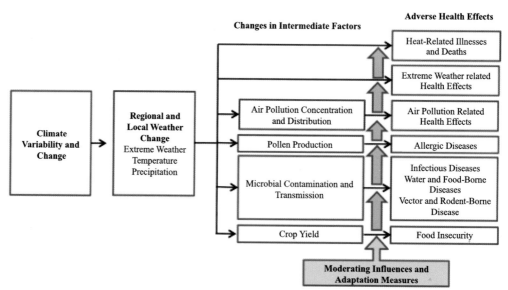

Figure 7.1 Potential health effects of climate variability and change adapted from Haines and Patz (2004).

adverse impacts, it is also acknowledged that climate change may positively affect human health. For example, warmer winters may result in decreased mortality from cold exposure (Huang et al. 2011), although hypothermia is not a large problem in the Northwest. However, most research concludes that the positive impacts of climate change on human health will be minor and outweighed by the adverse health impacts (Epstein and Mills 2005). The chapter also highlights how local and state health departments and other public health practitioners and researchers in the Northwest are addressing climate change through adaptation and mitigation planning, research, outreach and education, and other approaches. Finally, the chapter concludes with a summary of knowledge gaps and research needs.

7.2 Key Impacts of Climate Changes on Human Health

7.2.1 TEMPERATURE

The average temperature in the Northwest has increased by about 0.7 °C (1.3 °F) over the last century (Chapter 2). Average ambient temperatures and episodes of heat extremes in the Northwest are projected to continue increasing during the next century. In the Northwest, recent global and regional modeling suggests mean summer temperature increases of 1.9–5.2 °C (3.4–9.4 °F) by mid-century under a very high growth scenario (RCP8.5) (Chapter 2). Climate models are also unanimous that measures of heat extremes will increase and measure of cold extremes will decrease (Chapter 2, table 2.1). Specifically, the North American Regional Climate Change Assessment Program (NARCCAP) simulations (2041–2070 mean minus 1971–2000 mean, for an emissions scenario of continued growth [SRES-A2]) show an increase in the number of days maximum temperature is over 32 °C (90 °F), 35 °C (95 °F), and 38 °C (100 °F) as well as 134% and 163% increases in consecutive days with temperatures greater than 35 °C (95 °F) and 38 °C (100 °F), respectively, averaged over the region (Table 2.2). Increases in summer temperature are projected to be greatest in southern and eastern Oregon (east of Cascades) and most of Idaho (Chapter 2, fig. 2.8).

The Centers for Disease Control and Prevention (CDC) reported a total of 3,342 deaths resulting from exposure to extreme heat from 1999 to 2003 and 4,780 heat-related deaths in the United States from 1979 to 2002 (Luber et al. 2006; LoVecchio et al. 2005). During 1999–2003, cardiovascular disease was recorded as the underlying condition in 57% of the deaths in which hyperthermia was a contributing factor (Luber et al. 2006). Deaths among men accounted for 66% of all heat-related deaths and, for all groups, there was a greater number of heat-related deaths among men than women. Chicago, Illinois experienced two notable heat waves during the 1990s. In July 1995, 485 heat-related deaths and 739 excess deaths (i.e., attributed to an underlying condition such as cardiovascular disease) occurred as a result of a heat wave (Donoghue et al. 2003). Major risk factors for the heat-related deaths included advanced age and inability to care for oneself. During the 1999 heat wave in Chicago, 81 heat-related deaths occurred. Social isolation, advanced age, and medical and psychiatric conditions were major risk factors for a heat-related death (Naughton 2002). It is important to note the distinction between deaths specifically attributed to heat and excess deaths, which are largely attributed to underlying medical conditions, during hot days. Specifically, extreme heat worsens existing health conditions such as respiratory illness, cardiovascular disease, and renal failure, and may lead to increases in heart attacks, strokes, and all-cause mortality (Kaiser et al. 2007; Gosling et al. 2009; Knowlton et al. 2009).

Increased incidence of heat-related illness (HRI) can be expected in areas with increased temperature and increased occurrence of extreme heat events. HRI is a composite term referring to a broad spectrum of illness ranging from heat rash to death. Heat is created through a variety of the body's internal processes and absorbed when the ambient air temperature rises above body temperature (Lugo-Amador 2004). Body heat is dissipated via all four physical mechanisms of heat transfer (conduction, convection, radiation, and evaporation) with *convection* of heat from the body's core to the periphery via blood circulation and *evaporation* of sweat into ambient air (Lugo-Amador 2004). In addition to environmental temperature, direct sunlight, relative humidity, and wind velocity contribute significantly to heat-related illness.

Older adults, young children, persons with chronic medical conditions, and outdoor workers are particularly susceptible to HRI (Luber et al. 2006). Continual exposure to high temperature and heat extremes may cause several HRIs including heat rash, heat syncope (fainting), heat cramps, heat exhaustion, and heat stroke. Heat rash is caused by inflammation and obstruction of the sweat glands leading to sweat retention. Heat syncope is caused by disturbed blood distribution resulting in reduction of systolic blood pressure and cerebral hypotension with sudden unconsciousness. Heat cramps are typically preceded by large consumption of water without replacement salts in hot environmental conditions. Heat exhaustion is a result of prolonged exposure to heat in combination with inadequate intake of water and salt. Headache, thirst, extreme weakness, and confusion are common symptoms of heat exhaustion. Heat stroke is a life-threatening emergency characterized by high body temperature (> 40 °C [104 °F]) and a lost or severely reduced sweating capacity. Heat rash, heat syncope, heat cramps, and heat exhaustion are more common and less serious than heat stroke (Lugo-Amador 2004).

While outdoor workers have been identified as a population vulnerable to HRI (Balbus and Malina 2009), few studies have examined the morbidity associated with occupational HRI. Bonauto et al. (2007) examined the state workers' compensation fund in Washington State from 1995–2005 and found that the three highest HRI claim incidence rates by industry sector were Construction (12.1 per 100,000 FTE), Public Administration (12.0 per 100,000 FTE) and Agricultural, Forestry, Fishing (5.2 per 100,000 FTE). Of the 480 claims for HRI, 377 (78.5%) occurred as a result of working outdoors. Approximately 20% of all claims listed medication or an existing medical condition as a contributing factor to the HRI. In addition, Bonauto et al. (2007) reported that lack of acclimatization may have contributed to the HRI cases given that HRI claims occurring within one week of employment occurred more than four times as often as workers suffering injuries from all causes within the same time period. Schulte and Chun (2009) identified seven categories of climate-related hazards, including increased ambient air temperature, to develop a preliminary framework to explain how climate change may affect occupational health and safety. In addition to outdoor workers, indoor workers in workplaces such as greenhouses, manufacturing plants, and buildings without air conditioning are also at risk for HRI.

Data regarding heat-related deaths in the Northwest are limited. Jackson, Yost, et al. (2010) examined the historical relationship between age- and cause-specific mortality rates and heat events in the greater Seattle area, Spokane County, the Tri-Cities, and Yakima County in Washington from 1980 through 2006. They used the historical data in combination with climate projections for Washington State to predict the number of excess deaths by age group and cause during projected heat events in years 2025, 2045, and 2085. In the greater Seattle area, there are expected to be 96 (±12) excess deaths in 2025, 148 (±17) excess deaths in 2045, and 266 (±21) excess deaths in 2085 from all non-traumatic causes among persons 65 years and older under the middle warming scenario.[1]

The Oregon Health Authority (2012) analyzed warm season hospitalization and mortality data from 2000 to 2009 and found that each 5.6 °C (10 °F) increase in maximum temperature was associated with a 266% (95% confidence interval: 218–322%) increase in the rate of heat-related illness. This study also found that adults aged 65 and older were more vulnerable to adverse heat-related health outcomes than the general population. In addition, significant increases in other health effects such as electrolyte imbalance, nephritis, and acute renal failure hospitalizations, and all-cause and drowning mortality were associated with heat events in all regions of the state.

Greene et al. (2011) examined extreme heat events (EHE) and mortality in 40 large US cities including Seattle, Washington, and Portland, Oregon, assuming a very high growth emissions scenario (SRES-A1FI). The average number of projected EHE days in

---

1   Jackson, Yost, et al. (2010) define the middle scenario as the average of the PCM1 model forced with SRES-B1 (substantial reductions in emissions) and the HADCM model forced with SRES-A1B (continued growth peaking at mid-century).

Portland during 2020–2029, 2045–2055, and 2090–2099 is 37, 42, and 51, respectively. In Seattle, the average number of projected EHE days during the same periods is 51, 54, and 57, respectively. Similarly, the number of EHE-attributable deaths per summertime using the time periods 2020–2029, 2045–2055, and 2090–2099 in Portland is projected to be 8, 10, and 13, respectively. In Seattle, the number of EHE-attributable deaths per summertime over the same periods is projected to be 15, 14, and 18, respectively. These estimates are much lower than those calculated by Jackson, Yost, et al. (2010); however, the study area for Seattle included King, Pierce and Snohomish counties. Mortality was measured above the long-term adjusted baseline and was calculated using historical (1975–1999) statistical relationships between EHE-attributable mortality and meteorological conditions. Seattle was one of three cities that did not show the expected steady increase of EHE-attributable deaths over time.

7.2.2 EXTREME WEATHER EVENTS

7.2.2.1 Storms and Flooding

Another threat from climate change in the Northwest is extreme weather events including floods and storms. Projections of future changes in frequency and intensity of windstorms in the Northwest are ambiguous (Bengtsson et al. 2009), but projections of slight increases in extreme precipitation events are somewhat more robust, especially for western Washington (Chapter 2). Cyclones can create winds in excess of 161 kph (100 mph) and represent a direct health threat of injury or death. Windstorms represent a risk for direct injuries from flying debris, falling trees, downed power lines, and collapsed or damaged structures (Blackmore and Tsokri 2004). The Columbus Day storm of 1962 is an example of such a destructive storm. According to the National Oceanic and Atmospheric Administration (NOAA), the Columbus Day storm of 1962 was the strongest non-tropical wind storm ever to hit the continental United States, killing 46 people in Oregon, Washington, and British Columbia and injuring hundreds more (National Weather Service Forecast Office 2012). More recently, on December 1, 2007, an extra-tropical cyclone hit the Northwest, bringing winds in excess of 161 kph (100 mph) and leading to 18 deaths.

In the United States, flooding caused an average of 78 deaths per year during 2002–2011. In 2011, there were 113 fatalities in the United States due to floods, and 103 in 2010. Less than 1% of all US flood fatalities from 2002–2011 occurred in the Northwest (National Weather Service Office of Climate, Water, and Weather Services 2012). Flood risk during winter has increased in some rain dominant and near coastal mixed rain-snow basins in Washington and Oregon, especially those in which most floods already occur in winter and winter precipitation falls as a mix of rain and snow (Hamlet and Lettenmaier 2007). Future hydrologic modeling suggests that the largest increases in flooding will be in mixed rain-snow basins during winter (Mantua et al. 2010; see Chapter 3).

In addition to direct health consequences of windstorms and flooding (injuries and death), indirect health consequences are numerous. Like most natural disasters, increases in communicable diseases are a threat due to standing and contaminated water, crowded evacuation centers, and lack of clean water to wash and bathe with (Watson et al. 2007). Mold and microbial growth can increase risk of respiratory illness. Studies examining the impact of Hurricane Katrina showed that airborne mold concentrations in remediated homes of flooded areas were double that of non-flooded areas (Solomon et al. 2006). Recovery efforts from extreme weather events such as repairing homes and clearing debris also pose an increased risk for orthopedic injuries (Ohl and Tapsell 2000). Recovery efforts from extreme weather events like floods also expose clean-up workers and residents to contamination from hazardous and toxic substances that are released and spread as a result of the flooding.

7.2.2.2 Drought

Although a majority of climate models project increases in winter, spring, and fall precipitation, a majority also project decreased precipitation in the summer, which combined with temperature-driven loss of snow can lead to increased frequency of drought conditions (Chapter 2; Chapter 3). Potential health impacts of drought include mental health effects (see 7.2.6), various morbidities associated with drought-fueled wildfires (see 7.2.2.3), reduced water availability and quality (see Chapter 3), and food insecurity. This section will focus on drought's impact on food insecurity, as the other potential impacts of drought are covered elsewhere.

Food insecurity and food cost are related to droughts, as agriculture productivity is dependent on the availability of water (Rosegrant et al. 2009). The effects of drought are sometimes hard to capture, as the food commodity market is a global one. However, between 2005 and 2008 the cost of staple commodities (i.e., cereals, grain, and rice) more than tripled. While there were many economic reasons for this increase, such as fuel prices and political instability, global droughts in numerous countries were primary drivers of reduced supply (Webb 2010). In 2012, much of the United States experienced what the United States Department of Agriculture (USDA) declared as the worst and most extensive incidence of drought in 25 years. This resulted in drought disaster incidents being declared in 1,692 primary counties and natural disaster areas being declared in 276 contiguous counties across 36 states (USDA Farm Service Agency 2012). These included 4 primary and 8 contiguous counties in Oregon and 6 primary and 21 contiguous counties in Idaho. While the effects of this drought have yet to be fully realized, some early projections suggest a loss of at least $7–$20 billion, representing close to 10% total loss of the entire US crop and livestock production (Mühr et al. 2012).

Increased food prices pose a health threat in the form of malnutrition or food insecurity to vulnerable populations, primarily the poor and those living in rural areas of the Northwest. The main causes of mortality and morbidity associated with food and drought are reduced food intake and lack of varied diet leading to protein-energy malnutrition and micronutrient deficiency. Similarly, food insecurity typically leads to the consumption of calorie-rich, nutrient-poor foods, aggravating the obesity epidemic and all its health consequences.

Food-insecure households are, at times during the year, uncertain of having, or unable to acquire, enough food to meet the needs of all their members because they have insufficient money or other resources for food. The USDA examined household-level food insecurity and found that, during 2009–2011, 15.4%, 13.6%, and 13.7% of households in Washington, Oregon, and Idaho, respectively, had low or very low food security, compared to 14.7% of US households (Coleman-Jensen 2012a). In Idaho, the prevalence of food insecurity during 2009–2011 represented a statistically significant increase from

2006–2008 (+2.3 percentage points). In Washington, the food insecurity prevalence during 2009–2011 also represented a statistically significant increase from 2006–2008 (+4.3 percentage points) and 1999–2001 (+2.9 percentage points) (Coleman-Jensen 2012b).

7.2.2.3 Wildfires

Hot and dry conditions that bring on droughts also create an environment that is susceptible to wildfires and their associated health effects. In summer, the multi-model mean projected precipitation change is -9% by 2041–2070 compared with 1950–1999 while some models project decreases of 30%. Models that project the largest warming also project the largest decreases in precipitation (Chapter 2). In the Northwest, the risk of wildfires is greatest east of the Cascades where the increase in summertime temperatures is projected to be greatest. While the population density in the region is greatest west of the Cascades, all population centers are at risk due to wildfire smoke plumes drifting across the entire region.

Various studies show that much of the NW is at risk for increased future wildfire activity. Future projected changes in area burned range from less than 100% to greater than 500% increase in annual area burned compared to the 20th century and depend on vegetation type, region, time frame, method, and future climate scenario (Chapter 5). One study projected up to 40% more aerosol organic carbon and about 20% more elemental carbon present in the air largely due to increases in annual area burned by 2046–2055 compared to 1996–2005 in the western US (Spracklen et al. 2009). While the adverse effects of urban fine particulate matter (PM) air pollution (PM$_{2.5}$) on cardiovascular and respiratory health are known, the possible health effects associated with acute exposures to wildfire-generated PM are not well understood (Pope and Dockery 2006; Simkhovich et al. 2008; Willers et al. 2013). Wegesser et al. (2009) examined the health impact of exposure to elevated levels of wildfire-generated PM and the specific toxicity of PM arising from California wildfires in 2008. They found that PM concentrations were not only higher during the wildfires, but the PM was much more toxic to the lungs on an equal weight basis than was PM collected from normal ambient air in the region.

Elemental carbon, also known as black carbon, is well known for its harmful effect on the human respiratory system and is the primary component of fine particulate matter (PM$_{10}$ and PM$_{2.5}$) (Cavalli et al. 2010). Results from a study examining the impact of the large 2003 "Cedar Fire" in San Diego County show a 34% increase in asthma hospital admissions, concluding that wildfire-related increases in PM result in increased asthma incidents (Delfino et al. 2009). The study also found increased bronchitis and pneumonia hospital admissions associated with the fire. Similarly, a study following up the 2007 wildfires in California showed that combined area hospital admissions for respiratory complaints increased from 48.6 to 72.6 per day and asthma diagnosis increased from 21.7 to 40.4 per day during and immediately after the wildfires (Ginsberg et al. 2008). A similar increase in exacerbations of respiratory disease is possible given the projected increase in area burned in the Northwest.

7.2.3 AEROBIOLOGICAL ALLERGENS AND AIR POLLUTION

Climate change can have a direct negative impact on human health through respiratory disorders such as allergies, asthma, and other diseases of the lungs (Cecchi et al. 2010;

D'Amato et al. 2010; Reid and Gamble 2009). These mechanisms include the direct effects of an altered climate (e.g., temperature changes, humidity changes, more extreme weather) and effects from the agents of climate change (e.g., carbon dioxide, ground-level ozone, air pollution).

7.2.3.1 Aerobiological Allergens

Climate change has resulted in an alteration and extension of the time period during which plants pollinate. Increasing CO_2 concentrations, a primary driver of global climate change, and warmer temperatures are also major causes of changes in the allergenic proteins of plants. Experiments with ragweed have shown that plants grown in present era CO_2 concentrations (370 ppm) compared to pre-industrial era CO_2 concentrations (280 ppm) not only produce more pollen, but pollen with higher levels of the key allergy-causing protein (Singer et al. 2005). The effect was even greater when CO_2 levels were increased to projected future levels (600 ppm). Additional experimentation on ragweed showed that on average, high concentrations of CO_2 produced 32% more plant biomass, and 55% more pollen than plants grown in lower CO_2 concentrations (Rogers et al. 2006). Ziska et al. (2011) report that the duration of the ragweed pollen season has increased in recent decades as a function of latitude in North America, primarily a result of changes to the timing of the first frost of the fall season and duration of frost-free season. The risk to human health is evident given ragweed is found extensively throughout Idaho, eastern and southern Oregon, and in select areas of Washington, and approximately 26% of the US population has a positive skin test reaction to ragweed (Arbes et al. 2005; USDA National Resources Conservation Service).

Temperature also determines the concentration of various elements of plant pollen. A study involving birch trees, populous throughout the Northwest, showed that the trees flower earlier and also produce pollen with higher concentrations of allergenic proteins in warmer temperatures (Ahlholm et al. 1998). Temperature can have widespread effects on the reproductive phenology of all plants. In a large scale controlled experiment of multiple plant species, spring blooming plants advanced fruiting and blooming by 7.6 days in a warmer and wetter spring environment than average (Sherry et al. 2007). The same experiment showed that summer blooming plants delayed their blooming times by 4.7 days with the warmer and wetter lead season.

An effect of climate change on allergies is that plants can produce more pollen because of extended growing seasons and long fruiting times; pollen that has higher concentrations of allergenic proteins. Higher concentrations of pollen and allergenic proteins can result in more severe and longer-lasting allergy symptoms (Singer et al. 2005). Longer pollen seasons with pollen that contain higher concentrations of allergenic proteins increase the risk of triggering an attack among those suffering from asthma.

7.2.3.2 Air Pollution

Anthropogenic increases in particulate air pollution and tropospheric ozone concentration directly impact human health. Air pollution can exacerbate asthma symptoms among those that suffer from asthma, particularly in children (Epton et al. 2008; Liu et al. 2009). Especially vulnerable are those living in close proximity to heavy traffic, where negative effects have been more pronounced. Investigations using the 65,000 member

Women's Health Initiative cohort also showed increased risk of cardiovascular disease along with increased respiratory disorders in correlation to air pollution (Miller et al. 2007). In a review of multiple climate models, particulate matter and surface ozone was shown to increase during summertime months as a function of temperature and air stagnation (Jacob and Winner 2009). However, much of the Northwest will not be as strongly impacted as other regions of the country. Future changes in ozone pollution will vary across the country: larger increases are expected in the East, Southeast and Southwest regions, and lesser increases in the Northwest and Central states (Chen et al. 2009). Some urban areas are projected to experience increases of up to 20 ppbv of ozone mixing ratio, while increases in the Northwest are below 13 ppbv. Despite the relatively modest changes in ozone pollution in the Northwest as a whole (fig. 7.2), it is an important issue among people living in communities in central and eastern Oregon, eastern Washington, and the southern Idaho.

Health problems associated with ground-level ozone include chest pain, coughing, throat irritation, congestion, reduced lung function, worsening of bronchitis, emphysema, asthma, and repeated exposure may permanently scar lung tissue. A study of 448,850 participants in the American Cancer Society Cancer Prevention Study II linked long-term exposure to increased concentrations of ozone with higher risk of cardiopulmonary death (Jerrett et al. 2009). In addition, results from a study of 60 eastern US cities showed a positive interaction between increased temperature and ozone concentrations and mortality (Ren et al. 2008).

Data regarding the health effects of particulate air pollution or ground-level ozone in the Northwest are limited. Jackson, Yost, et al. (2010) used estimated relationships between ozone concentration and mortality to project the number of excess deaths due to ground-level ozone concentrations for mid-century (2045–2054) in King and Spokane counties in Washington based on the continued growth SRES-A2 emissions scenario.

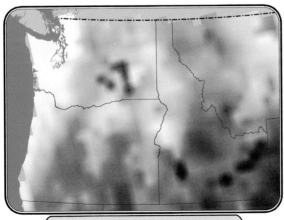

Change in summer averaged daily maximum 8h averaged (DM8H) ozone mixing ratios

+5 ppbv   +13 ppbv

Figure 7.2 Change in summer averaged daily maximum 8-hr ozone mixing ratios (ppbv) between a future case (2045–2054) and base case (1990–1999) based on future climate from a model forced with the continued growth emissions scenario (SRES-A2). Changes in ground-level ozone are due to global and local emissions, changes in environmental conditions and urbanization, and increasing summer temperatures. Adapted from Chen et al. (2009).

They estimated the annual number of May–September excess deaths due to ozone in King County to increase from 69 (95% confidence interval: 35–102) in 1997–2006 to 132 (95% confidence interval: 68–195) by mid-century. For Spokane County, the annual number of May–September excess deaths due to ozone was estimated to increase from 37 (95% confidence interval: 19–55) in 1997–2006 to 74 (95% confidence interval: 38–109) by mid-century.

7.2.4 INFECTIOUS DISEASES

Climatic factors such as temperature, humidity, rainfall, sea level rise, and ultraviolet radiation each impact the epidemiology of infectious diseases. In addition, interactions among vectors, animal reservoirs, microbes, and humans can result from changes to the physical environment, which also affect the transmission dynamics of infectious diseases (Institute of Medicine 2003). Specifically, climate can directly affect disease transmission through its effects on the microbial replication rate, microbial movement, movement and replication of vectors and animal hosts, and evolutionary biology (National Research Council 2001). Indirect effects of climate on disease transmission include effects operating through ecological changes and effects operating through changes in human activities. As such, climate change has the potential to impact the epidemiology of infectious diseases such as vector-borne diseases, water-borne diseases, and fungal diseases.

7.2.4.1 Vector-Borne Diseases

During the past 15 years, arboviruses (arthropod-borne viruses) such as West Nile virus, Chikungunya virus, Rift Valley fever virus, and Bluetongue virus have emerged and caused outbreaks in North America. Their emergence may be due to climate change, as climate is an important factor in determining the geographic and temporal distribution of arthropods (invertebrate animals having an exoskeleton), characteristics of the life cycles, dispersal patterns of associated arboviruses, evolution of arboviruses, and efficiency with which arboviruses are transmitted to vertebrate hosts (Gould and Higgs 2009). Specifically, longer, drier, and warmer summers in the Northwest may have a significant impact on the incidence of arboviruses in the Northwest.

West Nile virus (WNV) is a prime example of an emerging vector-borne disease that is now endemic in the Northwest. A mosquito-borne flavivirus, WNV debuted in North America in New York City in 1999 and moved across North America in subsequent years arriving in Idaho, Oregon, and Washington in 2003, 2004, and 2006, respectively (table 7.1). WNV was responsible for 712 known cases and 43 deaths in the United States in 2011 (CDC 2012b). During the last 10 years, the number of cases in the Northwest (Washington, Oregon, and Idaho) peaked in 2006 with 1,068 confirmed cases, of which 996 were in Idaho, which led the nation. There were six total cases in the three states during 2010 and 2011. Research has shown that increased temperatures influence WNV distribution in North America. Reisen et al. (2006) showed that above-average temperatures were associated with the original spread of WNV into western US states from the eastern US. Research regarding the association between precipitation and WNV spread has been inconsistent. While results from studies showed that prior drought contributed to

Table 7.1 Number of cases of select vector-borne diseases in the Northwest (2002–2010)

	2002	2003	2004	2005	2006	2007	2008	2009	2010
Hantavirus									
Washington	0	0	2	1	3	2	2	3	2
Oregon	0	2	0	0	0	0	0	2	3
Idaho	1	2	1	0	2	1	0	0	2
Lyme Disease									
Washington	11	7	14	13	8	12	22	15	12
Oregon	12	16	11	3	7	6	18	12	7
Idaho	4	3	6	2	7	9	5	4	6
West Nile Virus									
Washington	0	0	0	0	3	0	3	38	1
Oregon	0	0	3	7	69	26	16	11	0
Idaho	0	1	3	13	996	132	39	38	1

the initial outbreak of WNV, results from subsequent research have shown both positive and negative associations between precipitation and WNV (Shaman et al. 2005; Hubalek 2000). However, longer summers and higher temperatures may increase the incidence of WNV and other encephalitides.

The number of cases and incidences of other vector-borne diseases such as Lyme disease, hantavirus, malaria, and dengue are quite low in the Northwest (table 7.1) and the impact of climate change on these diseases is unknown. Doggett et al. (2008) examined Lyme disease in Oregon and calculated the incidence with exposure to ticks *only* in Oregon. Using field surveys, they adjusted the total number of reported cases of clinical Lyme disease cases during 1999–2004 based on probable tick exposure from 94 (total reported) to 23 (0.34 per 100,000 population).

7.2.4.2 Water-Borne Diseases

Higher ocean and estuarine temperature in the Northwest has the potential to increase the number of *Vibrio parahaemolyticus* infections from eating raw oysters or other shellfish. *V. parahaemolyticus* is a bacterium that lives in brackish saltwater and causes gastrointestinal illness in humans. The bacterium may require water temperatures 17.2 °C (63 °F) or higher to reach levels in shellfish which are sufficient to infect humans (Cieslak and Kohn 2008). *V. parahaemolyticus* naturally inhabits coastal waters in the United States and Canada and is present in higher concentrations during summer. Oysters harvested in Oregon and Washington have led to outbreaks of *V. parahaemolyticus* infection. During a 2006 outbreak, there were 8 confirmed and 8 probable cases in Oregon and 55 confirmed and 23 probable cases in Washington, much higher than the usual incidence during May–July of previous years (Balter et al. 2006). All confirmed and probable cases consumed shellfish from the same source, which were linked to harvest areas in Washington and British Columbia, Canada. Also, during July–August of 1997, an outbreak occurred in which 209 culture-confirmed *V. parahaemolyticus* infections (1 death) were associated with eating raw oysters harvested from California, Oregon, and Washington in the United States and British Columbia in Canada (Fyfe et al. 1998). The risk for *Vibrio* infection will increase with warming waters in the Northwest.

Anticipated increases in precipitation and subsequent flooding have the potential to wash animal intestinal pathogens into drinking water reservoirs and recreational waters. In particular, *Cryptosporidium*, protozoa found in cattle, can be transported by water and cause gastrointestinal illness in humans. During 2006–2010, the rate of cryptosporidiosis in Idaho was above the national rate in every year and the rate in Oregon was above the national rate in three of the five years (Yoder and Beach 2007; Yoder et al. 2010; Yoder et al. 2012). Of note, the rate of cryptosporidiosis in Idaho was nearly 10 times the national rate in 2007 during which there were 518 cases including 365 outbreak cases linked to recreational water. The Northwest will continue to be at risk of outbreaks of *Cryptosporidium* due to the abundance of recreational waters throughout the region.

7.2.4.3 Fungal Diseases

One fungal disease is of particular importance in the Northwest, as its presence in the region may be linked to climate change. The fungus *Cryptococcus gattii* was, until recently, considered to be restricted to tropical and subtropical climates and primarily affect persons with normal immunity (i.e., those uninfected with HIV) (Dixit et al. 2009). However, first recognized in the Northwest on Vancouver Island, British Columbia, Canada in 1999, *C. gattii* is now found in mainland British Columbia as well as Washington and Oregon (Datta et al. 2009). During January 1, 2004–July 1, 2010, 59 human cases of *C. gattii* were reported to the CDC from the Northwest including 15 from Washington, 43 from Oregon, and 1 from Idaho (fig. 7.3). Nine (20%) of the 45 patients for whom the outcome was known died because of the infection and 6 (13%) died with *C. gattii* infection (DeBess et al. 2010). The emergence of *C. gattii* in a temperate climate suggests the fungus may have adapted to a new climate or climate change might have created an environment in which the fungus can survive and propagate consistently (Dixit et al. 2009; Datta et al. 2009). Regardless, *C. gattii* is a notable example of an emerging infectious disease in the Northwest and raises the possibility of additional emerging diseases in the future.

7.2.5 HARMFUL ALGAL BLOOMS

Harmful algal blooms (HABs) occur when certain types of microscopic algae grow quickly in water, typically forming visible patches that may harm the health of the environment, plants, or animals. HABs can deplete the oxygen and block the sunlight that other organisms need to live. The natural toxins produced by some harmful algae can become concentrated in some filter feeding shellfish which, when eaten by humans, can cause illness or death (Moore et al. 2008). HABs can occur in marine, estuarine, and fresh waters and can impair drinking and recreational waters. On the US West Coast, the main toxin-producing algal species are dinoflagellates in the genus *Alexandrium* that cause paralytic shellfish poisoning and diatoms in the genus *Pseudo-nitzschia* that produce domoic acid and cause domoic acid poisoning (Horner et al. 1997).

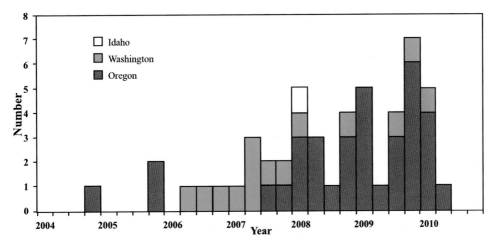

Figure 7.3 Cases of Cryptococcus gatti infection (n = 50) with known illness onset date, by quarter – Idaho, Washington, and Oregon, 2004–2010 (DeBess et al. 2010)

HABs appear to be increasing in frequency, intensity, and duration in all aquatic environments on a global scale, including along the coastlines and in the surface waters of the United States (Van Dolah 2000; Glibert et al. 2005) and may be partially due to climate change though the association remains poorly understood. Few studies have examined the association between climate change and HABs due to difficulty in separating the influence of climate variability and change from other known HAB contributing factors like eutrophication (Moore et al. 2008). Regardless, since many HAB species are currently limited by cold temperatures, warmer conditions in the future may lead to more outbreaks. For example, rising temperatures in the Northwest may promote earlier and longer lasting blooms of *A. catenella* in the Puget Sound region, and these blooms are correlated with shellfish toxicity (Moore et al. 2008; Horner et al. 1997). Climate change and changing ocean properties could cause more frequent HABs over greater geographic areas, increasing the risk for human illness and death from exposure to HAB toxins.

7.2.6 MENTAL HEALTH

The five preceding key impacts in this section focus on the climate impacts and resulting effect on human health. Because climate and weather events can affect mental health as well as physical health through several pathways, we devote a subsection to mental health. It is important to note that the impact of climate change on mental health will vary in different parts of the Northwest just as other impacts will vary across the region. Like physical health impacts, there are both direct and indirect mental health impacts of climate change. Direct impacts on mental health arise from traumatic weather events and their associated immediate impacts. Indirect impacts on mental health include persistent chronic stressors due to changes or the threat of changes to the physical, economic, and social environments among those affected.

One primary source of mental health distress following an extreme weather event is the potential destruction of one's home, which can affect mental health in two ways. First is the idea of losing what made the house a home, meaning the emotional impact of losing relics and items that were unique to the individual such as pictures, family heirlooms, and children's crafts (Carroll et al. 2009). Secondly, loss of home can lead to distress resulting from physical displacement (Davis et al. 2010). Distress may also come from witnessing a traumatic event. Studies have found that significant proportions of children, adults, and elderly individuals suffer from post-traumatic stress disorder (PTSD) symptoms following traumatic weather events. These mental health impacts can be long lasting. A survey conducted at five months, then again at one year, post Hurricane Katrina found an 89.2% increase in serious mental illness, a 31.9% increase in PTSD symptoms, and a 61.6% increase in suicidal thoughts and plans among those surveyed pre-hurricane (Kessler et al. 2008).

Mental health can be affected by a large array of secondary stressors as well. Financial concerns, recovery and rebuilding, family pressure, loss of leisure and recreation, and loss of security are examples of secondary stressors that can all accumulate into significant chronic stress (Lock et al. 2012). Living in a state of long-term stress can manifest as negative physical health outcomes, such as system-wide inflammation, high blood pressure, and unhealthy coping mechanisms such as smoking, excessive alcohol use, and poor dietary habits (Jackson, Knight, et al. 2010). For example, a longitudinal study found that experiencing a traumatic event increases the odds of tobacco use by 12% when controlling for other factors, including PTSD symptoms (Parslow and Jorm 2006).

Mental health impacts are not limited to experiencing a traumatic event. The threat of a climate event, the uncertainty of the future, or the loss of control over a situation can result in feelings of depression or helplessness (Sartore et al. 2008). Changes in the physical environment and social environment, as well as impacts from adaptation and mitigation measures can result in anxiety, depression, and feelings of loss of control (Berry et al. 2010). The weather itself can also have a direct impact on mental health. Seasonal affective disorder is a state of depression that varies with seasonal changes in environmental light levels and is more prevalent in northern latitudes with a deficiency in environmental light (Danilenko and Levitan 2012). A study in Australia revealed that hospital admissions during heat waves (temperatures exceeding 26.7 °C [80 °F]) are associated with a 7.3% increase in hospital admissions for mood or behavioral disorders (Hansen et al. 2008).

7.2.7 POTENTIAL HEALTH COSTS

The previous sections discuss the specific public health vulnerabilities and outcomes that have been linked to climate change. In this section we explore the connections between impact of climate change on human health and its economic dimensions: as a comparison of the costs and benefits (relative costs) of health policies developed to adapt to a changing climate, and the costs that will be incurred in spite of climate adaptation or mitigation policies. Research specific to the Northwest is lacking, so we draw upon the limited studies, which focus on valuing the impacts on a more aggregate (national) scale.

Economic research on the health costs is dependent on knowledge from other scientists and public health professionals regarding the likely magnitude of the health impacts on target populations or damage impacts, as well as the capacity of existing health care and public health systems and infrastructure to respond to increased demands. As Hutton (2011) notes, "the global evidence base on the economics of climate change is extremely weak, especially in the health sphere. Little is known still on the precise health impacts of climate change at a sub-national level, their economic costs, and the costs and benefits of measures to protect health from climate change." Valuing the health cost component of a changing climate is complex since not all health-related damages, such as loss of life and deteriorations in existing health conditions, can be easily monetized, and damages are dependent upon future climate scenarios. Also, using increased health care expenditures and household income approaches to isolate the increase in health care expenditures attributable to climate change is difficult due to current insurance markets and policy-induced responses, and to the inherent difficulty involved in attributing expenditure increases precisely to a climate-induced health event. As a result, the few economic valuation studies of impacts on public health focus on valuing only the most direct and quantifiable types of economic impacts, leading to a systematic undervaluation of the public health effects of climate change (Hutton 2011).

A conclusion reached earlier in this chapter is that for the Northwest, the potential health impact of climate change is still low, relative to other parts of the United States, with a few exceptions. Climate change is expected to worsen ozone pollution, heat waves, infectious disease outbreaks, river flooding, and wildfires. A recent study by Knowlton et al. (2011) estimated substantial healthcare costs associated with such climate change-related events in the United States (e.g., ozone air pollution, heat wave, hurricane, West Nile virus outbreak, flooding, and wildfires); these isolated climate-related events were estimated to cost about $14 billion in lost lives and health costs in the specific locations where they occurred. Though none of the events occurred in the Northwest per se, the Knowlton analysis provides a framework to quantify potential climate-related health costs to the Northwest region should similar well-defined events occur.

7.3 Northwest Adaptation Activities

With the assistance of academia, public health officials in the three states are taking a more active role in preparing for climate change and its associated hazards, improving their responses to them and working with vulnerable communities to help them adapt to these increased threats. As such, there are numerous adaptation activities taking place in the Northwest to combat the potential impact of climate change on human health.

As a starting point for adaptation activities in Oregon, the Oregon Coalition of Local Health Officials (CLHO) and the University of Oregon's Climate Leadership Initiative (CLI) conducted a study to determine the awareness, preparation, and resource needs for potential public health risks associated with climate change (Vynne and Doppelt 2009). In 2008, CLHO and CLI distributed a web-based survey to public health workers (N=84) in 35 county health departments in Oregon. Responses from 34 participants in 25 counties showed that, while most respondents view climate change as a serious or very serious problem, 97% of respondents do not consider climate change preparation as one of their top 5 priorities. Also, most respondents cited a need for additional resources such as training, funding, and staff to adequately prepare and adapt to climate change-related health risks.

The Oregon Health Authority (OHA) Public Health Division received funds from the CDC through a cooperative agreement in 2011 to address the health impacts of climate change. The overall goal of the project is to build local and state climate change capacity and increase awareness about climate change, the public health implications, and the mitigation and adaptation strategies that can improve public health. The Climate Change Capacity Building through Health Impact and Hazard Vulnerability project is ending its second year of funding and has achieved numerous milestones (Oregon Health Authority 2011). OHA recruited five local health jurisdictions to participate in the project and trained them on each step of the Building Resilience Against Climate Effects (BRACE) framework. The model consists of five steps and is designed for local health jurisdictions to be able to develop their own climate change adaptation plans. Local health jurisdictions have carried out climate and health assessments and have identified issues such as heat, vector-borne disease, and drought to focus on implementing adaptation strategies. OHA has also offered numerous trainings and resources to assist local health jurisdictions in communicating about climate and health.

OHA has also conducted climate and health analysis projects such as a 10-year retrospective analysis of the impact of heat events on health in Oregon. OHA has worked with the Oregon Climate Change Research Institute and the National Weather Service (NWS) to change inland Northwest NWS guidelines for heat advisories based on potential health effects. OHA also completed a health impact assessment of the Portland Metro region's Climate Smart Scenarios planning project to reduce greenhouse gas emissions. The health impact assessment identified the scenarios with the greatest health cobenefits and generated specific recommendations about including health in climate change planning efforts in the region and throughout Oregon.

The Resource Innovation Group (TRIG)/Climate Leadership Initiative (CLI) and Biositu, LLC conducted a similar collaborative effort designed to increase capacity among county, regional, and tribal public health departments and agencies in the Northwest. The two groups published *Public Health* and *Climate Change: A Guide for Increasing Capacity of Local Public Health Departments*, a guidebook designed to support efforts to initiate and integrate climate planning within county, regional, and tribal public health departments and agencies. The guidebook is meant to be worked through over six months to a year at the individual level or, ideally, in conjunction with or at least with support from, organization leadership. In 2010, TRIG/CLI also released two guidebooks for climate change in the public health sector. These two guidebooks describe specific activities that public health departments can take to reduce greenhouse gas emissions (mitigation) and prepare for health-related climate risks (adaptation), both internally and in their community.

The US Global Change Research Program (USGCRP) sponsored the National Climate Assessment Health Sector Workshop: Northwest Region in February of 2012. The purpose of the workshop was to provide a more nuanced representation of regional climate change impacts on human health. A report produced from the findings of the workshop highlighted adaptation efforts in Washington, Oregon, and Idaho (USGCRP 2012). Washington State began an effort that was co-chaired by the University of Washington and the Washington State Department of Health to examine climate change and health in 2007. In 2009, the State Agency Climate Leadership Act (SB5560) was adopted, bringing together multiple agencies to look at climate change issues affecting different sectors. While health was not mentioned as an explicit part of that legislation, it was included after the fact. As a result of SB5560, Washington drafted an integrated response strategy that includes a section on human health (Adelsman and Ekrem 2012). Also, researchers at the University of Washington have developed projections of impacts of excessive heat and are attempting to integrate findings into local planning and adaptation strategies (Jackson, Yost, et al. 2010). Public health practitioners in Seattle and King County have formed a climate change team that focuses on both mitigation and adaptation. Clark County is working on a county comprehensive plan update, which will include a health component for the first time. This update will recommend establishing a county climate action community to address local climate risks.

NW states are also taking steps to reduce the impact of HABs in the region. Oregon's Harmful Algal Bloom monitoring project was initiated in 2005, after a coast-wide shellfish harvesting closure. Oregon developed a monitoring program similar to Washington State's Olympic Region Harmful Algal Bloom (ORHAB) project. In 2006, Oregon Department of Fish and Wildlife (ODFW) in collaboration with Oregon State University, University of Oregon, and the NOAA Northwest Fisheries Science Center were awarded funds to develop an integrated HAB monitoring and event response program, Monitoring Oregon's Coastal Harmful Algae (MOCHA). Currently, ODFW staff in conjunction with Oregon Department of Agriculture is working to monitor 10 sites along the Oregon coast for any potential signs of the phytoplankton that cause domoic acid (*Pseudonitzschia sp.*) and paralytic shellfish (*Alexandrium sp.*) poisoning.

7.4 Knowledge Gaps and Research Needs

Looking forward, there are several knowledge gaps and research needs that must be filled in order to better understand the full impact of climate change on human health and to effectively adapt communities to the changing climate. Knowledge gaps and research needs relevant to the Northwest include:

Accurate surveillance data on climate-sensitive health outcomes. Increased and improved surveillance for climate-sensitive health outcomes are needed to develop and evaluate public health adaptation strategies and to better project the impacts of climate change on human health (English et al. 2009). Environmental and health outcome indicators for climate change have been proposed by the Council of State and Territorial Epidemiologists (CSTE) and are being tested in some states. Health outcome indicators include rate of heat-related deaths, hospitalizations, and emergency room visits during summer months; injuries and deaths due to extreme weather events; human cases of Lyme disease; human cases of West Nile virus; human cases of Valley fever (Coccidioidomycosis), Dengue fever, and Hantavirus; and allergic disease (English et al. 2009). While not all of the health outcome indicators are relevant to the Northwest (e.g., Coccidioidomycosis), public health officials in the Northwest can use the list proposed by CSTE as a framework for regional activities. In addition, heat-related illness is vastly underreported as a diagnosis and, instead, typically exacerbates an underlying condition. Additional coding for heat-related illness is needed to better identify the morbidity associated with increased temperatures.

Better matching of environmental monitoring with public health monitoring. Related to the need mentioned above regarding climate-sensitive health outcomes, improved linkages between environmental indicators and public health surveillance activities are also needed to develop and evaluate public health adaptation strategies and to better project the impacts of climate change on human health. Environmental indicators for climate change (e.g., pollen indicators, air mass stagnation events, greenhouse gas emissions, number of wildfires and percent of total acres impacted) have been developed and, coupled with health outcome indicators for climate change, will allow researchers to identify local and regional risks in the Northwest.

Increased regional-level modeling. As the impacts of climate change will vary across the Northwest, additional regional-level modeling of ozone and other air pollutants (Herron-Thorpe et al. 2010), temperature, precipitation, and wildfires is needed to allow researchers to develop corresponding regional-level mortality and morbidity projections. Again, these projections are needed to develop and evaluate public health adaptation strategies and to better project the impacts of climate change on human health.

Increased capacity building efforts among state health departments and local health jurisdictions throughout the Northwest. Health impact assessments (HIAs) and public health hazard vulnerability assessments (PH-HVAs) are important capacity building activities at any level of government and wider adoption of these activities throughout the region is needed. HIAs and PH-HVAs help to identify people, property, and resources that are at risk of injury, damage, or loss from natural or technological hazards as well as intentional threats. HIAs and PH-HVAs are generally conducted to help establish priorities for planning, capability development, and hazard mitigation; serve as a tool in the identification of hazard mitigation and adaptation measures; be a tool in a hazard-based needs analysis; serve to educate the public and public officials about hazards and vulnerabilities; and help communities make objective judgments about acceptable risk.

Improved understanding of the interaction between climate and vector-borne and zoonotic diseases (VBZD) in the Northwest and better understanding of how climate change will affect the epidemiology of VBZDs. Mills et al. (2010) identified several research studies needed to address this issue which can be applied to the Northwest. One important study is to establish baseline data on the geographic and habitat distribution of recognized zoonotic and vector-borne pathogens in the Northwest and their hosts and vectors. This would allow researchers to prospectively examine spatial and temporal changes. Another needed study is to establish longitudinal monitoring programs, which would identify factors associated with changes in the host, vector, and pathogen, as well as the interactions between the three. The goal is to identify changes in risk of VBZDs among humans. Finally, predictive models are needed to examine the changes in zoonotic disease risk and the projected distribution and abundance of major hosts and vectors using laboratory, field, and epidemiologic studies. A related example of this effort that is currently underway was the establishment of a surveillance system for *C. gattii* once it was first recognized in the Northwest.

Acknowledgments

The authors of the Health chapter wish to acknowledge and thank the Oregon Health Authority Public Health Division, Michael Heumann (HeumannHealth Consulting,

LLC), Stacy Vynne (Puget Sound Partnership), and an anonymous reviewer as well as Philip Mote and Meghan Dalton (Oregon State University) for their thoughtful review, comments, and feedback during the development of this chapter.

References

Adelsman, H., and J. Ekrem. 2012. *Preparing for a Changing Climate: Washington State's Integrated Climate Response Strategy.* Washington Department of Ecology, Publication No. 12-01-004, Olympia, Washington. http://www.ecy.wa.gov/climatechange/ipa_responsestrategy.htm.

Ahlholm, J. U., M. L. Helander, and J. Savolainen. 1998. "Genetic and Environmental Factors Affecting the Allergenicity of Birch (Betula pubescens ssp. czerepanovii [Orl.] Hämet-Ahti) Pollen." *Clinical and Experimental Allergy* 28 (11): 1384-1388. doi: 10.1046/j.1365-2222.1998. 00404.x.

Arbes, S. J., Jr., P. J. Gergen, L. Elliott, and D. C. Zeldin. 2005. "Prevalences of Positive Skin Test Responses to 10 Common Allergens in the US Population: Results from the Third National Health and Nutrition Examination Survey." *The Journal of Allergy and Clinical Immunology* 116 (2): 377-383. doi: 10.1016/j.jaci.2005.05.017.

Balbus, J. M., and C. Malina. 2009. "Identifying Vulnerable Subpopulations for Climate Change Health Effects in the United States." *Journal of Occupational and Environmental Medicine* 51 (1): 33-37. doi: 10.1097/JOM.0b013e318193e12e.

Balter, S., H. Hanson, L. Kornstein, L. Lee, V. Reddy, S. Sahl, F. Stavinsky, M. Fage, G. Johnson, J. Bancroft, W. Keene, M. Williams, K. MacDonald, N. Napolilli, J. Hofmann, C. Bopp, M. Lynch, K. Moore, J. Painter, N. Puhr, and P. Yu. 2006. "*Vibrio parahaemolyticus* Infections Associated with Consumption of Raw Shellfish--Three States, 2006." *Morbidity and Mortality Weekly Report 55* (31): 854-856.

Bengtsson, L., K. I. Hodges, and N. Keenlyside. 2009. "Will Extratropical Storms Intensify in a Warmer Climate?" *Journal of Climate* 22 (9): 2276-2301. doi: 10.1175/2008JCLI2678.1.

Berry, H. L., K. Bowen, and T. Kjellstrom. 2010. "Climate Change and Mental Health: A Causal Pathways Framework." *International Journal of Public Health* 55(2): 123-132. doi: 10.1007/ s00038-009-0112-0.

Blackmore, P., and E. Tsokri. 2004. "Windstorm Damage to Buildings and Structures in the UK during 2002." Weather 59 (12): 336-339. doi: 10.1256/wea.230.03.

Bonauto, D., R. Anderson, E. Rauser, and B. Burke. 2007. "Occupational Heat-Illness in Washington State, 1995-2005." *American Journal of Industrial Medicine* 50 (2): 940-950.

Carroll, B., H. Morbey, R. Balogh, and G. Araoz. 2009. "Flooded Homes, Broken Bonds, the Meaning of Home, Psychological Processes and Their Impact on Psychological Health in a Disaster." *Health & Place* 15 (2): 540-547. doi: 10.1016/j.healthplace.2008.08.009.

Cavalli, F., M. Viana, K. E. Yttri, J. Genberg, and J. P. Putaud. 2010. "Toward a Standardised Thermal-Optical Protocol for Measuring Atmospheric Organic and Elemental Carbon: The EUSAAR Protocol." *Atmospheric Measurement Techniques* 3 (1): 79-89. doi: 10.5194/ amt-3-79-2010.

Cecchi, L., G. D'Amato, J. G. Ayres, C. Galan, F. Forastiere, B. Forsberg, J. Gerritsen, C. Nunes, H. Behrendt, C. Akdis, R. Dahl, and I. Annesi-Maesano. 2010. "Projections of the Effects of Climate Change on Allergic Asthma: The Contribution of Aerobiology." *Allergy* 65 (9): 1073-1081. doi: 10.1111/j.1398-9995.2010.02423.x.

Centers for Disease Control and Prevention. 2011. "2010 Child Asthma Data: Prevalence Tables, Table L1." Last updated November 16. http://www.cdc.gov/asthma/brfss/2010/child/lifetime /tableL1.htm.

Centers for Disease Control and Prevention. 2012a. "2010 Adult Asthma Data: Prevalence Tables and Maps, Table L1." Last updated August 27. http://www.cdc.gov/asthma/brfss/2010 /lifetime/tableL1.htm.

Centers for Disease Control and Prevention. 2012b. "Final 2011 West Nile virus Human Infections in the United States." Last modified April 18. http://www.cdc.gov/ncidod/dvbid /westnile/surv&controlCaseCount11_detailed.htm

Chen, J., J. Advise, B. Lamb, E. Salathé, C. Mass, A. Guenther, C. Wiedinmyer, J.-F. Lamarque, S. O'Neill, D. McKenzie, and N. Larkin. 2009. "The Effects of Global Changes upon Regional Ozone Pollution in the United States." *Atmospheric Chemistry and Physics* 9: 1125-1141. doi: 10.5194/acp-9-1125-2009.

Cieslak, P., and P. Kohn. 2008. "Climate Change and Communicable Diseases in the Northwest." *Northwest Public Health* Fall/Winter: 16-17. http://www.nwpublichealth.org/docs/nph/f2008 /cieslak_kohn_fw2008.pdf.

Coleman-Jensen, A., M. Nord, M. Andrews, and S. Carlson. 2012a. *Household Food Security in the United States in 2011.* ERR-141, US Department of Agriculture, Economic Research Service, September 2012. http://www.ers.usda.gov/media/884525/err141.pdf.

Coleman-Jensen, A., M. Nord, M. Andrews, and S. Carlson. 2012b. *Statistical Supplement to Household Food Security in the United States in 2011,* AP-058. US Department of Agriculture, Economic Research Service, September 2012. http://www.ers.usda.gov/media/884603 /apn-058.pdf.

Colwell, R., P. Epstein, D. Gubler, M. Hall, P. Reiter, J. Shukla, W. Sprigg, E. Takafuji, and J. Trtanj. 1998. "Global Climate Change and Infectious Diseases." *Emerging Infectious Diseases* 4 (3): 451-452. doi: 10.3201/eid0403.980327.

Confalonieri, U., B. Menne, R. Akhtar, K. L. Ebi , M. Hauengue, R. S. Kovats, B. Revich, and A. Woodward. 2007. "Human Health." In *Climate Change 2007: Impacts, Adaptation and Vulnerability. Contribution of Working Group II to the Fourth Assessment Report of the Intergovernmental Panel on Climate Change,* edited by M. L. Parry, O. F. Canziani, J. P. Palutikof, P. J. van der Linden, and C. E. Hanson. 391–431. Cambridge University Press, Cambridge, UK.

D'Amato, G., L. Cecchi, M. D'Amato, and G. Liccardi. 2010. "Urban Air Pollution and Climate Change as Environmental Risk Factors of Respiratory Allergy: An Update." *Journal of Investigative Allergology and Clinical Immunology* 20 (2): 95-102. http://www.jiaci.org/issues /vol20issue2/vol20issue02-1.htm.

Danilenko, K. V., and R. D. Levitan. 2012. "Seasonal Affective Disorder." In *Neurobiology of Psychiatric Disorders,* edited by T. E. Schlaepfer and C. B. Nemeroff, 279-290. Elseveir, B.V. Amsterdam, The Netherlands.

Datta, K., K. H. Bartlett, R. Baer, E. Byrnes, E. Galanis, J. Heitman, L. Hoang M. J. Leslie, L. MacDougall, S. S. Magill, M. G. Morshed, K. A. Marr. 2009. "Spread of *Cryptococcus gattii* into Pacific Northwest Region of the United States." *Emerging Infectious Diseases* 15 (8): 1185-1191. doi: 10.3201/eid1508.081384.

Davis, T. E., A. E. Grills-Taquechel, and T. H. Ollendick. 2010. "The Psychological Impact From Hurricane Katrina: Effects of Displacement and Trauma Exposure on University Students." *Behavior Therapy* 41(3): 340-9. doi: 10.1016/j.beth.2009.09.004.

DeBess, E., P. R. Cieslak, N. Mardsen-Haug, M. Goldoft, R. Wohrle, C. Free, E. Dykstra, R. J. Nett, T. Chiller, S. R. Lockhart, J. Harris. 2010. "Emergence of *Cryptococcus gattii* – Pacific Northwest, 2004-2010." *Morbidity and Mortality Weekly Report* 59 (28): 865-868.

Delfino, R. J., S. Brummel, J. Wu, H. Stern, B. Ostro, M. Lipsett, A. Winer, D. H. Street, L. Zhang, T. Tjoa, and D. L. Gillen. 2009. "The Relationship of Respiratory and Cardiovascular

Hospital Admissions to the Southern California Wildfires of 2003." *Occupational and Environmental Medicine* 66 (3): 189-197. doi: 10.1136/oem.2008.041376.

Dixit, A., S. F. Carroll, S. T. Qureshi. 2009. "*Cryptococcus gattii*: An Emerging Cause of Fungal Disease in North America." *Interdisciplinary Perspective on Infectious Diseases* 2009: 840452. doi: 10.1155/2009/840452.

Doggett, J. S., S. Kohlhepp, R. Gresbrink, P. Metz, C. Gleaves, D. Gilbert. 2008. "Lyme Disease in Oregon." *Journal of Clinical Microbiology* 46 (6): 2115-2118. doi: 10.1128/JCM.00394-08.

Doherty, T. J., and S. Clayton. 2011. "The Psychological Impacts of Global Climate Change." *American Psychologist* 66 (4): 265-276. doi: 10.1037/a0023141.

Donoghue, E. R., M. Nelson, G. Rudis, J. T. Watson, G. Huhn, and G. Luber. 2003. "Heat-related deaths--Chicago, Illinois, 1996-2001, and United States, 1979-1999." *Morbidity and Mortality Weekly Report* 52 (26): 610-613.

English, P. B., A. H. Sinclair, Z. Ross, H. Anderson, V. Boothe, C. Davis, K. Ebi, B. Kagey, K. Malecki, R. Shultz, and E. Simms. 2009. "Environmental Health Indicators of Climate Change for the United States: Findings from the State Environmental Health Indicator Collaborative." *Environmental Health Perspectives* 117 (11): 1673-1681. doi: 10.1289/ehp.0900708.

Epstein, P. R., and E. Mills, eds. 2005. *Climate Change Futures: Health, Ecological and Economic Dimensions.* The Center for Health and the Global Environment, Boston, MA. http://coralreef.noaa.gov/aboutcrcp/strategy/reprioritization/wgroups/resources/climate/resources/cc_futures.pdf.

Epton, M. J., R. D. Dawson, W. M. Brooks, S. Kingham, T. Aberkane, J. A. E. Cavanagh, C. M. Frampton, T. Hewitt, J. M. Cook, S. McLeod, F. McCartin, K. Trought, and L. Brown. 2008. "The Effect of Ambient Air Pollution on Respiratory Health of School Children: A Panel Study." *Environmental Health* 7: 16. doi: 10.1186/1476-069X-7-16.

Fyfe, M., M. T. Kelley, S. T. Yeung, P. Daly, K. Schallie, S. Buchanan, P. Waller, J. Kobayashi, N. Therien, M. Guichard, S. Lankford, P. Stehr-Green, R. Harsch, E. DeBess, M. Cassidy, T. McGivern, S. Mauvais, D. Fleming, M. Lippmann, L. Pong, R. W. McKay, D. E. Cannon, S. B. Werner, S. Abbott, M. Hernandez, C. Wojee, J. Waddell, S. Waterman, J. Middaugh, D. Sasaki, P. Effler, C. Groves, N. Curtis, D. Dwyer, G. Gowdle, and C. Nichols. 1998. "Outbreak of *Vibrio parahaemolyticus* Infections Associated with Eating Raw Oysters – Pacific Northwest, 1997." *Morbidity and Mortality Weekly Report* 47 (22): 457-462.

Ginsberg, M., J. Johnson, J. Tokars, C. Martin, R. English, G. Rainisch, W. Lei, P. Hicks, J. Burkholder, M. Miller, K. Crosby, K. Akaka, A. Stock, and D. Sugerman. 2008. "Monitoring Health Effects of Wildfires Using the BioSense System--San Diego County, California, October 2007." *Morbidity and Mortality Weekly Report* 57 (27): 741-744.

Glibert, P. M., D. M. Anderson, P. Gentien, E. Graneli, K. G. Sellner. 2005 "The Global Complex Phenomena of Harmful Algae." *Oceanography* 18 (2): 136-147.

Gosling, S. N., J. A. Lowe, G. R. McGregor, M. Pelling, B. D. Malamud. 2009. "Assocations between Elevated Atmospheric Temperature and Human Mortality: A Critical Review of the Literature." *Climate Change* 92 (3-4): 299-341. doi:10.1007/s10584-008-9441-x.

Gould, E. A. and S. Higgs. 2009. "Impact of Climate Change and Other Factors on Emerging Arbovirus Diseases." *Transactions of the Royal Society of Tropical Medicine and Hygiene* 103 (2): 109-121. doi: 10.1016/j.trstmh.2008.07.025.

Greene, S., L. S. Kalkstein, D. M. Mills, and J. Samenow. 2011. "An Examination of Climate Change on Extreme Heat Events and Climate-Mortality Relationships in Large U.S. Cities." *Weather, Climate, and Society* 3: 281-292. doi: 10.1175/WCAS-D-11-00055.1.

Gubler, D. J., P. Reiter, K. L. Ebi, W. Yap, R. Nasci, and J. A. Patz. 2001. "Climate Variability and Change in the United States: Potential Impacts on Vector- and Rodent-Borne Diseases." *Environmental Health Perspectives* 109 (Supplement 2): 223-233. http://www.ncbi.nlm.nih.gov/pmc/articles/PMC1240669/.

Haines, A., and J. A. Patz. 2004. "Health Effects of Climate Change." *Journal of the American Medical Association* 291 (1): 99-103. doi: 10.1001/jama.291.1.99.

Hamlet, A. F. and D. P. Lettenmaier. 2007: "Effects of 20th Century Warming and Climate Variability on Flood Risk in the Western U.S." *Water Resources Research* 43 (W06427): 17. doi: 10.1029/2006WR005099.

Hansen, A., P. Bi, M. Nitschke, P. Ryan, D. Pisaniello, and G. Tucker. 2008. "The Effect of Heat Waves on Mental Health in a Temperate Australian City." *Environmental Health Perspectives* 116(10): 1369-1375. doi: 10.1289/ehp.11339.

Herron-Thorpe, F. L., B. K. Lamb, G. H. Mount, and J. K. Vaughan. 2010. "Evaluation of a Regional Air Quality Forecast Model for Tropospheric NO_2 Columns Using the OMI/Aura Satellite Tropospheric NO_2 Product." *Atmospheric Chemistry and Physics* 10: 8839-8854. doi:10.5194/acp-10-8839-2010.

Horner, R. A., D. L. Garrison, and F. G. Plumley. 1997. "Harmful Algal Blooms and Red Tide Problems on the U.S. West Coast." *Limnology and Oceanography* 42 (5): 1076-1088.

Huang, C., A. G. Barnett, X. Wang, P. Vaneckova, G. FitzGerald, and S. Tong. 2011. "Projecting Future Heat-Related Mortality under Climate Change Scenarios: A Systematic Review." *Environmental Health Perspectives* 119 (12): 1681-1690. doi: 10.1289/ehp.1103456.

Hubalek, Z. 2000. "European Experience with the West Nile Virus Ecology and Epidemiology: Could it Be Relevant for the New World?" *Viral Immunology* 13 (4): 415-426. doi: 10.1089/vim.2000.13.415.

Hutton, G. 2011. "The Economics of Health and Climate Change: Key Evidence for Decision Making." *Global Health* 7 (1): 18. doi: 10.1186/1744-8603-7-18.

Institute of Medicine. 2003. *Microbial Threats to Health: Emergence, Detection, and Response.* National Academy of Sciences, Washington, DC.

Jackson, J. S., K. M. Knight, and J. A. Rafferty. 2010. "Race and Unhealthy Behaviors: Chronic Stress, the HPA Axis, and Physical and Mental Health Disparities over the Life Course." *American Journal of Public Health* 100 (5): 933-939. doi: 10.2105/AJPH.2008.143446.

Jackson, J. E., M. G. Yost, C. Kar, C. Fitzpatrick, B. K. Lamb, S. H. Chung, J. Chen, J. Avise, R. A. Rosenblatt, and R. A. Fenske. 2010. "Public Health Impacts of Climate Change in Washington State: Projected Mortality Risks Due to Heat Events and Air Pollution." *Climatic Change* 102: 159-186. doi: 10.1007/s10584-010-9852-3.

Jacob, D. J., and D. A. Winner. 2009. "Effect of Climate Change on Air Quality." *Atmospheric Environment* 43 (1): 51-63. doi: 10.1016/j.atmosenv.2008.09.051.

Jerrett, M., R. T. Burnett, C. A. Pope, K. Ito, G. Thurston, D. Krewski, Y. Shi, E. Calle, and M. Thun. 2009. "Long-Term Ozone Exposure and Mortality." *The New England Journal of Medicine* 360 (11): 1085-1095. doi: 10.1056/NEJMoa0803894.

Kaiser, R., A. Le Tertre, J. Schwartz, C. A. Gotway, R. Daley, C. H. Rubin. 2007. "The Effect of the 1995 Heat Wave in Chicago on All-Cause and Cause-Specific Mortality." *American Journal of Public Health* 97 (Supplement 1): S158-162. doi: 10.2105/AJPH.2006.100081.

Kessler, R. C., S. Galea, M. J. Gruber, N. A. Sampson, R. J. Ursano, and S. Wessely. 2008. "Trends in Mental Illness and Suicidality after Hurricane Katrina." *Molecular Psychiatry* 13: 374-384. doi: 10.1038/sj.mp.4002119.

Knowlton, K., M. Rotkin-Ellman, L. Geballe, W. Max, and G. M. Solomon. 2011. "Six Climate Change-Related Events in the United States Accounted for about $14 Billion in Lost Lives and Health Costs." *Health Affairs (Millwood)* 30 (11): 2167-2176. doi: 10.1377/hlthaff.2011.0229.

Knowlton, K. M., M. Rotkin-Ellman, G. King, H. G. Margolis, D. Smith, G. Solomon, R. Trent, and P. English. 2009. "The 2006 California Heat Wave: Impacts on Hospitalizations and Emergency Department Visits." *Environmental Health Perspectives* 117 (1): 61-67. doi: 10.1289/ehp.11594.

Littell, J. S., E. E. Oneil, D. McKenzie, J. A. Hicke, J. Lutz, R. A. Norheim, and M. M. Elsner. 2010. "Forest Ecosystems, Disturbance, and Climatic Change in Washington State, USA." *Climatic Change* 102: 129-158. doi: 10.1007/s10584-010-9858-x.

Liu, L., R. Poon, L. Chen, A. M. Frescura, P. Montuschi, G. Ciabattoni, A. Wheeler, and R. Dales. 2009. "Acute Effects of Air Pollution on Pulmonary Function, Airway Inflammation, and Oxidative Stress in Asthmatic Children." *Environmental Health Perspectives* 117 (4): 668-674. doi: 10.1289/ehp11813.

Lock, S., G. J. Rubin, V. Murray, M. B. Rogers, R. Amlot, and R. Williams. 2012. "Secondary Stressors and Extreme Events and Disasters: a Systematic Review of Primary Research from 2010-2011." *PLOS Currents*, October 29. doi: 10.1371/currents.dis.a9b76fed1b2dd5c5bfcfc13c87a2f24f

LoVecchio, F., J. S. Stapczynski, J. Hill, J. A. Skindlov, D. Engelthaler, C. Mrela, G. E. Luber, M. Straetemans, and Z. Duprey. 2005. "Heat-Related Mortality--Arizona, 1993-2002, and United States, 1979-2002." *Morbidity and Mortality Weekly Report* 54 (25): 628-630.

Luber, G.E., C.A. Sanchez, and L.M. Conklin. 2006. "Heat-Related Deaths--United States, 1999-2003." *Morbidity and Mortality Weekly Report* 55 (29): 796-798.

Lugo-Amador, N.M., T. Rothenhaus, and P. Moyer. 2004. "Heat-Related Illness." *Emergency Medicine Clinics of North America* 22 (2): 315-327.

Mantua, N., I. Tohver, and A. Hamlet. 2010. "Climate Change Impacts on Streamflow Extremes and Summertime Stream Temperature and Their Possible Consequences for Freshwater Salmon Habitat in Washington State." *Climatic Change* 102 (1): 187-223. doi: 10.1007/s10584-010-9845-2.

McKenzie, D., Z. Gedalof, D. L. Peterson, and P. Mote. 2004. "Climatic Change, Wildfire, and Conservation." *Conservation Biology* 18 (4): 890-902. doi: 10.1111/j.1523-1739.2004.00492.x.

Miller, K. A., D. S. Siscovick, L. Sheppard, K. Shepherd, J. H. Sullivan, G. L. Anderson, and J. D. Kaufman. 2007. "Long-Term Exposure to Air Pollution and Incidence of Cardiovascular Events in Women." *The New England Journal of Medicine* 356 (5): 447-458. doi: 10.1056/NEJMoa054409.

Mills, J. N., K. L. Gage, and A. S. Khan. 2010. "Potential Influence of Climate Change on Vector-Borne and Zoonotic Diseases: A Review and Proposed Research Plan." *Environmental Health Perspectives* 118 (11): 1507-1514. doi: 10.1289/ehp.0901389.

Moore, S. K., V. L. Trainer, N. J. Mantua, M. S. Parker, E. A. Laws, L. C. Backer, and L. E. Fleming. 2008. "Impacts of Climate Variability and Future Climate Change on Harmful Algal Blooms and Human Health." *Environmental Health* 7 (Suppl2). doi: 10.1186/1476-069X-7-S2-S4.

Mote, P. W., and E. P. Salathé. 2010. "Future Climate in the Pacific Northwest." *Climatic Change* 102: 29-50. doi: 10.1007/s10584-010-9848-z.

Mühr, B., J. Daniell, B. Khazai, D. Köbele, M. Kunz, T. Kunz-Plapp, A. Leyser, M. Vannieuwenhuyse. 2012. *Analysis of U.S. Extreme Drought and Record Heat 2012*, 2nd Report. Center for Disaster Management and Risk Reduction Technology.

National Research Council. 2001. *Under the Weather: Climate, Ecosystems, and Infectious Disease.* National Academy of Sciences, Washington, DC.

National Weather Service Forecast Office. 2012. "Some of the Area's Windstorms." Accessed November 11. Portland, OR. http://www.wrh.noaa.gov/pqr/paststorms/wind.php.

National Weather Service Office of Climate, Water, and Weather Services. 2012. "Natural Hazards Statistics." Accessed November 11. http://www.nws.noaa.gov/om/hazstats.shtml.

Naughton, M. P., A. Henderson, M. C. Mirabelli, R. Kaiser, J. L. Wilhelm, S. M. Kieszak, C. H. Rubin, and M. A. McGeehin. 2002. "Heat-Related Mortality during a 1999 Heat Wave in Chicago." *American Journal of Preventive Medicine* 22 (4): 221-227. http://www.ajpmonline.org/article/S0749-3797(02)00421-X/abstract.

Ohl, C. A., and S. Tapsell. 2000. "Flooding and Human Health." *British Medical Journal (Clinical Research Edition)* 321 (7270): 1167-1168. http://www.ncbi.nlm.nih.gov/pmc/articles/PMC1118941/.

Oregon Health Authority, Public Health Division. 2011. *Increasing Public Health Capacity to Address Climate Change Adaptation through Health Impact and Hazard Vulnerability Assessment.* Cooperative Agreement with Centers for Disease Control and Prevention (unpublished).

Oregon Health Authority. 2012. *Extreme Heat and Health Impacts Project Summary Report,* November 2012 (unpublished).

Parslow, R. A., and A. F. Jorm. 2006. "Tobacco Use after Experiencing a Major Natural Disaster: Analysis of a Longitudinal Study of 2063 Young Adults." *Addiction* 101(7): 1044-50. doi:10.1111/j.1360-0443.2006.01481.x.

Pope III, C. A., and D. W. Dockery. 2006. "Health Effects of Fine Particulate Air Pollution: Lines that Connect." *Journal of the Air & Waste Management Association* 56 (6): 709-742.

Reid, C. E., and J. L. Gamble. 2009. "Aeroallergens, Allergic Disease, and Climate Change: Impacts and Adaptation." *Ecohealth* 6 (3): 458-470. doi: 10.1007/s10393-009-0261-x.

Reisen, W. K., Y. Fang, and V. M. Martinez. 2006. "Effects of Temperature on the Transmission of West Nile Virus by Culex Tarsalis (Diptera: Culicidae)." *Journal of Medical Entomology* 43 (2): 309-317. doi: 10.1603/0022-2585(2006)043[0309:EOTOTT]2.0.CO;2.

Reiter, P. 2001. "Climate Change and Mosquito-Borne Disease." *Environmental Health Perspectives* 109 (Supplement 1): 141-161. http://www.ncbi.nlm.nih.gov/pmc/articles/PMC1240549/.

Ren, C., G. M. Williams, K. I. Mengersen, L. Morawska, and S. Tong. 2008. "Does Temperature Modify Short-Term Effects of Ozone on Total Mortality in 60 Large Eastern US Communities? -- An Assessment Using the NMMAPS Data." *Environmental International* 34 (4): 451-458. doi: 10.1016/j.envint.2007.10.001.

Rogers, C. A., P. M. Wayne, E. A. Macklin, M. L. Muilenberg, C. J. Wagner, P. R. Epstein, and F. A. Bazzaz. 2006. "Interaction of the Onset of Spring and Elevated Atmospheric CO_2 on Ragweed (Ambrosia artemisiifolia L.) Pollen Production." *Environmental Health Perspectives* 114 (6): 865-869. doi: 10.1289/ehp.8549.

Rosegrant, M. W., C. Ringler, and T. J. Zhu. 2009. "Water for Agriculture: Maintaining Food Security under Growing Scarcity." *Annual Review of Environment and Resources* 34: 205-222. doi: 10.1146/annurev.environ.030308.090351.

Sartore, G. M., B. Kelly, H. Stain, G. Albrecht, and N. Higginbotham. 2008. "Control, Uncertainty, and Expectations for the Future: A Qualitative Study of the Impact of Drought on a Rural Australian Community." *Rural Remote Health* 8 (3): 950. http://www.rrh.org.au/articles/subviewnew.asp?ArticleID=950.

Schulte, P. A., and H. Chun. 2009. "Climate Change and Occupational Safety and Health: Establishing a Preliminary Framework." *Journal of Occupational and Environmental Hygiene* 6 (9): 542-554. doi: 10.1080/15459620903066008.

Shaman, J., J. F. Day, and M. Stieglitz. 2005. "Drought-Induced Amplification and Epidemic Transmission of West Nile Virus in Southern Florida." *Journal of Medical Entomology* 42 (2): 134-141. doi: 10.1603/0022-2585(2005)042[0134:DAAETO]2.0.CO;2.

Sherry, R. A., X. H. Zhou, S. L. Gu, J. A. Arnone, D. S. Schimel, P. S. Verburg, L. L. Wallace, and Y. Luo. 2007. "Divergence of Reproductive Phenology under Climate Warming." *Proceedings of the National Academy of Sciences* 104 (1): 198-202. doi: 10.1073/pnas.0605642104.

Simkhovich, B. Z., M. T. Kleinman, and R. A. Kloner. 2008. "Air Pollution and Cardiovascular Injury Epidemiology, Toxicology, and Mechanisms." *Journal of American College of Cardiology* 52 (9): 719-726. doi: 10.1016/j.jacc.2008.05.029.

Singer, B. D., L. H. Ziska, D. A. Frenz, D. E. Gebhard, and J. G. Straka. 2005. "Increasing Amb a 1 Content in Common Ragweed (*Ambrosia artemisiifolia*) Pollen as a Function of Rising Atmospheric CO_2 Concentration." *Functional Plant Biology* 32 (7): 667-670. doi: 10.1071/FP05039.

Solomon, G. M., M. Hjelmroos-Koski, M. Rotkin-Ellman, and S. K. Hammond. 2006. "Airborne Mold and Endotoxin Concentrations in New Orleans, Louisiana, after Flooding, October through November 2005." *Environmental Health Perspectives* 114 (9): 1381-1386. doi: 10.1289/ehp.9198.

Spracklen, D. V., L. J. Mickley, J. A. Logan, R. C. Hudman, R. Yevich, M. D. Flannigan, and A. L. Westerling. 2009. "Impacts of Climate Change from 2000 to 2050 on Wildfire Activity and Carbonaceous Aerosol Concentrations in the Western United States." *Journal of Geophysical Research-Atmospheres* 114: D20301. doi: 10.1029/2008JD010966.

The Resource Innovation Group and Biositu, LLC. *Public Health and Climate Change: A Guide for Increasing the Capacity of Local Public Health Departments.* http://www.theresourceinnovation group.org/storage/public-health-materials/Public%20Health%20Guide_FINAL.pdf

US Department of Agriculture Farm Service Agency. 2012. "2012 Secretarial Drought Designations – All Drought." October 17. http://www.usda.gov/documents/usda-drought-fast-track -designations-101712.pdf.

US Department of Agriculture Natural Resources Conservation Service. "PLANTS Profile, *Ambrosia L.* ragweed." Accessed November 13, 2012. http://plants.usda.gov/java/profile ?symbol=AMBRO.

US Global Change Research Program. 2012. National Climate Assessment Health Sector Workshop Report: Northwest Region. February 23-24. http://www.joss.ucar.edu/ohhi/nw_nca_health _sector_feb12/Health_and_CC_NW_Report.pdf.

Van Dolah, F.M. 2000. "Marine Algal Toxins: Origins, Health Effects, and Their Increased Occurrence." *Environmental Health Perspectives* 108 (Supplement 1): 133-141. http://www.jstor.org /stable/3454638.

Vynne, S., and B. Doppelt. 2009. *Climate Change Health Preparedness in Oregon: An Assessment of Awareness, Preparation, and Resource Needs for Potential Public Health Risks Associated with Climate Change.* Climate Leadership Initiative, University of Oregon, Eugene, Oregon. http://www.theresourceinnovationgroup.org/storage/ORPHSurveyReportFinal.pdf.

Watson, J. T., M. Gayer, and M. A. Connolly. 2007. "Epidemics after Natural Disasters." *Emerging Infectious Diseases* 13 (1): 1-5. doi: 10.3201/eid1301.060779.

Webb, P. 2010. "Medium- to Long-Run Implications of High Food Prices for Global Nutrition." *Journal of Nutrition* 140 (1): 143s-147s. doi: 10.3945/ jn.109.110536.

Wegesser, T. C., K. E. Pinkerton, and J. A. Last. 2009. "California Wildfires of 2008: Coarse and Fine Particulate Matter Toxicity." *Environmental Health Perspectives* 117 (6):893-897. doi: 10.1289/ehp.0800166.

Willers, S. M., C. Eriksson, L. Gidhagen, M. E. Nilsson, G. Pershagen, and T. Bellander. 2013. "Fine and Coarse Particulate Air Pollution in Relation to Respiratory Health in Sweden." *European Respiratory Journal*, January 11, doi: 10.1183/09031936.00088212.

Yoder, J. S. and M. J. Beach. 2007. "Cryptosporidiosis Surveillance--United States, 2003-2005." *Morbidity and Mortality Weekly Report Surveillance Summary* 56 (7): 1-10. http://www.cdc.gov /mmwr/preview/mmwrhtml/ss5607a1.htm.

Yoder, J. S., C. Harral, and M. J. Beach. 2010. "Cryptosporidiosis Surveillance--United States, 2006-2008." *Morbidity and Mortality Weekly Report Surveillance Summary* 59 (S6): 1-14. http:// www.cdc.gov/mmwr/preview/mmwrhtml/ss5906a1.htm.

Yoder, J. S., R. M. Wallace, S. A. Collier, M. J. Beach, and M. C. Hlavsa. 2012. "Cryptosporidiosis Surveillance--United States, 2009-2010." *Morbidity and Mortality Weekly Report Surveillance Summary* 61 (5): 1-12. http://www.cdc.gov/mmwr/preview/mmwrhtml/ss6105a1.htm?s_cid =ss6105a1_w.

Ziska, L., K. Knowlton, C. Rogers, D. Dalan, N. Tierney, M.A. Elder, W. Filley, J. Shropshire, L. B. Ford, C. Hedberg, P. Fleetwood, K. T. Hovanky, T. Kavanaugh, G. Fulford, R. F. Vrtis, J. A. Patz, J. Portnoy, F. Coates, L. Bielory, and D. Frenz. 2011. "Recent Warming by Latitude Associated with Increased Length of Ragweed Pollen Season in Central North America." *Proceedings of the National Academy of Sciences* 108 (10): 4248-4251. doi: 10.1073/ pnas.1014107108.

Chapter 8

Northwest Tribes
Cultural Impacts and Adaptation Responses

AUTHORS
Kathy Lynn, Oliver Grah, Preston Hardison, Jennie Hoffman, Ed Knight,
Amanda Rogerson, Patricia Tillmann, Carson Viles, Paul Williams

8.1 Introduction

Climate change will have complex and profound effects on tribal resources, cultures, and economies. Indigenous peoples have lived in the region for thousands of years, developing cultural and social customs that revolve around traditional foods and materials and a spiritual tradition that is inseparable from the environment. Projected changes in temperature, precipitation, sea level, hydrology, and ocean chemistry threaten not only the lands, resources, and economies of tribes, but also tribal homelands, ceremonial sites, burial sites, tribal traditions, and cultural practices that have relied on native plant and animal species since time immemorial (Williams and Hardison 2005, 2006, 2007, 2012).

The cultural, social, and economic integrity of tribal communities, as well as the physical and spiritual health of tribal members, are intertwined with the places, foods, medicines, and resources they adapted to over the centuries. Tribes are intimately connected to the Earth's resources, and this relationship of reciprocity is embedded in many indigenous cultures (Whyte 2013; Hardison and Williams, in review). Ecologically and culturally, tribes in the Northwest (NW) are embedded in the landscape (fig. 8.1). Climate change could lead to extirpation of some species, migration of species away from traditional gathering areas, and altered timing of resource availability relative to traditional practices, all of which could have disastrous impacts for tribes.

Tribes are responding to climate change in ways that make sense for their particular communities, cultural traditions, economies, and environments. This chapter synthesizes what is currently understood about key climate change vulnerabilities for tribes in the Northwest and potential consequences to a range of tribal cultural and natural resources, traditional foods, and economies. The chapter highlights how tribes in the Northwest are working with others to address climate change by contributing their knowledge about the environment, their staff and equipment on the ground and in the water, their resource management experience, and their authority as natural resource trustees. They are engaged in climate assessments, adaptation and mitigation planning, research, monitoring, outreach, and education. The chapter also cites examples of how tribes are beginning to pursue such efforts, driven by the recognition of changes and

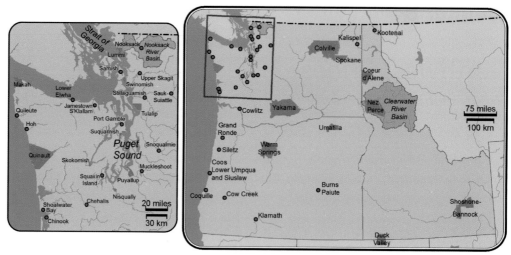

Figure 8.1 Federally recognized tribes in Washington, Oregon, and Idaho.

of the need to adapt, as an illustration of emerging means and methods for developing response to climate change. The chapter concludes with a summary of tribal needs to address the challenges of climate change, including strengthened government-to-government relationships, tribal capacity building and training, tribal participation in climate research and access to climate data, and development and deployment of resources for early detection and monitoring of fundamental ecosystem changes.

8.2 Tribal Culture and Sovereignty

Tribal culture, traditional ways of life, lands, and resources are increasingly vulnerable to the impacts from climate change (Lynn et al. 2011). Some tribes may experience physical loss of sovereign lands due to direct climate impacts such as inundation (Papiez 2009; Swinomish Indian Tribal Community 2009). Tribal ceremonies and traditional practices depend upon regular and seasonally specific access to certain species (Norgaard 2005; Swinomish Indian Tribal Community 2010). Changing species distribution may move important tribal resources to areas that are physically or legally less accessible to them. Altered riparian and freshwater health may impair the productivity and long-term viability of important resources, such as salmon. In the ocean, temperature change, sea level rise, and changing ocean chemistry may contribute to the extirpation of some species, the migration of others, and displacement by the arrival of new species (Washington State Blue Ribbon Panel 2012). Furthermore, reduced fish and shellfish populations pose a threat to tribal dietary health, spiritual health, and economic well-being (Swinomish Indian Tribal Community 2010).

These climate impacts occur within the framework of tribal sovereignty, and therefore it is helpful to review the tribal relationship with local and federal governments. At its establishment, the US legal system recognized in the Constitution that prior to colonization, tribes had the same status as any other nation state as sovereigns of their lands (Wilkins and Lomawaima 2001). In order to acquire legal title to tribal lands and resources, the United States acknowledged it had to acquire them through voluntary, negotiated cessions rather than by force (Wilkins and Lomawaima 2001). The Constitution of the United States provided this through the treaty-making process, with treaties being "the supreme law of the land" (Article VI US Constitution). The tribes ceded vast amounts of lands and resources in return for peaceful relations, guarantees of homelands and resources to continue their ways of life, and promises of protection from the United States (the *federal trust responsibility*; Morisset 1999; Pevar 2012).

In ceding lands and resources, tribes retained their sovereignty. There are two key issues to understand about this status and the bargains that were struck between nations. First, as is stated by the Reserved Rights Doctrine, treaties are not rights granted to the Indians, but rather "a reservation by the Indians of rights already possessed and not granted away by them" (established in *US v. Winans* [198 US 371 (1905)]). This means that tribes possess inherent rights as sovereigns under the supreme law of the land that have a status fundamentally different from rights granted by the federal government to citizens. The second issue is that treaties are to be primarily read as documents for ceded rights, not as listings of retained sovereign rights (Pevar 2012). In a sovereign-to-sovereign treaty process, any sovereign rights not explicitly negotiated are retained, so tribes have retained rights they have not explicitly given up (Pevar 2012).

The treaties establish the basis for the government-to-government relationship between tribes and the US government, which is grounded in the US Constitution, and elaborated through statutes, federal case law, regulations, and executive orders. Consultation is a core element of the government-to-government relationship and enables American Indian and Alaska Native Tribes and the United States to identify issues and concerns, and resolve differences through respectful dialogue between sovereigns. Federal agencies have a responsibility to consult with tribes when "engaging in policy making or undertaking initiatives that will affect the vital interests of tribes" (Parker 2009) under numerous federal statutes, regulations, and presidential executive orders (Galanda 2011). Federal agencies are required to develop consultation policies that "ensure meaningful and timely input by tribal officials in the development of regulatory policies that have tribal implications" (Executive Order No. 13175).

In the treaties, many tribes reserved rights to hunt, fish, and gather on lands outside of their reservations, because their ancestors knew their designated homelands would not be sufficient for them to maintain their ways of life (Bilodeau 2012). These areas have variously been referred to as "accustomed places," "accustomed grounds and stations," "accustomed stations," and "usual and accustomed places," each of which has been elaborated through court decisions and agreements (Bernholz and Weiner 2008). Through this history, tribes have maintained geographically bounded rights to natural resources and heritage that occur both on their reservations and on off-reservation lands (Gates 1955; Ovsak 1994).

Climate change thus has the potential to affect treaty-protected rights and rights as verified by court cases (including hunting, fishing, and gathering rights) through changes in tribally important species and habitat on a wide variety of lands. Figure 8.2 illustrates ceded lands for Washington State and illustrates the potential range of tribal territories that climate change may impact.

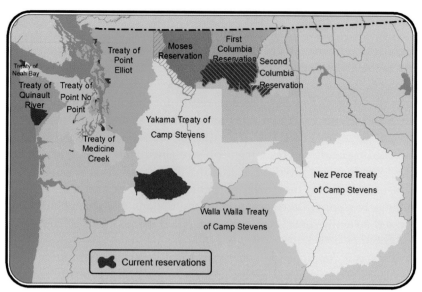

Figure 8.2 Treaty Ceded Lands. Washington State Historic Tribal Lands (Tribal Areas of Interest. Washington Department of Ecology)

The actual extent of these treaty tribal rights within reserved areas, often outside of reservation lands, has been the subject of one of the longest running cases in the US federal courts. Until the 1970s, Washington State controlled the fisheries outside of reservation boundaries, resulting in tribal harvest of only between two and five percent of the salmon. Through a series of protests and fish-ins, the tribes were able to get the federal government to recognize their trust responsibility and join in filing suit. In 1974, District Court Judge Boldt required the state of Washington to regulate its fisheries to ensure that treaty tribes have the opportunity to take up to 50% of the harvestable surplus of anadromous fish runs that originate in or pass through tribal usual and accustomed fishing grounds and stations (*United States v. Washington* [384 F. Supp. 312]). In 1994, District Court Judge Edward Rafeedee ruled that shellfish are fish, thus extending Boldt's ruling on salmon to shellfish (*United States v. Washington* [873 F. Supp. 1422]).

Treaty water rights also stand to be affected by climate change, and are intimately connected with treaty fishing rights. Changes to water quantity and quality, as well as surface and groundwater flows, are likely to affect salmon, particularly in the freshwater phase of their lifecycle (see Chapter 3). We focus here on Indian water rights conflicts related to the needs of salmon, in particular the need for sufficient instream flow, which may be affected by climate change. Generally, courts have held that tribes have rights to sufficient instream flow to support salmon on their reservation (Bilodeau 2012). Although the Winters Doctrine, established by the US Supreme Court in 1908, held that there are implied Indian water rights carried by reservations, these rights only extend to cover the primary purpose of the reservation (Bilodeau 2012). Thus the 9th Circuit Court of Appeals declared in *Skokomish Indian Tribe v. United States* (401 F.3d 979 (9th Cir. 2005)) that the Tribe did not have the right to water necessary to support fishing because they could not demonstrate that fishing was one of the primary purposes of the Tribe's treaty.

The decision was later amended to remove this finding; however, it left open the question of tribal reserved water rights for fishing.

As climate change intensifies competition for water, the issue of which water rights are guaranteed by treaty, and how to interpret those rights in light of changing conditions, will become increasingly important and intense. Tribal adaptation options will be strongly influenced by how the legal and regulatory systems address treaty rights in the face of climate change, and for tribes in the Northwest, tribal access to and control over both fish and water is a particularly critical issue. A major concern is the renegotiation of the Columbia River Treaty. The original Treaty focused on flood control and hydropower, and did not address the rights of the 15 federally-recognized tribes in the Columbia Basin, nor the importance of ecological function (Osborn 2012). Balancing the needs of hydropower, agriculture, fish, and others that rely on the Columbia is contentious now and will only become more so as climate change may result in lower and warmer river flows in the summer (Osborn 2012). This calls for meaningful tribal engagement in the renegotiation process.

8.3 Climatic Changes and Effects: Implications for Tribes in the Northwest

Climate change will affect the availability, processing, storage, and use of tribal cultural resources. Across the Northwest, tribes depend upon a wide variety of traditional foods to maintain a healthy diet, the loss of which has the potential to affect tribal health and economies (Norgaard 2005). These foods, including salmon, shellfish, and roots and berries, among others, are of significant cultural importance to tribes, as is illustrated in figure 8.3, which depicts the traditional preparation of salmon by the Coquille Indian Tribe. This section builds on the research and findings in the previous chapters of this report and highlights specific research and case studies describing the ways in which climate change may impact culture, traditional foods, and traditional ways of life among tribes in the Northwest. The focus here on specific climate change impacts connects this section with the companion chapters in this report. Within this section, the interconnected nature of tribal cultural-ecological systems is maintained through integrated discussion of tribal perspectives and risks to tribal resources. The information provided is not meant to be comprehensive, but rather an evolving understanding of climate change impacts on NW tribes.

8.3.1 WATER RESOURCES AND AVAILABILITY

Tribal communities are struggling to maintain healthy river systems in light of climate change (Kaufman 2011). Climate change is expected to reduce snowpack, shift the timing and magnitude of precipitation and runoff, increase stream temperatures, and magnify both drought conditions in the dry season and flooding in the wet season (Chapter 3), all of which could have significant effects on salmon and other aquatic and riparian species. Effects on fish threaten tribal fisheries and economies, and treaty rights to fish lose value if climate change reduces the availability of those fish (McNutt 2008; see also, section 8.2 in this chapter).

Figure 8.3 Salmon on Sticks. Photo Credit: Jon Ivy, Coquille Indian Tribe

Salmon fishing is an important economic pursuit for tribes throughout the Northwest (Papiez 2009), and not just those along the coast. Indeed, the Columbia River Inter-Tribal Fish Commission released a report on climate change and snowpack focusing on tribal lands in the Columbia Basin and pointed out that "The ceded lands of the Confederate Tribes of the Warm Springs Reservation of Oregon, the Confederated Tribes of the Umatilla Indian Reservation, the Confederated Tribes and Bands of the Yakama Nation, and the Nez Perce Tribe all host culturally precious natural resources, including salmon and steelhead, which benefit from the upland melting of snowpack to provide steady streamflows" (Graves 2008). Similarly, incorporating climate change into the renegotiation of the Columbia River Treaty will increase the likelihood of sustainable fish populations persisting into the future (see section 8.2).

8.3.2 WATER TEMPERATURE AND CHEMISTRY

Many tribes in the Northwest are people of the water. The locations and names of villages, even the names of the people themselves reflect the inseparable connection of tribes to water. Even tribes far from the sea depend on the oceans to rear the salmon that migrate hundreds of miles upriver. Fish and shellfish harvests are a primary source of income for tribal members in Puget Sound. The ex-vessel value of tribal shellfish harvests in Puget Sound, Washington, is around $50 million per year (Washington Department of Fish and Wildlife 2012). The health of these fisheries depends on how they are managed and on the health of the ecosystems they inhabit. In the face of multiple anthropogenic stressors impacting vital species, tribes and state agency managers are working to understand how to address these existing stressors as well as those that will result from climate change and changes to water chemistry.

BOX 8.1

Case Study: The Effect of Climate Change on Baseflow Support in the Nooksack River Basin and Implications on Pacific Salmon Species Protection and Recovery

The Nooksack Indian Tribe, a recognized Tribe under the Point Elliot treaty, inhabits the area around Deming, Washington, in the northwest corner of the state. The Tribe relies on various species of salmonids for ceremonial, commercial, and subsistence uses. The numbers of fish that return to spawn in the Nooksack River watershed have greatly diminished since Europeans colonized the area. Although direct counts are difficult, it appears that under today's conditions, native salmonid runs are less than 1.7–8% of the runs in the late 1800s (Lackey 2000). Many of these species and populations (i.e., evolutionarily significant units, or ESUs) are now protected under the federal Endangered Species Act. Of particular importance to the Tribe are spring Chinook salmon that return to, hold, spawn, and rear in the Nooksack River.

Historic and current land use practices have resulted in substantial loss of fish habitat in the Nooksack watershed, primarily due to removal of large woody debris from the river, removal of riparian forest, timber harvest, agriculture, alteration of stream channels, and transportation development (Coe 2001; Brown and Maudlin 2007). In the Northwest, observed trends in climate over the last century suggest increasing temperatures, increased winter flows, earlier snowmelt runoff, reduced summer flows, and variable precipitation amounts and timing as discussed in more detail in Chapters 2 and 3.

Climate change exacerbates existing stressors that also directly impact Pacific salmonids. These include a reduction in habitat function and services, increased stream temperatures, decreased summer flows, and increased duration of summer low flows due to land use and management. There is strong evidence climate change has contributed and will continue to contribute to a decrease in the summer baseflows that some salmonids depend upon for migration, holding, and spawning, as well as an increase in water temperatures that exceed the tolerance levels, and in some cases exceed lethal levels, of several Pacific salmon species in the Nooksack River (Mantua et al. 2010; Isaak et al. 2012). Further, higher peak flows are projected in the winter season, which would have exacerbating effects on other life stages of Pacific salmon.

Portions of the Nooksack River are listed as 303(d) Category 5 waters (under the federal Clean Water Act) that require a temperature Total Maximum Daily Load (TMDL) remediation project to bring those sections into compliance with the Act. Most traditional temperature TMDL projects do not address climate change. The Environmental Protection Agency (EPA) Office of Research and Development, in collaboration with the Nooksack Indian Tribe, has initiated a pilot research project focused on how to evaluate, design, and implement restoration tools in the South Fork Nooksack River as an extension of the South Fork Nooksack River TMDL Project that will address the projected increase of instream temperatures, loss of glacier melt contributions, decreased baseflows, and increased winter-time flows caused by continued climate change that adversely affect fish and fish habitat (Klein 2013). This is one of very few such projects currently being implemented in the United States. The outputs of the pilot research project will be a set of recommendations that will inform development of the TMDL, recommendations for updates to salmon recovery planning, and other land use and restoration planning efforts that can take climate change into direct consideration.

The dynamic response of glaciers to climate change in the Nooksack River watershed, particularly on Mount Baker, is vitally important to sustained flows and cool temperatures during the low-flow summer months when some salmonids are migrating and holding over to spawn (Grah and Beaulieu 2013). As such, it is of great

Box 8.1 (Continued)

importance to understand how climate change will alter melt contribution to, and stream temperatures in, the Nooksack River. (See Chapter 3 for a summary of observed and projected climate change effects on watersheds and glacier dynamics.) Approximately 22 km^2 (8.5 mi^2) of glaciated area drains into the Nooksack River basin. Pelto and Brown (2012) showed that the average retreat of Mt. Baker glacier termini has been approximately 370 m (1214 ft) over the period of 1979–2009. Although the glaciers on Mt Baker are experiencing significant terminal retreat, their accumulation zones are experiencing minimal thinning, suggesting they have the capacity to survive the current climate conditions (Pelto 2010). It is unlikely that this behavior applies to future climatic conditions with increased temperature and decreased snow accumulation (Pelto 2010; Pelto and Brown 2012). With increasingly negative mass balances, the current terminal retreat rate is insufficient to approach equilibrium (Pelto and Brown 2012). There are minimal field observations of the glaciers that directly feed the Nooksack River, emphasizing the need for more thorough investigation (Grah and Beaulieu 2013). The Tribe is implementing pilot studies on Mt. Baker glaciers to better determine their behavior in response to recent climatic trends and subsequent changes in glacier melt contributions to the Nooksack River (Grah and Beaulieu 2013).

For the Nooksack watershed in particular, stream temperatures are projected to increase up to 1 °C (1.8 °F) by 2020, 1–2 °C (1.8–3.6 °F) by 2040, and 2–3 °C (3.6–5.4 °F) by 2080 under the SRES-A1B emissions scenario, which is characterized by continued growth peaking at mid-century (Nakićenović et al. 2000). These temperature increases will pose substantive challenges to salmonid survival in the Nooksack River. Projected decreases in summertime low flows

of 13% by 2025, 27% by 2050, and 40% by 2075 (Dickerson-Lange and Mitchell, in review) could further reduce available habitat by approximately 10% for spring Chinook during the summer holding period (Grah and Beaulieu 2013).

As a result of these current and projected changes, stream temperatures may transition from the current favorable status to a stressful status in the Nooksack River watershed (Mantua et al. 2010). However, total loss of salmonid populations is not expected and thus, effective restoration will be needed to limit the harmful effects of climate change on top of existing stressors in the watershed.

Floodplain reconnection, as well as riparian and instream restoration (Mantua et al. 2010), have been demonstrated to be effective and imperative in protecting and sustaining salmonid species in the Northwest (Beechie et al. 2012) and in the Nooksack River, particularly in light of existing stressors resulting from land management. Because the impacts of climate change occur over time, the Nooksack Indian Tribe is currently implementing an aggressive restoration program that addresses not only the effects of land management, but also the potential changes in quantity and quality of the Nooksack River flows. Implementing climate-smart restoration is imperative to the survival and perpetuation of salmonids that the Nooksack Indian Tribe relies on.

Beechie et al. (2012) state that restoring floodplain connectivity, restoring streamflow regimes, re-planting deforested riparian zones, and re-aggrading incised channels are the restoration techniques most likely to ameliorate adverse streamflow and temperature changes caused by climate change and increase habitat diversity and population resilience. A large portion of the riparian zone of the Nooksack River has been deforested, leading to a substantial loss of protective

Box 8.1 (Continued)

shading (Coe 2001). Restoring riparian shading along important salmon-bearing streams will be essential to ameliorate increasing stream temperatures. Further, current restoration plans may need to be updated or re-written to address climate change. Beechie et al. (2012) suggest adapting salmon recovery plans to account for climate change by using a decision support process that includes local habitat factors limiting salmon persistence and recovery. The Nooksack Indian Tribe is particularly focused on recovery of salmon to sustainable populations that can support harvest for all Tribal uses. The possible extinction of salmonids, particularly spring Chinook salmon, from the Nooksack River is unacceptable because the Tribe is dependent on these species, and being place-based, the Tribe cannot move its geographic base or homeland to where salmon will be located under future climatic conditions.

Beechie et al. (2012) also suggest using scenarios of climate change effects on streamflow and temperature, assessments of the ability of restoration actions to address climate change effects, and assessments of the ability of restoration actions to increase habitat diversity and salmon population resilience. These measures will help ensure the survival of Nooksack River salmonids, given continued climate change and the adverse effects of past and present land management, and will promote the re-establishment of populations capable of sustainable harvest for all Tribal uses. The Nooksack Tribe has been actively implementing instream and riparian restoration to remediate the effects of land management, and now is also doing so to address climate change and ensure the survival of salmonids.

Few, if any, local strategies can slow the shrinking of glaciers in the Nooksack River watershed and the subsequent reduction in baseflows so critical to Pacific salmonids. The Tribe believes that watershed, riparian, and in-channel restoration strategies, as well as floodplain reconnection restoration will be effective at promoting the survival of salmonids in the face of climate change to ensure that the Tribe's ceremonial, cultural, and subsistence needs are met in the future.

As highlighted in Chapter 4, increasing ocean acidification and hypoxia pose significant but poorly understood threats to many species at the base of the food web, including fish and shellfish species widely used by tribes and others. For example, decreasing pH is associated with observed declines in the abundance and mean size of mussels from Tatoosh Island on the Makah Reservation in Washington, and has been linked to mass larval mortality of Pacific oyster larvae in shellfish hatcheries (Feely et al. 2012; Wootton et al. 2008). Researchers are currently assessing the contribution of nutrients from stormwater runoff and other sources of pollution to the acidification of Puget Sound (Washington State Blue Ribbon Panel 2012). Local solutions, including reducing the nutrient content of wastewater input into Puget Sound or using seaweed culture to take up carbon dioxide, as proposed by the Tulalip Tribes and the Washington State Blue Ribbon Panel on Ocean Acidification (2012), may reduce acidity in localized areas. These efforts could help address acidification while providing other widespread benefits, including increasing habitat and reducing water pollution.

Increasing water temperature averages and extremes can also have major effects on fish and shellfish that are important to tribes. The occurrence of harmful algal blooms (HABs) such as *Alexandrium catenalla*, the dinoflagellate responsible for paralytic shellfish poisoning, and *Dinophysis*, which causes diarrhetic shellfish poisoning, is a health concern for tribes, and could disrupt commercial harvests leading to economic losses (Moore et al. 2011; see also Chapters 4 and 7). Although there is much yet to understand about the dynamics of phytoplankton and microbial communities in relation to climate change, anomalous climate conditions have been linked with unusual blooms of dinoflagellates, cyanobacteria, and other contributors to HABs (e.g., Cloern et al. 2005, Dale et al. 2006, Paerl and Huisman 2009). In response to possible health effects linked to increases in harmful algal blooms, some tribes have initiated regular water testing at key shellfish beaches to limit risk, and participate in Sound Toxins, a HAB monitoring program (Papiez 2009; www.soundtoxins.org). Such local monitoring programs could help increase our understanding of how climate change may influence the dynamics of HABs in this oceanographically complex region.

Increasing air and water temperatures can also affect shellfish populations that tribes depend upon. Over the past 50 years, warmer summers have led to a 51% decrease in the vertical extent of the California mussel (*Mytilus californianus*) along the Strait of Juan de Fuca and southern Vancouver Island (Harley 2011). Because the ochre star (*Pisaster ochraceous*), a major predator and a primary determinant of the lower limit of the mussels, is relatively unaffected by changes in air temperature, mussel populations have entirely disappeared in the warmer parts of this region (Harley 2011), squeezed out from above by warmer air and from below by the sea stars. Because the interaction of tides and topography create a range of microclimates along the Pacific coast (Helmuth et al. 2006; Helmuth 1998), one adaptation option is to map out areas more likely to remain below critical temperature thresholds and prioritize those areas for tribal aquaculture or shellfish harvest.

8.3.3 SEA LEVEL RISE

Tribal coastal infrastructure and ecosystems are threatened by sea level rise and associated increases in the frequency and severity of coastal tidal surges. The unpredictability of the El Niño-Southern Oscillation, decadal ocean variability, tectonic plate movement, and weather systems complicate projections. However, sea levels are expected to rise by about 60 cm (2 ft) by the end of the century, although some models and projections do not rule out sea level rise of as much as 1.4 m (4.6 ft) by 2100, compared with the year 2000 (National Research Council 2012). This will seriously affect coastal tribal communities, as well as coastal ecosystems (see Chapter 4).

The homelands of the Quileute, Hoh, Swinomish, and others may face inundation from rising seas, storm surge, and rapid erosion, leading to a loss of sovereign land (Papiez 2009; Swinomish Indian Tribal Community 2009). These lands include culturally important spaces such as burial grounds and traditional fishing and shellfish gathering areas (Papiez 2009; Swinomish Indian Tribal Community 2009).

The projected impacts from sea level rise could also affect the economies and infrastructure of coastal tribes, which are often, by circumstance or necessity, located in vulnerable low-lying areas subject to potential inundation. The economic livelihood and vitality of such communities is directly threatened by both incremental sea level rise and by the combination of sea level rise, storm surges, and increasing wave heights (NRC 2012; Ruggiero et al. 2010). For example, economic development on Swinomish lands is located primarily in low-lying areas at risk from the indirect impacts of rising seas and storm surge (Swinomish Indian Tribal Community 2009; see also section 8.4.2).

On small coastal reservations there is a tension between allowing coastal habitat shift and maintaining space for land-based needs. In the past, coastal habitat shifted landward with rising seas, limiting habitat loss. But modern infrastructure may prevent such shifts. Further, rising sea levels may induce changes in the type and distribution of existing estuarine habitat throughout the Northwest (Glick et al. 2007). Estuarine ecosystems are some of the most biologically rich in the Northwest and play vital roles in maintaining water quality while providing habitat for young salmon, shellfish, native birds, and countless other species (Parker et al. 2006; Kaufman 2011). Sea level rise will inundate some existing shellfish beds, making them less accessible for harvest; in some cases, shellfish habitat may be lost (Swinomish Indian Tribal Community 2009). This is also an issue for species that spend key life cycle stages in near shore areas, including Dungeness crab and salmon. Historically, many coastal peoples along the Pacific coast actively maintained "clam gardens," by moving rocks to create berms and terraces; it may be that a return to this active manipulation of intertidal habitat will enable tribes to maintain high and sustainable populations of shellfish with rising sea levels.

8.3.4 FORESTS AND WILDFIRE

Large-scale changes in tree distribution across the Northwest are projected to occur over time as individual tree species adapt to changing CO_2 levels and changing climate (Littell et al. 2010; DellaSala et al., in review; see also Chapter 5). Changes in climate will affect the resources and habitats that tribes depend upon for cultural, medicinal, economic, and community health (Voggesser et al. 2013). As cited in Voggesser et al. (2013), species losses and shifts in species ranges are already being observed (Rose 2010; Swinomish Indian Tribal Community 2010), including northward or elevational migration of temperate forests, contraction or expansion of other plant species, and changes in the distribution and density of wildlife species (Trainor et al. 2009).

Direct climate change impacts on forest ecosystems are projected to be considerable, and wildfires and other climate-related forest disturbances are considered the most significant threats (Chapter 5; Littell et. al 2010). The compounding impacts from wildfire, invasive species and insect outbreaks, and other impacts to forests and ecosystems, pose a threat to the traditional foods, including berries, plants used for traditional basket weaving, and wildlife that tribes depend upon for their traditional ways of life (Voggesser et al. 2013).

8.4 Tribal Initiatives in the Northwest

Tribes respond to climate change as social, political, cultural, economic, and legal entities. Many NW tribes are engaged in climate change impacts and vulnerability assessments, developing climate change adaptation plans and highly localized solutions to climate change. These include ecosystem-based approaches to climate change, pursuing research and education efforts, and reducing greenhouse gas emissions. This section illustrates various ways that tribes in the Northwest are addressing climate change.

8.4.1 CLIMATE CHANGE IMPACTS AND VULNERABILITY ASSESSMENTS

Climate change vulnerability assessments are often the first step in understanding climate projections and identifying potential climate impacts. In 2012, the Port Gamble S'Klallam and Jamestown S'Klallam Tribes received funds through the EPA Indian Environmental General Assistance Program fund to initiate climate change assessments and adaptation planning. As neighboring tribes, the Port Gamble S'Klallam and Jamestown S'Klallam Tribes share concerns about climate change impacts and are coordinating their efforts to share information and leverage resources where appropriate. The Port Gamble S'Klallam Tribe is conducting an impacts assessment to identify and document major areas of concern, including ocean acidification, increasing temperatures, and sea level rise.

The Jamestown S'Klallam Tribe has begun work on an impacts assessment as part of a longer-term effort to develop a tribal adaptation plan. Key impacts of concern in their initial efforts include inundation and erosion of infrastructure, loss of usual and accustomed fishing and gathering areas, exposure of burial sites from extreme weather events, decline in estuary health, and increases in wildfire. The project team has divided impacts based on specific planning areas within their reservation and ancestral territory. This allows them to utilize their resources most efficiently, as the tribal project team is able to identify areas that are most likely to experience severe impact such as shellfish gathering areas. In addition to a written impacts assessment, the team is creating extensive maps of tribal assets and infrastructure, as well as high-risk zones.

8.4.2 CLIMATE CHANGE ADAPTATION PLANS

Tribal climate change adaptation planning is bringing together the understanding of potential impacts of climate change on tribal culture, resources, and economy. The Swinomish Indian Tribal Community developed one of the first climate change adaptation plans in the region. The Swinomish Tribe's historic and cultural reliance on salmon fishing, shellfish harvesting, and other traditional marine resources, as well as their economy and infrastructure, is threatened by a number of climate change impacts (Swinomish Indian Tribal Community 2009). These potential threats, along with regional identification of the lower Skagit River area as a high-risk area for sea level rise, and local extreme weather events such as severe storms and tidal surges, served as a catalyst for efforts by the Tribe to examine climate change issues (Bauman et al. 2006). In response, the Tribe assembled an interdisciplinary team to document and plan for these potential impacts. As their first major task, the tribal project team and science advisors reviewed multiple climate models and mapped risk zones for potential impact areas, then completed a detailed vulnerability assessment and risk analysis of projected impacts. In 2009, the tribe released an impacts assessment reporting on a broad range of potentially significant impacts. Potential inundation threatens 445 hectares (1,100 acres) on the north end of the

Reservation, an area containing critical tribal enterprises in the Tribe's primary development lands that currently provide the bulk of the Tribe's revenues, including a $50 million resort complex and a $2 million gas station, in addition to a $4 million wastewater treatment plant and other supporting infrastructure. In addition, the Tribe's only agricultural lands lie within this threatened low-lying area, and two bridges provide the sole means of access to the Reservation through similar at-risk low-lying lands. Culturally important shellfish beds and fish habitat around the perimeter of the Reservation are equally threatened. Impacts also included risks such as inundation of gathering areas and the effects of increasing temperature on the health of tribal members (Swinomish Indian Tribal Community 2009).

Following the impacts assessment, the tribe released a Climate Adaptation Action Plan in the fall of 2010. Initial recommendations included a starter list of follow-up projects, with emphasis on the need for ongoing monitoring to evaluate effectiveness. Project participants understand they are working for the future, that results of these efforts may not be seen by this generation, and that it will require continued focus on opportunities to engage the federal government on addressing impacts on trust and treaty rights.

The Nez Perce Tribe, in collaboration with government agencies and scholars, created the Clearwater River Subbasin Climate Change Adaptation Plan in 2012. The subbasin is located within the traditional Nez Perce homeland, and the majority of Nez Perce tribal members continue to live there. The Nez Perce Tribe developed this adaptation plan to better understand the projected local impacts of climate change and to identify some key adaptation strategies to help preserve the natural and economic resources of the Clearwater River subbasin. As a living document, the Tribe recognizes that the plan is the first step to assessing the vulnerability of resources in the region and developing strategies to adapt to regional changes in climate. The plan examines different climate change scenarios in the region to identify how potential changes may affect the region's forests, water, and economy and to address potential solutions for these regional changes and risks (Clark and Harris 2011). Because the subbasin is a mix of tribal, public, private, and wilderness land, the plan emphasizes the importance of partnership in researching and implementing responses to climate change impacts in the region.

Other tribes in the Northwest are also beginning to consider climate change adaptation planning within existing government and departmental constructs. The Coquille Indian Tribe in Oregon is integrating climate change considerations in their Tribal Strategic Plan update to begin planning for impacts to tribal infrastructure, natural resources, culture, economy, health, and safety. The Karuk Tribe of California (whose territories extend into Southern Oregon) developed an Eco-Cultural Resources Management Plan consisting of a long-term adaptation strategy for the protection, enhancement, and utilization of cultural and natural resources. The plan establishes a framework for considering a wide range of human and environmental stressors to the Karuk Tribe, including climate change (Karuk Tribe 2010).

8.4.3 ECOSYSTEM-BASED APPROACHES TO ADDRESSING CLIMATE CHANGE

Some tribes in the Northwest are taking a holistic and ecosystem-based approach to understanding and addressing climate change. The Nisqually Tribe, for example, is seeking to address the threats that climate change may pose to the Nisqually River. These threats include reduced snowpack and shrinking glacial melts leading to reductions in river flow and resulting effects on salmon habitat (Kaufman 2011). The Nisqually Tribe is a partner in the Nisqually River Delta Restoration Project, which promotes "system resiliency to loss of habitats and biodiversity, climate change effects such as increased winter storms, rainfall, and flooding, and rise in sea levels resulting in loss of shoreline areas" (Tillmann and Siemann 2011).

For the Tulalip Tribes, adapting to climate change has meant focusing on interrelated changes in local ecosystems. While changes in marine ecosystems such as ocean acidification, sea level rise, and warming oceans are a definite concern, the Tulalip Tribes emphasize that no ecosystem can be looked at in isolation. Therefore, they are focusing ongoing research with both the University of Washington and the University of Colorado on what changes are likely to occur in riparian ecosystems, surrounding areas, and estuaries. Understanding how climate change affects those areas is especially important because of the immense cultural resources that estuaries and riparian habitats supply for Tulalip people (e.g., shellfish and salmon). Researching changes to these ecosystems will help the Tulalip Tribes better understand how to protect these cultural resources in the future, and how to adapt to what may be inevitable alterations to local landscapes and ecosystems. For example, early spring snowmelts are causing irregular channelization of rivers, and damaging juvenile salmon habitat (Beechie et al. 2012; Yarnell et al. 2010); working to protect salmon habitat requires a holistic understanding of what changes are occurring, and what potential responses the Tulalip can enact. One response being explored by the Tribes is the construction of artificial wetlands in the uplands that can help slow runoff, provide salmon refuges, and increase infiltration (US EPA 1993, 2005). Another approach is through the Sustainable Land Strategy, where the Tulalip are working with farmers and municipal and county land managers to develop measures to better control runoff from the land into the ocean. Such an approach can reduce carbon losses that exacerbate climate-related coastal acidification, and non-acidification stressors on marine ecosystems and species, and links the health of Puget Sound to the health of the land (Washington State Blue Ribbon Panel 2012). Tribal climate change initiatives have stressed that some changes to local ecosystem health will occur regardless of successful mitigation efforts, and therefore the Tulalip Tribes must prepare and adapt in order to ensure the continued health of the Tulalip culture and people.

8.4.4. RESEARCH AND EDUCATION

Research partnerships have provided another mechanism for tribes to engage in climate change initiatives in the Northwest. The Suquamish Tribe is partnering with other entities to understand carbon driven impacts, while at the same time continuing to work to reduce the rate of pollution increase and habitat degradation in Puget Sound ecosystems. Current projects include a research partnership with University of Washington and NOAA's Northwest Fisheries Science Center to study the effects of pH on crab larvae, work with agencies who monitor water quality to share data and address identified gaps, and development of a low cost, autonomous plankton image recognition system to study how changes in water quality impact zooplankton in the environment. The

Suquamish tribal community has guided engagement in climate issues; a 2012 survey of tribal members found that 80% of respondents suggested that all groups need to work together to address ocean acidification, and the tribes and federal government should take the lead. The Suquamish Tribe has also acted on the recognition that the next generations will face ever-increasing challenges to protect natural resources and adapt to climate change. They will be creating an online database of links to high quality materials teachers can use to supplement the core educational standards.

8.4.5 REDUCING GREENHOUSE GAS EMISSIONS

A number of tribes in the Northwest have begun taking action to address climate change through the reduction of greenhouse gas emissions. The Confederated Tribes of Siletz Indians created the Siletz Tribal Energy Program to reduce energy waste in homes through weatherization programs and work towards energy independence by retrofitting buildings with solar energy (University of Oregon 2011a). The Lummi Nation Strategic Energy Plan is aimed at reducing emissions through renewable energy development, including the potential use of wind energy, and the use of a geothermal heat pump in a new tribal administrative building (University of Oregon 2011b). The Coquille Tribe has developed a range of conservation measures to reduce greenhouse gas emissions, including implementing more renewable energy, reducing waste, and increasing recycling at tribal buildings and businesses (University of Oregon 2011c). The Nez Perce Tribe carbon sequestration program has developed a mitigation strategy that resulted in the restoration of forest habitat on Nez Perce land, increased water supply and quality, reduced erosion, and increased area for cultural activities such as root digging (University of Oregon 2011d).

8.5 Tribal Research and Capacity Needs and Considerations for the Future

To address the needs of tribes in regards to climate change, future climate research, policies, and programs should examine how reserved rights, treaty rights, and tribal access to cultural resources will be affected by climate change and potential species and habitat migration. Understanding of climate change impacts can be strengthened through tribally-led research on how to preserve access to traditional foods and gathering areas where possible, and how to prepare areas likely to experience change so they continue to be gathering sites. Finally, consideration of the potential impacts resulting from the implementation of adaptation and mitigation strategies on tribal resources is an important part of understanding how climate change will affect tribes.

8.5.1. TRIBAL RESEARCH AND CAPACITY NEEDS

In the Northwest, assessments of physical and ecological climate change impacts (see Chapters 2-6) outpace assessments related to health, social, cultural, and economic impacts (Bauman et al. 2011; MacArthur et al. 2012; see Chapter 7 for health impacts), as well as assessments of the ability of tribes to respond effectively to climate change (for non-tribal assessments, see Finzi Hart 2012 and Washington Coastal Training Program

2008). A recent assessment for the Northwest synthesized the climate change-related challenges, needs, and opportunities of resource managers, conservation practitioners, and researchers working in the North Pacific Landscape Conservation Cooperative (NPLCC) region (Tillmann and Siemann 2012). The assessment included targeted American Indian and Alaska Native tribes, as well as Canadian First Nations in the NPLCC region, which extends from southeast Alaska to northern California west of the major mountain ranges. Note that research needs specific to tribes located east of the Cascade Mountain range (outside of the NPLCC region) were not assessed and need further investigation.

Of the 195 project participants, 29 (14.9%) represented Tribes, First Nations, or Alaska Native Communities (indigenous communities were represented third most commonly), and approximately one-third of tribal participants were from Oregon and Washington. Findings from the assessment include several needs and activities specific to tribes and managing valued ecosystems, habitats, species, and resources in light of current and projected climate change effects. The identified needs and activities focus on species and habitats that are culturally and economically significant, identifying whether and how to incorporate traditional ecological knowledge and western science in the NPLCC's work, and increasing collaboration, both within formal government-to-government relationships and through less formal means. Following are summaries of the key needs and activities identified by tribal representatives during this process:

- *Support efforts to identify whether and how to incorporate traditional ecological knowledge and western science in climate change work.* Even though tribal representatives noted traditional ecological knowledge and western science may be incompatible, several tribal and non-tribal participants in the NPLCC assessment suggested incorporating traditional and local knowledge through an explicit, tribally-led process in order to understand which changes are most important at local and human scales and to make decisions that meet cross-cultural needs.

- *Support training and other capacity-building within tribes.* Support for tribal capacity should be targeted to meet tribal needs along the spectrum of climate change knowledge and preparedness because capacity to address climate change varies by tribe, with some tribes in the NPLCC region just beginning to address climate change and others leading the way with innovative approaches to climate change adaptation. Participants requested the NPLCC hold conferences and workshops, provide access to data and tools, and develop or disseminate guidance to support decision-making.

- *Facilitate collaboration and communication between tribes and resource agencies.* Participants stated enhanced collaboration and communication would facilitate sharing, with the free, prior, and informed consent from indigenous communities, of traditional knowledge as well as local, on-the-ground information. Shared information would assist with landscape conservation decisions and identification of baselines, dispersal corridors, and migration corridors.

- *Facilitate science communication with the public and educators.* Participants suggested facilitating and coordinating local-level testimonies from tribal communities to communicate evidence of climate change observed by indigenous communities.

- *Protect tribal lands, trust resources, and tribal rights.* Climate change effects on territory, cultural resource viability, and the ability to sustain traditional ways of life were key issues identified by participants. For example, as described earlier in this chapter, declining pH due to ocean acidification may negatively impact tribal trust resources in Puget Sound, but it may be possible to alleviate these effects by managing non-climate stressors also influencing pH.

- *Research to assess and identify the most vulnerable cultural and natural resources.* To identify where and when species are likely to be vulnerable to climate change effects, participants suggested modeling potential impacts and supplementing with other research methods when model results are uncertain. Project participants emphasized baseline data on the health of species such as fish and shellfish and suggested focusing on vulnerable communities and resources, especially where traditional foods, cultural and natural resources, and harvest areas are located. This information is critical to identifying what foods are not healthy for people to consume.

- *Produce data and research that enables tribal and non-tribal decision makers to work together to address climate change effects.* Participants suggested focusing on the outputs that assist the "highest level agency staff in understanding how to work together" (quote from April 20, 2012 workshop in Juneau, Alaska) and on making data, studies, and research from science organizations easily accessible to tribes, indigenous entities, and other agencies. Improved collaboration between tribal and non-tribal decision makers would assist tribes in identifying which issues and policies to address. For example, improved collaboration through increased access to data, studies, and research would provide knowledge about when gathering of traditional foods and resources is appropriate. It would also assist with a more streamlined and integrated approach to consultation as a result of improved communication between tribal and non-tribal agencies.

8.5.2 CONSIDERATIONS FOR THE FUTURE

As highlighted throughout this chapter, it is important to explore and address the effects of climate change on tribal health, culture, economies, and infrastructure. The social, legal, and regulatory context for tribes is sufficiently distinct that tribal vulnerability and adaptation require explicit attention. Because climate affects the abundance and health of cultural resources and traditional foods, there must be an understanding of the consequences to a range of tribal sectors and traditional ways of life. Furthermore, unintended consequences that may have implications for tribal resources and rights must be fully explored, such as species shifts that may result in species moving out of usual and accustomed areas.

Traditional knowledge plays a key role in tribal approaches to understanding climate change impacts and identifying strategies for adaptation. Traditional knowledge has developed since time immemorial, and reflects a highly complex understanding of how to live within specific social and ecological systems, and how to adapt to changes within those systems (Williams and Hardison 2012). Tribal people have experience using traditional knowledge, the connections between language, place and cultural resources, and the strength of their social and cultural ties to deal with sudden, violent changes to their lifeways, such as European colonization or the complete loss of access to species central to their culture. Traditional knowledge has long recognized the necessity of adapting to changing conditions for tribal cultural survival, and many tribal communities are relying on these knowledge systems to provide for their continued cultural well-being. Traditional knowledge can inform tribal and non-tribal understanding of how climate change may impact tribal resources and traditional ways of life. However, its use in climate change initiatives should ensure respect for the values associated with these knowledge systems and protection of sacred knowledge and wisdom, and give tribes decision-making authority over how traditional knowledge is used.

There are numerous efforts in the region and around the country bringing tribes together with non-tribal entities to be pro-active in addressing climate change. This includes the Pacific Northwest Tribal Climate Change Network (tribalclimate.uoregon. edu), which is a collaboration between the University of Oregon Environmental Studies Program and the United States Department of Agriculture Forest Service Pacific Northwest Research Station. The network plays an important role in bringing tribal voices into regional and national climate change-related activities, as well as sharing information about resources and activities among tribes and other organizations along the West Coast. Also in the Northwest, the NPLCC funded several tribes and First Nations to integrate traditional ecological knowledge into landscape management and identify tribal and First Nations information needs related to conservation and land management (University of Oregon 2013).

In the United States, federally-recognized tribes are tied to their homelands by law as well as by culture (Williams and Hardison 2005, 2006, 2007, 2012), yet the impacts of climate change will not recognize geographic or political boundaries. Within government-to-government interactions, there is an opportunity (in addition to formal consultation processes) to foster effective communication, partnerships and collaboration, and mechanisms to exchange knowledge while protecting culturally sensitive information. Initiating dialogue and interactions about climate change through the government-to-government relationship shows recognition and respect for tribal sovereignty. Collaboration, shared resources, and an exchange of knowledge will be critical in leveraging limited resources and fostering an understanding of how climate change will impact communities and landscapes. For example, a report from the Treaty Indian Tribes in Western Washington (2011) examined the risks salmon face from habitat loss and lists recommendations for the federal government to remedy these losses. Primary recommendations call for increased leadership to improve federal agency coordination to achieve salmon recovery goals and an end to federal funding of state actions that either do not contribute to, or actually impede, recovery of salmon habitat. While not directly about climate change, these recommendations may set an example for federal agency coordination and government-to-government interactions on climate issues.

In July 2012, the US Senate Committee on Indian Affairs held a Congressional oversight hearing to examine the disproportionate impact that climatic changes have on tribal homelands and the resources available to mitigate and adapt to the changing environment. This hearing highlighted the importance of strengthening the government-to-government relationship to protect tribal rights and resources in the face of climate change. In his testimony at the hearing, Billy Frank Jr., chairman of the Northwest Indian Fisheries Commission, stated, "In the end, our treaty fishing rights are based on abundance, and it is that abundance that must be restored for those rights to have meaning. That abundance must come from a combination of improved habitat and hatchery production. The federal government must honor its treaties and exert its authority by exercising its trust obligation to the tribes to protect those resources" (Frank 2012).

Acknowledgments

The authors wish to thank three reviewers for their important comments and contributions to a previous version of the manuscript. The authors would also like to thank Amy Snover (University of Washington, Climate Impacts Group), and Philip Mote and Meghan Dalton (Oregon State University) for the opportunity to include this chapter and for their helpful review and feedback.

References

Bauman, Y., B. Doppelt, S. Mazze, and E. C. Wolf. 2006. "Impacts of Climate Change on Washington's Economy: A Preliminary Assessment of Risks and Opportunities." Washington Department of Ecology, Publication 07-01-010. https://fortress.wa.gov/ecy/publications/publications /0701010.pdf.

Beechie, T., H. Imaki, J. Greene, A. Wade, H. Wu, G. Pess, P. Roni, J. Kimball, J. Stanford, P. Kiffney, and N. Mantua. 2012. "Restoring Salmon Habitat for a Changing Climate." *River Research and Applications*. doi: 10.1002/rra.2590.

Bernholz, C. D., and R. Weiner Jr., R. J. 2008. "The Palmer and Stevens "Usual and Accustomed Places" Treaties in the Opinions of the Courts." *Government Information Quarterly* 25 (4): 778-795. doi: 10.1016/j.giq.2007.10.004.

Bilodeau, K. A. 2012. "The Elusive Implied Water Right for Fish: Do Off-Reservation Instream Water Rights Exist to Support Indian Treaty Fishing Rights?" *Idaho Law Review* 48 (3): 515–551. http://www.uidaho.edu/~/media/Files/orgs/Law/law-review/2012-symposium /Bilodeau.ashx.

Brown, M., and M. Maudlin. 2007. "Upper South Fork Nooksack River Habitat Assessment." Lummi Nation Natural Resources Department. 156 pp. http://lnnr.lummi-nsn.gov/Lummi Website/userfiles/1_Upper%20South%20Fork%20Nooksack%20River%20Habitat%20 Assessment%202007.pdf.

Clark, K., and J. Harris. 2011. "Clearwater River (ID) Subbasin Climate Change Adaptation Plan." Nez Perce Tribe Water Resources Division. 58 pp. http://www.mfpp.org/wp-content/uploads /2012/03/ClearwaterRiver-Subbasin_ID_Forest-and-Water-Climate-Adaptation-Plan_2011 .pdf.

Cloern, J. E., T. S. Schraga, C. B. Lopez, N. Knowles, R. G. Labiosa, and R. Dugdale. 2005. "Climate Anomalies Generate an Exceptional Dinoflagellate Bloom in San Francisco Bay." *Geophysical Research Letters* 32 (14): L14608. doi: 10.1029/2005GL023321.

Coe, T. 2001. "Nooksack River Watershed Riparian Function Assessment." Natural Resources Department, Nooksack Indian Tribe. Report #2001-001. 114 pp. http://www.whatcom-mrc.what comcounty.org/documents/NooksackRiparianSummary.pdf.

Dale, B., M. Edwards, and P. C. Reid. 2006. "Climate Change and Harmful Algal Blooms." In *Ecology of Harmful Algae,* edited by E. Granéli, and J. T. Turner. 367-378. Ecological Studies Vol. 189. Dordrecht, The Netherlands: Springer-Verlag.

DellaSala, D. A., P. Brandt, M. Koopman, J. Leonard, C. Meisch, P. Herzog, P. Alaback, M. I. Goldstein, S. Jovan, A. MacKinnon, and H. Von Wehrden. "Climate Change May Trigger Broad Shifts in North America's Pacific Coastal Temperate Rainforests." *Conservation Biology.* In review.

Dickerson-Lange, S. E., and R. Mitchell. "Modeling the Effects of Climate Change Projections on Streamflow in the Nooksack River Basin, Northwest Washington." *Hydrological Processes.* In review.

Executive Order No. 13175. "Consultation and Coordination with Indian Tribal Governments" November 6. 2000. http://ceq.hss.doe.gov/nepa/regs/eos/eo13175.html.

Feely, R. A., T. Klinger, J. A. Newton, and M. Chadsey. 2012 "Scientific Summary of Ocean Acidification in Washington State Marine Waters." NOAA OAR Special Report Publication No. 12-01-016. 176 pp. https://fortress.wa.gov/ecy/publications/summarypages/1201016.html.

Finzi Hart, J. A., P. M. Grifman, S. C. Moser, A. Abeles, M. R. Myers, S. C. Schlosser, and J. A. Ekstrom. 2012. "Rising to the Challenge: Results of the 2011 Coastal California Adaptation Needs Assessment." USCSG-TR-01-2012.

Frank, B., Jr. 2012. "Testimony of Billy Frank, Jr. Chairman Northwest Indian Fisheries Commission before the Senate Committee on Indian Affairs oversight hearing on the impacts of environmental changes on treaty rights, traditional lifestyles and tribal homelands." Northwest Indian Fisheries Commission. http://www.indian.senate.gov/hearings/upload/Billy-Frank-testimony071912.pdf.

Galanda, G. 2011. "The Federal Indian Consultation Right: A Frontline Defense Against Tribal Sovereignty Incursion." *Government Affairs Practice Newsletter* 2 (1): 1-14. Winter 2010–11. American Bar Association Business Law Section. http://apps.americanbar.org/buslaw/committees/CL121000pub/newsletter/201101/galanda.pdf.

Gates, C. 1955. "The Indian Treaty of Point No Point," *Pacific Northwest Quarterly* 46 (2): 52-58.

Glick, P., J. Clough, and B. Nunley. 2007. "Sea-Level Rise and Coastal Habitats in the Pacific Northwest: An Analysis for Puget Sound, Southwestern Washington, and Northwestern Oregon." Seattle, WA: National Wildlife Federation.

Grah, O., and J. Beaulieu. 2013. "The Effect of Climate Change on Glacier Ablation and Baseflow Support in the Nooksack River Basin and Implications on Pacific Salmon Species Protection and Recovery." *Climatic Change.* doi: 10.1007/s10584-013-0747-y.

Graves, D. 2008. "A GIS Analysis of Climate Change and Snowpack on Columbia Basin Tribal Lands." Columbia River Inter-Tribal Fish Commission Report #08-05, Portland, Oregon.

Hardison, P. D., and T. Williams. "Culture, Law, Risk and Governance: The Ecology of Traditional Knowledge in Climate Change Adaptation." *Climatic Change.* In review.

Harley, C. D. G. 2011. "Climate Change, Keystone Predation, and Biodiversity Loss." *Science* 334: 1124-1127. doi: 10.1126/science.1210199.

Helmuth, B. 1998. "Intertidal Mussel Microclimates: Predicting the Body Temperature of a Sessile Invertebrate." *Ecological Monographs* 68 (1): 51-74. doi: 10.1890/0012-9615(1998)068[0051:IMMPTB]2.0.CO;2.

Helmuth, B., B. R. Broitman, C. A. Blanchette, S. Gilman, P. Halpin, C. D. G. Harley, M. J. O'Donnell, G. E. Hofmann, B. Menge, and D. Strickland. 2006. "Mosaic Patterns of Thermal Stress in the Rocky Intertidal Zone: Implications for Climate Change." *Ecological Monographs* 76 (4): 461-479. doi: 10.1890/0012-9615(2006)076[0461:MPOTSI]2.0.CO;2.

Isaak, D., S. Wollrab, D. Horan, and G. Chandler. 2012. "Climate Change Effects on Stream and River Temperatures across the Northwest U.S. from 1980–2009 and Implications for Salmonid Fishes." *Climatic Change* 113 (2): 499-524. doi: 10.1007/s10584-011-0326-z.

Karuk Tribe. 2010. "Eco-Cultural Resource Management Plan." Karuk Tribe Department of Natural Resources. http://www.karuk.us/karuk2/images/docs/dnr/ECRMP_6-15-10_doc.pdf.

Kaufman, L. 2011 "Seeing Trends, Coalition Works to Help a River Adapt." *New York Times*, July 20. www.nytimes.com/2011/07/21/science/earth/21river.html?_r=2.

Klein, S., J. Butcher, B. Duncan, and H. Herron. 2013. "EPA Region 10 Climate Change and TMDL Pilot Project Research Plan." EPA/600/R/13/028. US Environmental Protection Agency, Office of Research and Development, Corvallis, OR.

Lackey, R. T. 2000. "Restoring Wild Salmon to the Pacific Northwest: Chasing an Illusion?" In *What We Don't Know about Pacific Northwest Fish Runs: An Inquiry into Decision-Making under Uncertainty*, edited by P. Koss and M. Katz, 91-143. Portland, Oregon: Portland State University.

Littell, J. S., E. E. Oneil, D. McKenzie, J. A. Hicke, J. A. Lutz, R. A. Norheim, and M. M. Elsner. 2010. "Forest Ecosystems, Disturbance, and Climatic Change in Washington State, USA." *Climatic Change* 102: 129-158. doi: 10.1007/s10584-010-9858-x.

Lynn, K., K. MacKendrick, and E. Donoghue. 2011. "Social Vulnerability and Climate Change: Synthesis of Literature." Gen. Tech. Rep. PNW-GTR-838. Portland, OR: US Department of Agriculture, Forest Service, Pacific Northwest Research Station. 70 pp.

MacArthur, J., P. Mote, J. Ideker, M. Figliozzi, and M. Lee. 2012. "Climate Change Impact Assessment for Surface Transportation in the Pacific Northwest and Alaska." Washington State Department of Transportation, Report WA-RD 772.1. http://www.wsdot.wa.gov/research/reports/fullreports/772.1.pdf.

Mantua, N. J., I. Tohver, and A. F. Hamlet. 2010. "Climate Change Impacts on Streamflow Extremes and Summertime Stream Temperature and Their Possible Consequences for Freshwater Salmon Habitat in Washington State." *Climatic Change* 102: 187-223. doi: 10.1007/s10584-010-9845-2.

McNutt, D. ed. 2008. "Native Peoples: The Miners Canary on Climate Change." Olympia, WA: Evergreen State College. http://nwindian.evergreen.edu/pdf/climatechangereport.pdf

Moore, S. K., N. J. Mantua, and E. P. Salathé. 2011. "Past Trends and Future Scenarios for Environmental Conditions Favoring the Accumulation of Paralytic Shellfish Toxins in Puget Sound Shellfish." *Harmful Algae* 10: 521-529. doi: 10.1016/j.hal.2011.04.004.

Morisset, M. D. 1999. "Recent Developments in Defining the Federal Trust Responsibility (The Case of the Reluctant Guardian)." In *The Implementation of Indian Law: Handling Interactions Among Three Sovereigns*, edited by Washington State Bar Association. Seattle: Washington State Bar Association.

Nakićenović, N., O. Davidson, G. Davis, A. Grübler, T. Kram, E. Lebre La Rovere, B. Metz, T. Morita, W. Pepper, H. Pitcher, A. Sankovshi, P. Shukla, R. Swart, R. Watson, and Z. Dadi. 2000. *Special Report on Emissions Scenarios: A Special Report of Working Group III of the Intergovernmental Panel on Climate Change*, Cambridge University Press, Cambridge, UK, 599 pp. http://www.grida.no/climate/ipcc/emission/index.htm.

National Research Council. 2012. "Sea-Level Rise for the Coasts of California, Oregon, and Washington: Past, Present, and Future." Committee on Sea Level Rise in California, Oregon, and Washington, Board on Earth Sciences and Resources, and Ocean Studies Board Division on Earth and Life Sciences National Research Council. The National Academies Press, Washington, DC.

Norgaard, K. M. 2005. "The Effects of Altered Diet on the Health of the Karuk People." http://ejcw.org/documents/Kari%20Norgaard%20Karuk%20Altered%20Diet%20Nov2005.pdf.

Osborn, R. P. 2012. "Climate Change and the Columbia River Treaty." *Washington Journal of Environmental Law & Policy* 2 (1): 75-123. http://digital.law.washington.edu/dspace-law/bitstream/handle/1773.1/1148/2WJELP075.pdf?sequence=1.

Ovsak, C. M. 1994. "Reaffirming the Guarantee: Indian Treaty Rights to Hunt and Fish Off-Reservation in Minnesota." *William Mitchell Law Review* 20 (4): 1177.

Paerl, H. W., and J. Huisman. 2009. "Climate Change: A Catalyst for Global Expansion of Harmful Cyanobacterial Blooms." *Environmental Microbiology Reports* 1 (1): 27–37. doi: 10.1111/j.1758-2229.2008.00004.x.

Papiez, C. 2009. "Climate Change Implications for Quileute and Hoh Tribes of Washington." MS thesis, Evergreen State College of Washington. http://academic.evergreen.edu/g/grossmaz/papiez.html.

Parker, A. 2009. "Government-to Government Consultation." A Presentation to the Affiliated Tribes of Northwest Indians 2009 Mid-year Conference, Tacoma, WA. http://files.ncai.org/governance/Parker_Consultation_Policies_2009.10.15.pdf.

Parker, A., Z. Grossman, E. Whitsell, B. Stephenson, T. Williams, P. Hardison, L. Ballew, B. Burnham, J. Bushnell, and R. Klosterman. 2006. "Climate Change and Pacific Rim Nations." Northwest Indian Applied Research Institute. Evergreen State College, WA. http://academic.evergreen.edu/g/grossmaz/IndigClimate2.pdf.

Pelto, M. S. 2010. "Forecasting Temperate Alpine Glacier Survival from Accumulation Zone Observations." *The Cryosphere* 4: 67-75. doi: 10.5194/tc-4-67-2010.

Pelto, M. S., and C. Brown. 2012. "Mass Balance Loss of Mount Baker, Washington Glaciers 1990-2010." *Hydrological Processes* 26 (17): 2601-2607. doi: 10.1002/hyp.9453.

Pevar, S. 2012. *The Rights of Indians and Tribes*. New York: Oxford University Press.

Rose, K. A. 2010. "Tribal Adaptation Options: A Review of the Scientific Literature." US Environmental Protection Agency Region 10. Seattle, WA.

Ruggiero, P., P. D. Komar, and J. C. Allan. 2010. "Increasing Wave Heights and Extreme Value Projections: The Wave Climate of the U.S. Pacific Northwest." *Coastal Engineering* 57 (5): 539–552. doi: 10.1016/j.coastaleng.2009.12.005.

Swinomish Indian Tribal Community. 2009. "Swinomish Climate Change Initiative Impact Assessment Technical Report." Swinomish Indian Tribal Community, Office of Planning and Community Development, La Conner, WA. http://www.swinomish.org/climate_change/Docs/SITC_CC_ImpactAssessmentTechnicalReport_complete.pdf.

Swinomish Indian Tribal Community. 2010. "Swinomish Climate Change Initiative Climate Adaption Action Plan." Swinomish Indian Tribal Community, Office of Planning and Community Development, La Conner, WA. www.swinomish.org/climate_change/Docs/SITC_CC_AdaptationActionPlan_complete.pdf.

Tillmann, P., and D. Siemann. 2011. "Climate Change Effects and Adaptation Approaches in Marine and Coastal Ecosystems of the North Pacific Landscape Conservation Cooperative Region." National Wildlife Federation. 257 pp. http://www.fws.gov/pacific/Climatechange/nplcc/pdf/NPLCC_Marine_Climate%20Effects_Draft%20Final.pdf.

Tillmann, P., and D. Siemann. 2012. "Advancing Landscape-Scale Conservation: An Assessment of Climate Change-Related Challenges, Needs, and Opportunities for the North Pacific Landscape Conservation Cooperative." National Wildlife Federation – Pacific Region, Seattle, WA.

Trainor, S., S. Chapin, A. McGuire, M. Calef, N. Fresco, M. Kwart, P. Duffy, A. Lovecraft, T. Rupp, L. DeWilde, O. Huntington, and D. Natcher. 2009. "Vulnerability and Adaptation to Climate-Related Fire Impacts in Rural and Urban Interior Alaska." *Polar Research* 28 (1): 100-118. doi: 10.1111/j.1751-8369.2009.00101.x.

Treaty Indian Tribes in Western Washington. 2011. "Treaty Rights At Risk: Ongoing Habitat Loss, the Decline of the Salmon Resource, and Recommendations for Change." http://www.orca network.org/habitat/whitepaper628finalpdf.pdf.

United States v. State of Washington, 384 F. Supp. 312, Western District of Washington, 1974.

United States of America, et al., Plaintiffs, vs. State of Washington et al., Defendants. No. CV 9213, Sub-proceeding No. 89-3 United States District Court For The Western District Of Washington 873 F. Supp. 1422, 1994 US Dist. Decision December 20, 1994, Decided.

US Environmental Protection Agency. 1993."Constructed Wetlands for Wastewater Treatment and Wildlife Habitat: 17 Case Studies." EPA832-R-93-005. Washington: United States Environmental Protection Agency.

US Environmental Protection Agency. 2005. "National Management Measures to Protect and Restore Wetlands and Riparian Areas for the Abatement of Nonpoint Source Pollution." EPA-841-B-05-003. Washington: Office of Water, United States Environmental Protection Agency.

University of Oregon. 2011a. "Siletz Tribal Energy Program." Pacific Northwest Tribal Climate Change Profile Project. http://tribalclimate.uoregon.edu/tribal-profiles/.

University of Oregon. 2011b. "The Lummi Nation: Pursuing Clean Renewable Energy." Pacific Northwest Tribal Climate Change Profile Project. http://tribalclimate.uoregon.edu /tribal-profiles/.

University of Oregon. 2011c. "Climate Change and the Coquille Indian Tribe: Planning for the Effects of Climate Change and Reducing Greenhouse Gas Emissions." Pacific Northwest Tribal Climate Change Profile Project. http://tribalclimate.uoregon.edu/tribal-profiles/.

University of Oregon. 2011d. "Nez Perce Tribe: Carbon Sequestration Program." Pacific Northwest Tribal Climate Change Profile Project. http://tribalclimate.uoregon.edu/tribal-profiles/.

University of Oregon. 2013. "Indigenous Peoples and Northwest Climate Initiatives: Exploring the Role of Traditional Ecological Knowledge in Resource Management." Pacific Northwest Tribal Climate Change Profile Project. http://tribalclimate.uoregon.edu/tribal-profiles/.

Voggesser, G., K. Lynn, J. Daigle, F. K. Lake, and D. Ranco. 2013. "Cultural Impacts to Tribes from Climate Change Influences on Forests." *Climatic Change*. doi: 10.1007/s10584-013-0733-4.

Washington Coastal Training Program. 2008. "Needs Assessment Data Summary – Climate Training Topics." http://nerrs.noaa.gov/Doc/pdf/training/needs_assessment_data_summary. pdf.

Washington Department of Ecology. 2006. "Impacts of Climate Change on Washington's Economy. A Preliminary Assessment of Risks and Opportunities." Washington Department of Ecology, Olympia, Washington. Publication No. 07-01-010.

Washington Department of Fish and Wildlife. 2012. Fish Ticket Database (unpublished report)

Washington State Blue Ribbon Panel on Ocean Acidification. 2012. *Ocean Acidification: From Knowledge to Action, Washington State's Strategic Response,* edited by H. Adelsman, and L. Whitely Binder. Washington Department of Ecology, Olympia, Washington. Publication No. 12-01-015.

Whyte, K. P. 2013. "Justice Forward: Tribes, Climate Adaptation and Responsibility in Indian Country" *Climatic Change*. doi: 10.1007/s10584-013-0743-2.

Wilkins, D. E., and K. T. Lomawaima. 2001. *Uneven Ground: American Indian Sovereignty and Federal Law.* Norman: University of Oklahoma Press.

Williams T., and P. D. Hardison. 2005. "Global Climate Change, Environmental Change and Water Law." Paper presented at the Law Seminars International Conference "What is Next for Washington Water Law?" May 20. Tulalip Tribes of Washington, Tulalip, Washington (unpublished, available from the Author).

Williams T., and P. D. Hardison. 2006. "Impacts on Indigenous Peoples." In *Climate Change and Pacific Rim Indigenous Nations,* edited by A. Parker, Z. Grossman, E. Whitesell, B. Stephenson, T. Williams, P. Hardison, L. Ballew, B. Burnham, J. Bushnell, R. Klosterman, Northwest Indian Applied Research Institute (NIARI), The Evergreen State College, Olympia, Washington, USA. http://academic.evergreen.edu/g/grossmaz/IndigClimate.pdf.

Williams T., and P. D. Hardison. 2007. "Global Climate Change: Justice, Security and Economy." Paper Presented at the Salmon Homecoming Forum, September 17. Tulalip Tribes of Washington, Tulalip, Washington (unpublished, available from the Author).

Williams T., and P. D. Hardison. 2012. "Climate Threats to Pacific Northwest Tribes and the Great Ecological Removal: Keeping Traditions Alive." In *Asserting Native Resilience: Pacific Rim Indigenous Nations Face the Climate Crisis,* edited by Z. Grossman and A. Parker. Corvallis, Oregon: Oregon State University Press.

Wootton, J. T., C. A. Pfister, and J. D. Forester. 2008. "Dynamic Patterns and Ecological Impacts of Declining Ocean pH in a High-Resolution Multi-Year Dataset." *Proceedings of the National Academy of Sciences* 105 (48): 18848-18853. doi: 10.1073/pnas.0810079105.

Yarnell, S. M., J. H. Viers, and J. F. Mount. 2010. "Ecology and Management of the Spring Snowmelt Recession." *BioScience* 60 (2): 114-127. doi: 10.1525/bio.2010.60.2.6.